Hybrid Nature

Urban and Industrial Environments

Series editor: Robert Gottlieb, Henry R. Luce Professor of Urban and Environmental Policy, Occidental College

Hybrid Nature

Sewage Treatment and the Contradictions of the Industrial Ecosystem

Daniel Schneider

The MIT Press
Cambridge, Massachusetts
London, England

For information about special quantity discounts, please email special_sales@mitpress.mit.edu.
This book was set in Sabon by Graphic Composition, Inc., Bogart, Georgia. Printed and bound in the United States of America.

Library of Congress Cataloging-in-Publication Data

Schneider, Daniel, 1959–
Hybrid nature : sewage treatment and the contradictions of the industrial ecosystem / Daniel Schneider.
    p.   cm. — (Urban and industrial environments)
Includes bibliographical references and index.
ISBN 978-0-262-01644-5 (hardcover : alk. paper)—ISBN 978-0-262-51638-9 (pbk. : alk. paper)
1. Sewage—Purification—Biological treatment—History. 2. Industrial ecology—Philosophy.
3. Contradiction. I. Title.
TD755.S287   2011
628.3—dc22

                                                                                2011009079

10   9   8   7   6   5   4   3   2   1

For Jacob and Rosie

Contents

Acknowledgments

I'd first like to thank Mike Guthrie, former chief operator at the Northeast Plant of the Urbana-Champaign Sanitary District. While giving my watershed planning class tours of the plant, he first got me interested in how operators manage the ecosystems of the sewage treatment plant, which started me on this project. I'd like to thank him for his enthusiasm and insights.

Christopher Hamlin has been a great supporter and font of knowledge since I first contacted him out of the blue about his work on the history of biological sewage treatment. He encouraged me to pursue this research, agreed to help organize panels with me, read the entire manuscript, and gave me extensive feedback. His comments have greatly improved my arguments and saved me from several errors. Needless to say, any remaining errors are very much my own.

I would particularly like to thank Rip Sparks, Jim Kitchell, Mary Power, Dianna Padilla, John Lyons, and Eduardo Santana for encouraging me in my historical pursuits. These friends and ecologists all were important in reassuring me that my historical research had value for ecologists and ecology. I'd also like to thank Chip Burkhardt for his enthusiasm, interest, and numerous conversations about the history of science. My advisor Tom Frost died as I was beginning this research. His advice and insights would have made this a much better book, and I wish I could have shared it with him. Lew Hopkins, longtime head of the Department of Urban and Regional Planning at the University of Illinois built a department that valued and supported interdisciplinary research. Likewise, Dave Philipp and John Epifanio, directors of the Center for Aquatic Ecology at the Illinois Natural History Survey, enthusiastically supported my seemingly nonbiological efforts. I'd like to thank all my colleagues in both departments for making such interesting places to work. Helen Bustamante, my graduate student and native Mancunian, graciously offered to do research for me on her trips back home. I shared many conversations about environmental history with my other Mancunian graduate student, Glenn Sandiford. My brother Norman Schneider helped me figure out a variety of legal questions and read the manuscript as well.

Sherri Brown and family provided a wonderful home while I did research in New York City. I also got to stay with my parents and brothers in Washington, D.C., and Philadelphia. Julie Anderson sponsored my visit to the Center for the History of Science, Technology and Medicine at the University of Manchester. CHSTM was an incredibly welcoming place and made my research visit to the Manchester City Archives much more productive and enjoyable, for which I'd also like to thank John Pickstone and Mick Worboys.

I began research for this book while on one sabbatical in the Department of Integrative Biology at the University of California–Berkeley and neared completion while on another at Berkeley's Science Technology and Society Center and Office of the History of Science and Technology. I thank Mary Power, Charis Thompson, and Cathryn Carson for sponsoring my visits there.

I'd like to particularly thank Denise Whelan and the other staff in the Records Office at the Milwaukee Metropolitan Sewerage District, who helped find the records, made them available, and provided me a place to work in their crowded office. Tisa Overman and Jamie Staufenbeil with the Milorganite division of MMSD arranged a tour of the Jones Island Plant with process control engineer Sid Arora. Pete Cookingham at Michigan State's Turfgrass Information Center was incredibly helpful in guiding me through their materials on Milorganite and its relation to the turf grass industry. Faith Ward at the Water Environment Federation guided me through their archives. Archivists at the National Archives in New York, Philadelphia, and College Park were also very helpful. I'd also like to thank the staff of the Manchester Archives, Grand Rapids City Archives, Milwaukee Public Library, Milwaukee County Historical Society, University of Illinois Archives, University of California Water Resources Center Archives, and Rockefeller Archives. I'd also like to thank David Jenkins for sharing his father's papers and his encyclopedic knowledge of the activated sludge process. The MIT Press and my editors Clay Morgan and Robert Gottlieb have been very helpful and have made excellent suggestions for improving the book. Kathy Caruso did an excellent job copyediting, and her careful reading saved me from a number of embarrassing mistakes. This research has been supported by grants from the National Science Foundation (SES-0551858), University of Illinois Center for Advanced Study, University of Illinois Campus Research Board, and the Rockefeller Archives Foundation.

I'd like to thank my family for all their help and encouragement, my brothers Peter and Norman, and my parents, Irving and Zola. I've been most fortunate to be married to one of the best historians of women, law, or medicine there is. I've learned most of what I know about doing history from the remarkable example of Leslie J. Reagan. She is also one of the best readers I know of and made many excellent comments on the manuscript. I would like to also thank her for making it possible for me to go on my many research trips. I thank my children, Jacob and Rosie, for, among everything else, not minding that I've dedicated a book on sewage to them, although Rosie did say, "Poop??? Ewww."

Jacob has also been a willing companion on several trips to sewage treatment plants, and he told me, "You should write a book, Papa," for which I thank him.

Abbreviations

AMSDW	*Association of Managers of Sewage Disposal Works, List of Members and Proceedings*
ASM-Dep	*Activated Sludge Inc. v. The City of Milwaukee,* Depositions, Box 8303, MMSD
BP	Edward Bartow Papers, Record Series 15/5/35, University Archives, University of Illinois at Urbana-Champaign
CDT	*Chicago Daily Tribune*
CT	*Chicago Tribune*
DHP	Daniel Hoan Papers, Milwaukee County Historical Society
EN	*Engineering News*
ENR	*Engineering News-Record*
ER	*Engineering Record*
GSD-Dep	*Edgar C. Guthard and Activated Sludge Ltd. v. The Sanitary District of Chicago,* Depositions, Box 8303, MMSD
JRSI	*Journal of the Royal Sanitary Institute*
JSCI	*Journal of the Society of Chemical Industry*
MJ	*Milwaukee Journal*
MJS	*Milwaukee Journal Sentinel*
MS	*Milwaukee Sentinel*
MMSD	Records Office, Milwaukee Metropolitan Sewerage District
N/MC	Noer/Milorganite® MMSD Collection, Turfgrass Information Center, Michigan State University
NYT	*New York Times*
Surveyor	*Surveyor and Municipal and County Engineer*

SWJ   *Sewage Works Journal*

USPO   Patented Case Files, United States Patent Office, Record Group 241,
           National Archives and Records Administration, College Park

WEF   Water Environment Federation

Introduction

This book is an environmental history of the biological sewage treatment plant. Biological sewage treatment, like electricity, power generation, telephones, or mass transit, is one of the key technologies of the late-nineteenth- and twentieth-century city (figure 0.1). Present in almost every town and city in developed nations, sewage treatment plants are a major part of their urban infrastructure, responsible for protecting not only public health, but also the ecology of rivers, lakes, and oceans. In the United States alone, there were over 16,500 sewage treatment plants in 2004. The estimated capital stock of public sewerage facilities in 1997 was $274 billion, with annual spending for construction and operation and maintenance of almost $50 billion.[1]

Besides being a ubiquitous engineering component of the modern city, biological sewage treatment plants are also ecosystems. As such, they rely on the ability of microorganisms and other plants and animals to degrade sewage and produce a pure effluent. Ecosystems can be defined as biophysical systems in which communities of organisms—bacteria, fungi, plants, animals—consume food, energy, and nutrients. In turn, these communities transform the energy and resources, cycling various substances back to the environment. In a biological sewage treatment plant, bacteria convert the organic matter in waste to carbon dioxide and methane and proteins and organic nitrogen to ammonia, nitrate, and nitrogen gas. The cells of the bacteria grown on the sewage, along with any remaining organic matter, make up the solid waste of the treatment plant, the sludge or so-called biosolids. The sewage treatment plant differs from natural ecosystems, though, in the extent of human intervention in its creation and management. This book documents and explores these complex relations between society and nature that were involved in the establishment and operation of the sewage treatment plant ecosystem, from the mid-nineteenth century to the present.

Figure 0.1
Jones Island Treatment Plant, Milwaukee Sewerage Commission, c. 1927. When this plant,
located in Milwaukee's harbor, began operation in 1925, it was the largest sewage treatment
plant in the world. Milwaukee had established a testing station there in 1915 that was instrumental in the development of the activated sludge process. The plant is still operating today.
Source: Milwaukee Metropolitan Sewerage District

The Industrial Ecosystem

The sewage treatment plant was the most important example of a novel kind of
ecosystem that proliferated in the late nineteenth century, what I call the *industrial
ecosystem*. In the industrial ecosystem, the metabolic processes of an ecosystem are
exploited to extract resources such as food, fabrics, pharmaceuticals, or fuel.[2] The
biological sewage treatment plant was critical to the development of the industrial
ecosystem more broadly. It is now the most common industrial ecosystem, and as
a critical tool for protecting public health and the environment, it has occupied
the attentions of an extremely large group of scientists, engineers, city and public

health officials, and workers in the plants themselves. Advances in the understanding of microbial biology and ecology made in studies of sewage treatment spread to public health, microbiology, agriculture, and industry. Further, the sewage treatment plant did important cultural work in changing ideas about the relation between the natural and industrial that has had profound impacts on the history of biotechnology and genetic engineering today.

Like the sewage treatment plant, the industrial ecosystem had its origins in practices that relied on the power of microorganisms. But even before microorganisms were identified or their functions understood, people had been using them to produce beverages like wine, sake, and beer; foods such as bread, sauerkraut, and vinegar; and industrial products like flax and saltpeter. With the rise of microbiology as a science, however, and the coalescing of Darwin's theory of evolution with physiology and botany into the science of ecology, scientists developed the theoretical background and understanding to deliberately intensify and simplify these biological processes and turn them to mass production. Traditional practices were thus transformed into industrial ecosystems. The industrial ecosystem was just one aspect of the industrialization of nature, a part of the broader shift in the mode of production toward mechanization, fossil fuels, and the factory. As Edmund Russell notes, "Industrialization was a biological process as well as a mechanical process."[3]

Biological sewage treatment, scientific brewing using pure yeast culture techniques, and the use of bacterial fermentation to produce industrial chemicals all came about simultaneously in the late nineteenth century.[4] Emil Christian Hansen in Denmark and Max Delbrück in Germany applied the ideas of natural selection and the struggle for existence to the culture and use of yeast in beer making. In the United States, sanitary scientists William Sedgwick and E. O. Jordan used Darwin's theory to understand the role of bacteria in purifying sewage. Massachusetts chemist Charles Avery manipulated environmental conditions of bacterial fermentation to produce lactic acid, the first explicit application of bacteria to the commercial production of chemicals.

These theories and techniques quickly coalesced into a general program for the management of industrial fermentation that focused on the relation between organisms and environment. "A very thorough knowledge of the nature of those organisms and of the influence of environment on their chemical activities is essential to efficient and successful factory work," argued an early advocate for industrial microbiology. "For every organism," he continued, "there is a particular set of conditions, the observance of which is absolutely essential to successful working."[5] These conditions were to be based on a knowledge of the distribution of the organisms in nature, their "probable habitat." Microbiology was thus connected to ecological science and the theoretical underpinnings of the industrial ecosystem were established.

From its origins in the late nineteenth century, the industrial ecosystem quickly spread. Craft processes like brewing or vinegar making were industrialized while fermentation scientists discovered new microbial pathways for chemical production. Hansen's pure yeast culture in brewing expanded across Europe and the United States.[6] Microbiologists at the Pasteur Institute in Paris identified novel fermentation products. As scientists worked out the biology and chemistry of fermentation, they created other industrial ecosystems that provided bulk chemicals, like acetone or butyl alcohol (figure 0.2), enzymes like cellulase or protease, pharmaceuticals like antibiotics and vitamins. The scale of these industrial ecosystems could be enormous. The largest industrial ecosystem in the early twentieth century was probably the Curtis Bay acetic acid factory in Baltimore. In response to the huge need for acetic acid as a feedstock for explosives in World War I, the factory had scaled up the traditional method of making vinegar in which wine was dripped through a wooden cask containing beech or birch shavings. The Curtis Bay plant consisted of over one thousand fermentation tanks, each eighteen feet tall and over ten feet in diameter, covering over twenty-seven acres of ground. The fermenters were filled with birch shavings over which alcohol from an adjacent factory was dripped in a constant stream. The bacteria that grew on the wood shavings converted the alcohol to acetic acid.[7]

The industrial ecosystem now pervades modern food, chemical, and pharmaceutical manufacturing. The bacterium *Corynebacterium glutamicum*, for instance, is used to produce the amino acids L-glutamic acid and L-lysine. These two amino acids are the basis of a $4 billion global industry, L-glutamic acid in the flavor enhancer MSG, and L-lysine as an animal feed additive.[8] The bacterium is grown in huge stainless steel fermentation tanks, and fed either molasses or corn syrup. Production plant operators induce the bacteria to produce the amino acids in excess of their own metabolic requirements by manipulating the environmental conditions in the tanks. Mining companies use bacteria to extract gold, copper, uranium, and other metals from low-grade ore. These bacteria use the minerals in the ore for energy, producing sulphuric acid that leaches the precious metals from the crushed rock. In this "biological smelting," hundreds of thousands of tons of ore are piled over thirty feet high, and water that has been inoculated with bacteria is trickled through the pile, collected at the base, and pumped back to the top. After about six months, the bacteria have completely leached the available metals.[9] The cephalosporin antibiotics, a $6.5 billion worldwide market, are produced by the fungus *Acremonium chrysogenum*, which is grown industrially on sugars, oils, and oilseed meal. By changing the food source, the industrial operator can induce the fungus to change its growth pattern and begin making antibiotics.[10]

Industrial ecosystems like these are an increasingly dominant mode of production. As the price of oil that fuels the petrochemical industry climbs, and advances in

Figure 0.2
Fermenter Room at the Terre Haute plant of Commercial Solvents Corporation, 1929. In these tanks a strain of *Clostridium* bacteria first identified by Chaim Weizmann fermented corn starch to produce butyl alcohol, acetone, and ethanol. Each of the fifty-two tanks at the plant would be filled with 40,000 gallons of corn starch paste and 800 gallons of bacterial culture. After about twenty-four hours, "the whole content of the great tank is seething and foaming," and after forty-eight hours, the fermentation would be complete. *Source: Commercial Solvents Corporation*, 1929, Community Archives, Vigo County Public Library

genetic engineering provide more efficient biological production, the share of global production from industrial ecosystems is rapidly increasing. One study suggested that by 2010, industrial microbes would be responsible for one-fifth of all global chemical production, totaling $1.6 trillion.[11] This rapid increase is creating conflict in its wake. As biofuels become an increasingly important source of energy, people worry that their production will usurp a large proportion of the food grain market, driving up prices and fueling conversion of forests to cropland. As biotechnology generates novel organisms for producing pharmaceuticals and other products, activists are concerned about the patenting and privatization of fundamental products of nature, including the human genome. As sewage treatment plants apply their waste sludge as fertilizer and soil conditioner to agricultural fields, they also apply the trace quantities of heavy metals and other toxics concentrated into the sludge. As the industrial ecosystem expands to include much of the wild rivers, forests, and oceans, the logic of the industrial ecosystem comes to dominate the natural world as well.

In this book, I use the history of the biological sewage treatment plant to trace the origin and growth of the industrial ecosystem and to explore its logic. I examine the forces that shaped the ecosystems in the plant itself. Given that societies decided that sewage should be treated, why did sewage treatment take the particular form it did? To answer that question, I examine the sewage scientists, engineers, and sanitarians who theorized, designed, and built sewage treatment plants, as well as the operators who managed them.

"Like Sailing on Top of a Cesspool"

Sewage treatment arose in nineteenth-century England, where the twin processes of urbanization and industrialization first accelerated. London, Birmingham, Manchester, and other cities increased in size exponentially during the nineteenth century. Industrial discharge and human excrement overwhelmed the technical, legal, and administrative systems already in place for dealing with waste and nuisance. At the same time, advances in chemistry and microbiology and the birth of the science of ecology led to the scientific understanding of decomposition as a biological process. The biological sewage treatment plant, as an explicitly biological entity, or ecosystem, came out of these parallel developments.

By polluting water supplies, sedimenting rivers, and creating both a foul nuisance and a public health problem, sewage disposal became one of the most taxing problems facing the industrial city, first in nineteenth-century England, and later across the Continent and in the United States.[12] In the first part of the nineteenth century, human waste was collected in each household in a privy vault or cesspool, to be periodically carried away by scavengers to either be used as "night soil" or fertilizer or dumped on empty lands or in nearby water bodies (figure 0.3). Even water closets, introduced in the early nineteenth century, often emptied into cesspools rather than sewers. Sewers, where they existed, were often covered-over stream channels or other natural drainage ways, and were primarily for carrying away storm drainage, not waste. Indeed, it was often illegal to place household or sanitary waste in the sewers.

The broad acceptance of the water closet had to await the widespread availability of municipal water systems. But the combination of piped water and the flush toilet soon overwhelmed the capacity of the cesspools and privy vaults, as well as the early sewer drains. In response to the sanitary problems of cesspools and other disposal methods that kept waste in the vicinity of households, sanitarians began to advocate for a water carriage system, in which wastes would be transported by water from water closets and sinks through networks of self-cleansing sewers. By the mid-nineteenth century, cities began to expand their systems of sewers to efficiently drain both storm water and waste. London built its intercepting sewers in 1865, Brooklyn, New York, in 1855, Chicago in 1859. By the end of the century, the great majority of cities in Europe, Great Britain, and North America had sewerage systems.

But by combining waste with the water used to convey it through the sewer system, Victorian sanitary reformers created a new problem: sewage. Sewage was a new and distinct substance from the wastes that emptied into privy vaults and cesspools. That material was primarily human feces and urine. In contrast, sewage was a highly variable substance, containing the "solid and liquid excrements of the population," but also "the ingredients of soap, the refuse from kitchens, the drainings and washings from markets, stables, cow-houses, pigsties, slaughterhouses, etc., the refuse drainage from many factories and trading establishments, the washings of streets and other open surfaces." What before had been an often barely manageable volume of waste was now mixed with two hundred times its weight of water.[13]

Most cities took the expedient solution of simply dumping the liquid waste in the nearest body of water. When the fouled rivers became intolerable, sewers were extended to move the sewage farther away. In London, sewage flowed through an outdated network of ancient and more recently built sewers to the Thames River. With the increasing flow of sewage, concern over the quality of the Thames began increasing in the 1820s, reaching a crescendo following the "Great Stink" of 1858. The solution for Londoners was to build intercepting sewers, large sewers that captured the flow of all of the smaller sewers draining toward the river, and carrying that flow along the banks of the Thames to points downstream of the city. Completed in 1865, this system greatly improved the quality of the Thames in the city. But the problem of pollution was simply concentrated and transferred downstream.[14]

Nineteenth-century documents describe the horrific state of the rivers downstream of large cities. Boating on the Thames was "like sailing on top of a cesspool,"

FIG. 2

MANCHESTER CORPORATION DRY ASH CLOSET
a.a. *Screens to separate cinders from ashes and to direct the latter into the excrement pail.*

Figure 0.3
Dry ash closet. Ashes were used to help deodorize excrement in the dry earth system of waste disposal. Before the common use of water closets and sewers created the sewage problem, households used various privies, cesspools, or ashpits for collecting and removing excrement. The dry earth closets were an improvement on these early disposal systems, and competed with the water carriage system. In some cities, like Manchester, the dry earth system persisted long after sewerage became widespread. These systems have been recently revived in composting toilets and advocated in books like *The Humanure Handbook*. *Source*: Samuel M. Gray, Proposed Plan for a Sewerage System, and for the Disposal of the Sewage of the City of Providence, 1884

wrote engineer John Baldwin Latham evocatively. As one riverman on the Thames put it, "You could see lumps of stuff rise from the bed of the river, and then they would break open, and a fearful stench came from them when they broke open." Thames harbor masters described the river as "nearly black at half tide" and "little better than an open sewer." The nuisance was "spoken of on all sides" of the river. As the pollution killed the fish in rivers, fishermen had to find new livelihoods. One London fisherman, Henry Jones, was able to continue making a living off the river. He became a scavenger, collecting the fat and grease that floated on the river and making what some called "Thames mud butter." He and a handful of other scavengers steamed the grease in barges on the river and sold it for lubricating oil, or more fearfully perhaps, "Dutch Butter," reputed to be the "slimy sewage of the Thames . . . sent to Holland and from thence imported back to London markets" for human consumption.[15]

Industrialization and urbanization came later to the United States, and the resulting sewage as well. But soon these problems led to the widespread pollution of rivers and lakes. In 1904, Upton Sinclair described one arm of the Chicago River, known as "Bubbly Creek," downstream of the meatpacking district (figure 0.4). "One long arm of it is blind, and the filth stays there forever and a day. The grease and chemicals that are poured into it undergo all sorts of strange transformations, which are the cause of its name; it is constantly in motion, as if huge fish were feeding in it, or great leviathans disporting themselves in its depths. Bubbles of carbonic acid gas will rise to the surface and burst, and make rings two or three feet wide. Here and there the grease and filth have caked solid, and the creek looks like a bed of lava; chickens walk about on it, feeding, and many times an unwary stranger has started to stroll across, and vanished temporarily."[16]

"The New Problem of Sewage Disposal"

The creation of this new substance called "sewage" with all of its attendant problems led to "the new problem of Sewage Disposal," explained C.-E. A. Winslow, a leading American sanitary engineer, in 1915.[17] The water carriage system had removed wastes from households and neighborhoods, but had transferred the problem to "the end of the pipe," as noted by Martin Melosi. Because of its enormous volume and highly variable makeup, the disposal of sewage was far more complex than the disposal of privy waste had been.

The first comprehensive examination of the sewage disposal problem was initiated in 1857, with the establishment of a British Royal Commission to investigate the ways British cities might dispose of their sewage. The commission concluded that "the present state of sewage outfalls in many towns give rise to nuisance and danger of a formidable character."[18] Other commissions followed that investigated

Figure 0.4
Bubbly Creek at Morgan St., Chicago, 1911. The water carriage system, coupled with industrialization and urbanization, led to severe pollution of rivers, lakes, and estuaries. Bubbly
Creek, near the stockyards of Chicago, was so named because the gases of putrefaction from
sewage and industrial waste would constantly rise to the surface. Conditions like this led to
the demand for sewage treatment. *Source*: Chicago History Museum, *Chicago Daily News*
negatives collection, DN-0056839

the state of rivers in the United Kingdom, the efficacy of the various sewage disposal
technologies, and potential laws and administrative structures that could solve the
problem of pollution caused by sewage.

The reports of these commissions were highly influential, both in Great Britain
and abroad. Sanitarians in the United States paid close attention to the so-called
Blue Books and developments in Britain. U.S. cities sent delegations to Great Britain
to investigate the advances in treatment technology. In the United States, recently
established state boards of health began to investigate sewage problems, and individual cities convened commissions and conducted studies to find solutions to their
pollution crises. American cities hired British sanitary engineers to conduct studies

and draft reports on their sewage situations. In both countries, constructing sewage works was primarily left to individual cities. As a result, cities' engineers began conducting research into sewage treatment, and many of the most important developments were the result of this publicly funded and conducted research.

Despite the reports of the commissions and other investigative bodies, it was not clear that any single best means of treating sewage existed.[19] Various municipalities, based on their local situation and political and legal pressure to treat sewage, tried many different techniques, none with perfect success. When the 1857 Commission on Town Sewage issued its final report in 1865, it declared that the best method for treating sewage was to apply it to land, and land treatment or sewage irrigation became established as the primary means for treating sewage in England. Irrigation mirrored previous uses of human waste as fertilizer, and it involved spreading the sewage water over land, irrigating and fertilizing the soil simultaneously. There, the soil was thought to act as a filter that removed the solid particles in sewage and purified the effluent. The city of Edinburgh, regarded as the first city to use irrigation extensively, had disposed of its sewage in the Craigentinny Meadows since the beginning of the century, where it fertilized lush crops of grasses. Many other towns and cities in England and on the Continent followed. Cities like Berlin and Paris managed large farms that used and purified those cities' wastes. In a rapidly urbanizing world, however, sewage farms also required a large amount of expensive land. In humid climates like England, the water in sewage often presented a further problem. Most times of the year, there was plenty of rain, and the additional water in sewage simply flooded farm fields, leaving the land "sick," with sewage pooling on its surface.

The second major royal commission on sewage, the 1868 Commission on River Pollution, was dominated by the work of Edward Frankland, one of the most prominent chemists of his era.[20] Frankland conducted influential experiments on the ability of soils to filter sewage. Using glass tubes 6 ft. long and 1 ft. in diameter, filled with mixtures of sand and soil, he passed sewage taken from the London sewers through the tubes. Frankland found that sewage was only purified effectively if air were allowed to enter the soil. He concluded that the purification process was one of chemical oxidation.[21] Engineer John Bailey Denton built on this suggestion to develop a practical system of intermittent downward filtration in which extensive beds of soil would be allowed to rest between applications of sewage, allowing air to enter the soil and replenish the oxygen used to oxidize the sewage. Denton also abandoned the idea that growing crops was necessary for purification. Intermittent filtration thus severed the connection between farming and land treatment.

As an alternative to land treatment, chemists developed a myriad of techniques for separating out the solid particles in sewage by adding various kinds of chemicals, like aluminum or manganese salts, that would precipitate out the materials in suspension. The water, cleared of these solids, could be discharged into rivers, leaving what chemists (and investors) hoped would be a valuable fertilizer in the precipitated solids. Chemical precipitation was an industrial solution to the problem of industrialization and urbanization. Precipitation, however, was severely criticized, both because it failed to purify the effluent of the dissolved contaminants, and because it was often too expensive, costing more than the value of the fertilizer produced.

Beginning in the late 1880s, sanitary scientists began to understand the role of microorganisms in purifying sewage in soil. Engineers and scientists at the Lawrence Experiment Station of the Massachusetts State Board of Health were critical in this development. They built on the ideas of Frankland and Denton on downward filtration, but they also incorporated the recent advances of Pasteur and Darwin in microbiology and evolution. They began to understand the role of microorganisms in purifying sewage and developed explicitly biological processes of sewage treatment. Intermittent downward filtration soon morphed into the biological filter (figure 0.5). At the same time, London chemist William Dibdin began experimenting with biological treatment using what he called the contact bed. Both the biological filter (also called a trickling filter or percolating filter) and contact bed mimicked the treatment of sewage on land but concentrated it in a smaller area. In these systems, engineers built artificial beds of sand, gravel, clinker, slate, or other material and poured the sewage into the tank. Bacteria and other organisms grew on the material and purified the sewage.

These methods all relied on aerobic bacteria that required oxygen. Another set of processes developed in the 1890s and first decade of the twentieth century utilized anaerobic bacteria that were able to grow in the absence of oxygen. The septic tank, invented by Donald Cameron, the cultivation filter bed of W. Scott Moncrieff, Arthur Travis's hydrolytic tank, and the Imhoff tank were all closed tanks in which anaerobic bacteria could thrive. These bacteria tended to digest or liquify the sewage solids, reducing the amount of solid material in the waste stream, which could then be more efficiently treated in the various aerobic sewage filters.

In 1914, chemists Gilbert Fowler, Edward Ardern, and William Lockett of Manchester, UK, introduced the last major innovation in biological sewage treatment: the activated sludge process. In this process, sewage flowed into tanks which were bubbled with air to maintain aerobic conditions. Large populations of bacteria grew in the roiling tanks of sewage, feeding on the ammonia and organic matter. When the air was turned off or the sewage pumped to a quiescent clarifying tank, the bacteria would settle, cleansing the sewage of solids as they settled to the bottom of the tank, and leaving a clear effluent. Like a sourdough starter or the "mother of vinegar" used to make vinegar, the collected bacteria at the bottom of the clarifier would be recycled back into the activated sludge tank to keep the process going.

BACTERIAL SEWAGE PURIFICATION
ADAMS' PAT. " CRESSET " REVOLVING DISTRIBUTORS

The Broxburn and Uphall Sewage Purification Works

Figure 0.5
A trickling filter, Broxburn and Uphall Sewage Treatment Works. These circular beds would be filled with stone, slate, clinker, or other material. Sewage was sprayed over the top in a constant stream by the rotating arms. Bacteria that grew on the surfaces would purify the sewage as it trickled through the bed. Other microscopic organisms as well as insects and snails would also live on the filter, creating a complex ecosystem. *Source*: *Bacterial Sewage Treatment*, catalog from Adams Hydraulics, 1921, Milwaukee Metropolitan Sewerage District

The Contradictions of Biological Sewage Treatment

By 1914, then, the main processes of sewage treatment—the septic tank, the biological filter, and activated sludge—were established. These processes quickly dominated sewage treatment and remain the dominant treatment processes to this day. As Christopher Hamlin points out, though, the application of biological theories to sewage treatment was not a simple case of applying scientific knowledge to sanitary "problems" or of improvements in scientific understanding yielding corresponding improvements in treatment. Rather, these new treatment methods were adopted in a complex social context in which supporters of a variety of purification methods battled on political, cultural, economic, and scientific grounds. Sewage science was a highly contentious arena.[22] Physicians, scientists, engineers, public health officials, and treatment plant workers all argued over almost every aspect of the design and operation of treatment plants.

As these participants fought over the natural or artificial nature of the biological sewage plant, whether sewage processes could be patented and natural processes made private, whether sewage plants should be managed using scientific theory or industrial craft, or if sewage treatment should value potential profit from fertilizer over the greater purification of sewage, their disagreements highlighted the fundamentally contradictory nature of the biological sewage treatment plant and the industrial ecosystem more broadly. The resolution of these multiple contradictions occupied scientists, physicians, engineers, and industrial workers over the following century of development, construction, and operation of sewage treatment plants and other industrial applications of microorganisms. How these contradictions played out had enormous implications not only for the practice of sewage treatment but for other industrial ecosystems, including modern biotechnology. In attempting to resolve these contradictions, sewage workers created a new, hybrid form of nature, the industrial ecosystem.

Hybrid Nature is organized around the multiple contradictions that define the industrial ecosystem. Each chapter focuses on one of these contradictions, to produce overlapping narratives of the history of the sewage treatment plant. The sewage treatment plant was the most important industrial ecosystem, but its history paralleled and influenced other industrial ecosystems like brewing, chemical manufacture, and biotechnology. I follow these other systems as well in this narrative. Throughout I focus on the connected histories of sewage treatment in England and the United States. Most of the techniques for treating sewage were first developed in England. As industrialization spread from England across the Atlantic, so too did industrialization's environmental problems. Theories of microbial action in biological sewage treatment also crossed back and forth across the Atlantic as did many individual sewage scientists and engineers. Just after English researchers introduced the activated sludge process, World War I interrupted the progress England was making in sewage treatment, and the United States moved to the fore in research and development. For the period after the war, my focus shifts more heavily to the United States.[23]

The industrial ecosystem combined the natural and industrial in new, dynamic ways. In chapter 1, I examine the central contradiction of the industrial ecosystem, whether it was or should be considered a natural or artificial environment. Sanitarians of the nineteenth century placed great rhetorical importance on the "natural," and much of the justification for the treatment of sewage on land came from this understanding. In contrast, biological sewage treatment was originally characterized as an artificial process. Only through the naturalization of biological sewage treatment was it able to compete politically with land treatment. Coupled with this naturalization, however, was a drive to industrialize the biological processes. Scientists and engineers sought to accelerate, intensify, and regulate the biological activity of the treatment plant. The development of biological sewage treatment from land treatment to biological filters to activated sludge was thus a simultaneous process of naturalization and denaturing.

Chapter 2 examines the long-running conflict among engineers over the ethics of patenting and the public or private nature of fundamental biological processes. Both the septic tank and activated sludge were patented by English engineers who sought to enforce their patents in the United States, and many sanitary engineers organized nationally to resist the sewage syndicates and block the patents. These engineers viewed the patenting of sewage treatment processes as a transfer from public to private hands of a technology critical for the public health. This conflict exposed deep professional divisions between engineers. Municipal engineers viewed their work from the perspective of the public interest and criticized their colleagues who had patented advances in sewage treatment. For engineers working for industrial firms, in contrast, patent monopoly was a key business strategy. The challenges to the sewage patents were eventually defeated in the courts, with important implications for industrial microbiology in general, as well as modern biotechnology.

In chapter 3, I move into the plant itself and examine how it was managed by both engineers and workers. The category "worker" was fundamentally ambiguous, used to refer to both the human labor operating the plant as well as the bacteria responsible for the biological transformations. In a professional struggle for jurisdiction, sanitary engineers, using scientific process control, and operators, using factory craft, competed to exert control over the new space of the sewage treatment plant. Sewage professionals were from a diverse mix of disciplines, including chemistry, agriculture, medicine, public health, biochemistry, ecology, and engineering. A history of sewage treatment is also a history of professional identity and conflict. Through professionalization, training, and certification, engineers sought to control the unruliness of both bacterial and human labor by applying the principles of scientific management and laboratory control. At the same time, operators developed their own techniques for managing the plants that relied more on experience, close observation, sight, and smell.

The production and marketing of sewage fertilizer is the focus of chapter 4. Cities often hoped to profit from sewage treatment by recycling the sewage as fertilizer. In the mid-nineteenth century, a number of commercial firms were established to profit from sewage purification. These hopes, however, were repeatedly dashed as cities confronted what many thought to be a fundamental contradiction between purification and profit. Cities could either purify sewage, but at great cost, or produce a marketable fertilizer, but at the expense of adequate treatment. With the invention of the activated sludge process, hopes were renewed that sewage treatment could be profitable. Most notably, Milwaukee, Wisconsin, embarked on an almost century-long effort to produce and market sewage sludge. But by entering the marketplace,

Milwaukee has been subject to the fundamental contradictions of purification and profit as well as the contradictions of capitalism.

Chapter 5 examines the sewage treatment plant ecosystem roughly from the passage of the 1972 Clean Water Act in the United States to the present. Despite a century of biological sewage treatment, the contradictions of the industrial ecosystem have persisted and continue to dramatically shape society's response to sewage pollution. Privatization, scientific control, profit, and ideas of nature all became intertwined as sewage treatment plants struggled with the increasing demands placed on them. What has remained constant, however, is the continued importance of the living organism in sewage treatment. But as the pressures placed on the sewage treatment plant to deal with newer and more pollutants have increased, the industrial ecosystem is teetering. Despite the diversity and adaptability of the bacteria responsible for treatment, the living organism is not infinitely malleable.

In chapter 6, I broaden the focus and trace the importance of sewage treatment to modern biotechnology and explore how the contradictions of the industrial ecosystem persist in and structure industries based on genetic technology. The sewage patent cases played a crucial role in the landmark Supreme Court case *Diamond v. Chakrabarty* that laid the intellectual property framework for biotechnology. This case ruled that for the first time living organisms themselves could be patented. The decision was based on legal precedents first established in the sewage patent cases, but also on the cultural work that both naturalized and denatured microbial processes. As the "natural" became incorporated into the industrial ecosystem, the industrial moved into the wild. I trace how genetic material from activated sludge plants has become incorporated into wild species. The industrial ecosystem has literally hybridized with the wild.

Finally, in the conclusion, I consider the hybrid nature of the industrial ecosystem more explicitly. The biological sewage treatment plant and the industrial ecosystem are hybrids, systems composed of elements of both the natural and artificial. As sanitary engineers and industrial microbiologists have been creating these hybrids, society has often been unable to recognize and come to terms with their hybrid nature. As the logic of the industrial ecosystem expands to include much of the larger biosphere, we are hybridizing the industrial and the wild and extending the contradictions of the industrial ecosystem to include much of the "natural" world of rivers, forests, and oceans. Coming to terms with these hybrid ecosystems will be essential to prevent their uncontrolled proliferation.

1

Natural vs. Artificial: "The Right Way to Dispose of Town Sewage"[1]

At the close of the nineteenth century, prominent physician and sanitarian George Vivian Poore spoke to a London medical society on urban sanitation: "We see the pipes, the engines, the ventilators, the hospitals, and the smoke of the destructor; we hear the incessant thud of steam machinery." "But," he continued, contrasting this industrial scene to the healing powers of nature, "we never get a glimpse of the bright side of the matter, the return which Nature inevitably makes to nourish our bodies, gladden our senses, and freshen the air."[2] Describing a scene that might have come out of Charles Dickens's *Hard Times*, Poore was drawing upon a well-established literature contrasting the nineteenth-century industrial and pastoral landscapes of England.[3] However, Poore was not describing, like Dickens, the steam engines, smoke, and polluted rivers of the woolen mills, dye factories, and machine shops of England's industrialized cities. Rather, he was criticizing the industrial nature of the sanitary apparatus itself. Sanitarians were responding to the health problems and river and air pollution caused by industrialization by building their own industrial apparatus: vast networks of sewers and pumping plants, huge furnaces for incinerating the sewage of cities, giant schemes to treat sewage with the products of England's expanding chemical industry (figure 1.1).

Other sanitary scientists and engineers, however, saw industrialization of the sanitary apparatus as a necessary response to the impact of urbanization and industrialization itself. "The requirements of civilized man have created certain artificial conditions which can only be met by corresponding artificial treatment," wrote a correspondent to the *Times*, also in 1898. Unlike Poore, though, these sanitarians did not see the solution as pitting the industrial against the natural. Rather, they proposed a hybrid solution, in which the natural processes of purification advocated by Poore would be put to work in concentrated form. In creating their solution to the sanitation problem, these writers advocated for the recently developed processes of biological sewage treatment, in which "the law of nature need not be transgressed or departed from."[4] Rather nature would be used, improved upon, sped up, and intensified. Biological sewage treatment, as envisioned by writers like these, combined the

Figure 1.1
"The smoke of the destructor." This 1898 advertisement appeared in a sanitary engineering textbook, showing the industrial nature of the sanitary apparatus at the close of the nineteenth century. Destructors were used to incinerate both refuse and sewage sludge. *Source*: William H. Maxwell, *The Removal and Disposal of Town Refuse* (London: Sanitary Publishing Company, 1898)

artificial and the natural to create a new form of nature, the *industrial ecosystem*. From the sewage farms and irrigation of the mid-nineteenth century, to the septic tanks and bacterial filters of the 1890s, and the activated sludge process of 1914, sanitary scientists and engineers created ecosystems where the microbial purification of a "natural" soil became progressively intensified and simplified, in a word, industrialized.

The new sewage treatment schemes of the late nineteenth century relied on the biological action of bacteria and other organisms. But sanitary scientists and engineers of the period did not, at first, necessarily consider these biological processes to be natural. In contrast to what was regarded as "natural" treatment of sewage by the soil, biological sewage treatment was originally considered "artificial." Because of the perceived importance of naturalness to sewage treatment schemes, biology had to undergo a cultural and scientific process of naturalization.[5] During the late 1880s, scientific research and sanitary rhetoric had so naturalized biological treatment that by the early 1900s it was widely considered natural. Critical to this transformation was the application of Darwin's theory of natural selection to understanding the ecology of sewage treatment processes.

At the same time, however, sewage scientists denatured biological treatment through a drive for increasing industrialization of the process, characterized by biological intensification, simplification, and control. The shift from a natural to an industrial ecosystem was a process of denaturing—of removing the natural elements of an ecosystem and replacing them with elements that would allow the simplification and intensification of the biological process.

Denaturing was an essential characteristic of industrialization in general, because nature, with all its variability, was an obstacle to mechanization. To take advantage of the increased speed and power made available by steam, the unevenness of nature and natural processes needed to be removed. Variability was an obstacle to industrialization, and key inventions during the Industrial Revolution were designed to overcome this variability. Handspun threads were strong in some places, weak in others, and would break during the stresses placed on them by mechanical looms. The spinning mule produced a more even thread, amenable to mechanization. Iron and copper ores, roasted in large heaps during smelting, produced an irregular product "as the interior of the heaps is usually heated to excess whilst the outer parts are sometimes underheated." These open heaps were replaced by kilns and reverberatory furnaces that allowed "a more perfect control of the temperature, thus yielding a more uniform product."[6] Simplification and control went hand in hand with the intensification and acceleration of production processes made possible by mechanization. All of these processes of denaturing were designed to remove variability over both space and time. This "annihilation of space and time" is a central part of capitalism, industrialization, and modernity.[7] Denaturing was such an essential part of industrialization that it became an ideological goal in its own right, pursued for its own sake.

Because of this simultaneous denaturing and naturalization in the industrial ecosystem, the artificial and natural were in constant tension in the biological sewage plants, breweries, and bacterial fermentation plants created during this period. This tension helped drive the development of sewage treatment and the industrial ecosystem over the next century and to the present. The balance between the industrial and the natural was constantly shifting, fought over by scientists and engineers, ruled on by government commissions and the courts, and ultimately limited by the living organisms themselves.

Supporters and critics of modern biological treatment methods all sought to marshal arguments from nature to support their favored processes, meanwhile denigrating others' as artificial. But the categories "natural" and "artificial" were complex and highly charged, exposing conflicting ideals of nature and industry, rural and urban, moral and immoral, sacred and profane. The strength of the conflict between opposing views toward sewage stemmed from a variety of sources. Christopher Hamlin notes that questions of water and sewage treatment were often heard in the adversarial venues of the courtroom and parliamentary inquiry. The adversarial nature of these proceedings was then carried over into meetings of scientific and professional societies where "scientists refused to look upon one another's data as faithful representations of nature or to regard one another's conclusions as honest, impartial generalizations."[8] Underlying these controversies were also questions of old-fashioned greed. Many proponents of various treatment schemes had a pecuniary interest; some were patentees, others had financial interests in particular plans, others consulted for sewage treatment companies. Nevertheless, even when pecuniary interests were not involved, the conflict between natural and artificial processes was intense.

These debates, I argue, reflected different ideological stands on the questions of modernity and industrialization. Opponents of bacterial treatment or pure culture techniques argued for the preservation of traditional systems of agriculture, labor, and craft. The "artificial" techniques associated with industrialization represented attacks on these traditional modes of production. Proponents also had an ideological commitment to the project of industrialization and denaturing, seeing the rationalization of the process as a goal in itself, despite the failure of many efforts of simplification and control to improve treatment.

As each new biological treatment process was introduced, there was a simultaneous naturalizing and denaturing of the purification process. The new processes were naturalized by connecting them to a continuum of older, natural processes, and progressively denatured by simplifying and concentrating the biological activity. Activated sludge represented the most industrialized form of bacterial treatment;

yet engineers sought to denature and control the process even further. In the end, engineers were unable to simplify and denature the ecosystems at will. The systems themselves resisted the ultimate denaturing.[9]

Land Treatment as the Right and Natural Way to Dispose of Sewage

Chadwick's industrial vision for sanitation, the water carriage system so decried by Poore, was only partially implemented. Many cities and towns in England built the water distribution system, built the water closets and sewers, but neglected to purify the resultant sewage, instead dumping it in the nearest river or estuary. With the twin processes of industrialization and urbanization, the concentration of human and manufacturing waste created "an evil of immense magnitude."[10] British government, responding to the pollution crisis, repeatedly established a number of investigatory commissions to propose remedies for the destruction of the rivers of England. The Sewage of Towns Commission's 1865 report established land treatment as the sole effective method for purifying sewage in England. "The right way to dispose of town sewage is to apply it continuously to land, and it is only by such application that the pollution of rivers can be avoided," the commission asserted.[11] To control pollution, the commission advocated large sewage farms that would be irrigated with town sewage. As the sewage filtered through the soil, it would be purified by nature, through a mechanism that, while not entirely understood, was well grounded in observation as well as both theory and theology (figure 1.2).

Much of the justification for land treatment as the "right way to dispose of town sewage" was an understanding of sewage irrigation as the "natural" way to dispose of town sewage. "The 'law of nature' has made no provision whatever for the disposal of the sewage, otherwise than that of restoring it to the soil," declared a correspondent to the *Times* in 1868. "It is idle in the extreme to attempt otherwise to deal with this great question."[12] As Christopher Hamlin put it, commentators of the period sought "to demonstrate that the laws of Nature were indeed consistent with the goals of Victorian civilization and the sanitary options available to Victorian cities."[13]

Sanitarians used a variety of natural metaphors to justify this practice. Even before the importance of soil microorganisms in purifying sewage was understood, the soil was imbued with living characteristics, and its ability to purify sewage was characterized as a natural property of a living soil. Chadwick's vision of circulating the waste of cities to farms and then back underlay the idea of sewage farming, and the organismal metaphors of arteries, veins, and circulation they used helped to naturalize that ideal. If the pump that moved water and sewage through Chadwick's circulatory system was the heart, then the soil was the lung. In a highly influential study in the report of the Rivers Pollution Commission in 1870, Edward Frankland, one of England's leading chemists, invoked these bodily metaphors: "A field of

Figure 1.2
Sewage farm, near Barking, 1868. This sewage farm was established at the northern outfall of London's drainage scheme by the Metropolis Sewage & Essex Reclamation Company when their attempts to raise capital for a more ambitious scheme failed. In sewage farming, the sewage was first settled in a tank to remove the larger mineral and organic particles and then distributed via channels to different parts of the farm. Sewage farms ranged from relatively small farms on the scale of a single estate, like this one, to tens of thousands of acres treating the sewage of large cities like Berlin and Paris. *Source*: *Illustrated London News*

porous soil irrigated intermittently virtually performs an act of respiration, copying on an enormous scale the lung action of a breathing animal." Although Frankland used this organic metaphor, he saw the purification process as one of chemical oxidation, and not, in fact, the action of living organisms.[14]

The practice also had its roots in the disposal methods for waste prior to the construction of the water carriage system. Excrement had typically been disposed of using dry conservancy methods—the carting away of feces from privies as "night soil" to be disposed of on farms surrounding towns and cities. This disposal of feces on land had long roots in traditions that helped justify it as a natural practice. It mimicked the use of livestock manure as a fertilizer, and was thus rooted in the farming practices of rural England. Disposal methods were thus considered natural to the extent that they mimicked traditions of rural England.[15] Furthermore, dry conservancy had its analog in wild and domesticated animals that buried their own waste. Adopting the use of land to purify water-carried sewage thus seemed like a natural extension of both previous human practices as well as examples from the natural world. In addition to these metaphors of nature, supporters also drew on biblical exhortations and appeals to natural theology and the providential recycling of the nutrients in sewage. Through sewage recycling were connected the themes of morality, agricultural productivity, social advancement, and health. Ignoring recycling meant "pestilence, filth and drunkenness."[16] Arguing against disposal schemes that did not return waste to the earth, one sanitarian declared them a "flagrant violation of a fundamental law of Nature when we neglect to restore to Mother Earth what she has so liberally given us."[17]

By the 1860s, land treatment was so widely characterized as natural that sanitarians did not have to explicitly claim that status. Rather, they simply referred to the alternatives as artificial. In 1864, the Select Committee on the Metropolis Sewage stated that "no efficient artificial method has been discovered to purify . . . water which has once been infected by town sewage." "Soils," in contrast, the report continued, "have a great and rapid power of abstracting impurities from sewage water, and rendering it again innocuous and free from contamination."[18]

Engineers in the United States, relying heavily on the reports of the various British commissions, also adopted the distinction between natural and artificial methods. William Hall, California's state engineer, classified sewage treatment processes into "the application of sewage to land," on the one hand, and the "artificial treatment of sewage" on the other.[19] This contrast of "artificial" methods with "soils" or "land" reveals the unstated but powerful notion that treatment on soils was the natural means of purifying sewage. As sewage treatment technologies evolved, however, and biological processes laid the claim to a "natural" status, this unstated assumption about land treatment would become explicit, and the focus of escalating conflict among sanitarians.

Naturalizing Biology

Despite its official sanction, sewage irrigation often failed, creating more nuisance than it solved. Cities produced more sewage than could be assimilated by the land, and sewage farmers were forced to flood their fields with sewage even during heavy rains. If farmers applied sewage at too high a rate, the soil frequently became clogged or "sick," and residents near sewage farms complained of smells and feared disease. With continued pollution of England's rivers, cities were under increasing pressure to develop new methods for treating their waste. The recent findings of bacteriologists in the 1870s suggested new ways of thinking about sewage purification. As scientists began to uncover the biological basis for the purification of sewage in soil, "microbe agency" began to compete with land in the development of treatment schemes. At the same time, these new biological schemes began to compete with land for the appellation "natural." Many sanitarians at first considered the explicitly bacteriological methods of sewage treatment to be artificial because they were not accomplished using a natural soil. For the new microbial methods to be accepted, biology itself needed to be naturalized.

Scientists considered nitrification, or the conversion of ammonia to nitrate, to be the chief measure of sewage purification because well-nitrified sewage would not putrefy. Initially thought to be a process of chemical oxidation, nitrification in soils had been attributed to the direct action of air in the soil. It was not until the demonstration that nitrification was a biological process caused by "organized ferments" that sewage treatment came to be regarded as a biological process. In 1877, scientists Theophile Schloesing and Achille Muntz, pursuing ideas from Pasteur, studied the nitrification of sewage in soils. They slowly filtered sewage through a glass tube filled with sterile sand and limestone. Only after twenty days of filtration did they see nitrates in the effluent from the tube, indicating nitrification. They then sterilized the tube with chloroform vapors and saw the disappearance of the nitrates. With the subsequent withdrawal of the chloroform, nitrates did not appear until washings from fresh soil were added to the tubes. Schloesing and Muntz interpreted these results as proof that an "organized ferment" or microorganism in the soil was responsible for converting ammonia to nitrate. The delay in initial nitrification was due to the time needed for the population of the microbes to increase. Robert Warington, a British agricultural chemist, quickly recognized the importance of these results and publicized the French research to the English scientific community. He also began a series of experiments himself on the biological process of nitrification, establishing its relation to light, temperature, and other environmental conditions.[20] Warington was more interested in agricultural soils than in sewage, but Henry Robinson, a civil engineer interested in sewerage, publicized the findings of Warington to the sanitation community and explained their significance to the treatment of sewage using land.[21]

Despite the work of Warington and Robinson, as Christopher Hamlin observes, the implications of these findings for sewage treatment were mostly ignored by the mainstream sanitary community in England.[22] Instead, it fell to a new public health laboratory in the United States, the Lawrence, Massachusetts, experiment station. Sewage treatment first rose to importance in the United States in the Northeast, where urbanization and industrialization had first occurred. By the late 1870s, Massachusetts' rivers had become progressively polluted from increasing urbanization and industrial wastes.[23] The Massachusetts State Board of Health established an experiment station at Lawrence to investigate how sewage might best be disposed of.[24] The first director of the experiment station was Hiram Mills, an engineer who first rose to prominence as the hydraulic engineer for the Essex Company, which owned a dam and water power rights to the Merrimack River. Mills assembled a group of scientists from various fields, including biology, chemistry, and engineering to build on the advances in sewage biology and chemistry occurring in Europe.

Mills had rejected sewage irrigation as impractical in Massachusetts: even a small city like Lawrence would require a sewage farm one-fourth the area of the city itself. Rather, building on the experiments of Frankland, he thought that intermittent filtration might be a feasible alternative. According to his rough calculations, the area of land needed would be only one tenth that needed for a sewage farm. But where Frankland considered the process to be purely chemical, the Lawrence researchers, using the insights of Warington, Schloesing, and Muntz, examined the bacterial flora of sewage, the filters, and the effluent.[25] Mills built a number of indoor and outdoor tanks and filled them with sands, gravels, and soils typical of various regions of the state. Starting in 1888, using sewage from the city of Lawrence, the station began detailed biological and chemical studies of the sewage treatment processes in these tanks.[26] These studies were enormously influential in the United States as well as Europe and provided guidance on the design of what became known as biological filters. These studies were also instrumental in confirming the developing idea that sewage treatment on land was essentially a biological process, accomplished by microorganisms living in the soil.

At the same time, the manner in which the studies were conceived and carried out helped to make this newly understood process, carried out in outdoor tanks of prepared soil seem entirely natural. The Massachusetts experimenters emphasized the natural characteristics of bacterial treatment methods by comparing bacteria in sewage filters to bacteria in nature and by applying to intermittent filters the same ecological and evolutionary theories used to understand natural systems. These ecological approaches emphasized an understanding of organisms and environment that was rooted in ideas of adaptation and Darwin's evolutionary theory. Understanding sewage treatment as an ecological process was critical in naturalizing the artificial bacterial methods and creating the industrial ecosystem.

The biological studies were conducted chiefly by William T. Sedgwick, professor at MIT, and his former student Edwin O. Jordan. These bacteriologists took the critical step of connecting the bacterial organisms to their environment in studying sewage treatment. This approach, characteristic of the new direction in biogeography and physiology that was beginning to coalesce as ecology during this period, was critical to naturalizing biological sewage treatment and establishing sewage treatment plants as industrial ecosystems. The link between organism and environment was Darwin's theory of natural selection. By connecting the distribution of bacteria with studies of bacteria in their natural environment and ideas of natural selection, Sedgwick and Jordan naturalized the processes taking place in their experimental sewage treatment tanks. Both Sedgwick and Jordan were Darwinists. Sedgwick had graduated from the Johns Hopkins University in 1881, where he studied with Henry Newell Martin, himself a student of the noted anatomist and early Darwin supporter Thomas Henry Huxley. Sedgwick wrote an influential textbook on biology that connected the form and function of an organism to its environment.[27] Jordan was later to become one of the leading bacteriologists of the period and professor at the University of Chicago. His approach to microbiology was notable during the period for its emphasis on ecological relations in natural habitats of the bacteria, and adaptation and selection rather than simply medical applications.[28]

The relations of the organisms to their environment were key considerations for Sedgwick and Jordan in understanding nature, and they attributed them to evolution by natural selection. Jordan argued that one needed to understand the forces affecting the distribution of bacteria in their natural habitats of water, soil and air. "The species in these several classes are different," he argued, "because their conditions of life—food, temperature, etc.—are different in the different habitats."[29] Jordan viewed adaptation, the fit of the organism to its environment, as arising from the explicitly Darwinian process of natural selection. Bacteria in sewage were derived from the natural environment, Jordan argued, but not all species would be found. As bacteria differed in their various natural habitats, they would differ in sewage. Here, Jordan adopted an explicitly Darwinian perspective: "The sewage itself—a nutritive medium of varying composition and richness—will contain only those species capable of living and holding their own in the continual struggle for existence. So far as the conditions of life in sewage differ from conditions of life elsewhere, so far will the sewage be inhabited by species peculiarly adapted to those conditions . . . The chemical composition of the sewage undoubtedly debars some species from taking part in the contest." However, it was not only environmental conditions that could affect bacterial distribution. Competition among bacteria could also determine which species could survive. "Many species also which would perhaps grow in sterilized sewage are not able to exist in the presence of other and more powerful forms."[30] These phrases, "conditions of life," "struggle for existence," "contest," and "adapted," all derive from Darwin's *Origin of Species* and linked Lawrence's study of sewage treatment to Darwin's theory of evolution by natural selection.

Darwin's central metaphor, as argued by Robert Young, was "natural" selection. Darwin emphasized this term to explicitly compare selection by nature with the artificial breeding of the farmer.[31] Darwin used the well-known examples of "sports" and breeding to argue for a similar, but infinitely greater, capacity of nature to select forms or variants. By invoking the effect of the environment in selecting for adapted varieties in Jordan's studies of bacteria in sewage treatment, the Lawrence studies helped associate sewage treatment with the natural rather than artificial process of selection and helped naturalize biological sewage treatment. This connection between Darwinian selection and environment characterized the newly emerging field of ecology. Haeckel, in establishing the foundation of ecology, had explicitly connected it to Darwin and defined it as the "science of the struggle for existence."[32] Jordan's studies, examining the struggle for existence of bacteria in sewage, naturalized sewage treatment processes; the incorporation of ecological thinking established the sewage treatment plant as an ecosystem.

But sewage was not the only arena in which Darwin's theory was important in naturalizing the industrial ecosystem. The conflict between natural and artificial and the use of the metaphor of natural selection were also explicit during this same period in the brewing industry. Following Pasteur's research on fermentation in beer, Emil Christian Hansen, a Danish brewery scientist, had identified wild yeasts as the culprits in spoiled beer and developed a method of yeast culture in which, through laborious and meticulous laboratory techniques, the brewer could isolate and grow pure cultures consisting of only a single desirable variety of yeast. In these studies, however, Hansen emphasized not natural but artificial selection. As a result, other brewing scientists criticized his methods as artificial, in contrast to more natural ways of producing beer. The contest between natural and artificial appeared in almost identical terms as it had in sewage treatment.

Artificial selection was indeed at the heart of Hansen's program.[33] He began his book expounding his technique of yeast culture with an epigram from Darwin: "Man selects varying individuals, sows their seeds, and again selects their varying offspring." Hansen explicitly compared his work with that of the plant breeder. "It did not merely suffice to prepare a pure culture," he emphasized. Rather, "a systematic selection of the most suitable species or race must be made."[34]

With the perfection of these techniques in 1883, Hansen promoted pure yeast culture throughout the brewing world (figure 1.3). While brewers in many regions quickly adopted pure yeast culture, it met stiff resistance in others.[35] In northern Germany in particular, brewing scientists considered these pure yeast methods, based as they were on metaphors of breeding and artificial selection, to be "artificial" ("künstliche"). In its place, they argued for a "natural" ("natürliche") technique to develop pure yeast cultures, based not on artificial selection but on ecology and natural selection.[36] German chemist Max Delbrück criticized the pure yeast culture as an artificial manner of brewing and sought to produce infection-free cultures using methods based on the brewers' traditional techniques and not the artificial techniques of Hansen. Delbrück, like Sedgwick and Jordan, emphasized the conditions of life for the microorganisms in the industrial ecosystem. What naturalized Delbrück's process, similar to the naturalization of the sewage filters at Lawrence, was ecology and natural selection. By manipulating the ecological conditions of the wort, the acidity, aeration, temperature, and alcohol content—what he termed the "climate" —Delbrück could favor the growth of some organisms over others, producing a "natural pure culture" that avoided infections of yeasts and other microorganisms. Delbrück also explicitly invoked Darwin's language. Through the "struggle for existence"[37] among the yeasts and other microorganisms, Delbrück argued, the brewer could achieve a pure yeast culture in a more natural way.

Figure 1.3
Pabst Brewing Co., Milwaukee, Wisconsin. Breweries were perhaps the most common industrial ecosystem at the beginning of the twentieth century. In the 1880s, brewers began using
microbiological techniques to cultivate cultures of pure yeast rather than using wild strains.
Pabst was one of the first American breweries to use Hansen's pure yeast culture techniques,
beginning in 1887. *Source*: Jno T. Faber Publisher, Milwaukee, Wisconsin, postmarked 1908

Having established the framework for understanding the species composition of
sewage, soil, and effluent, Sedgwick hoped to elucidate the biological mechanism
of purification. He correlated the chemical changes taking place in sewage to the
populations of bacteria in the filter, from the time it was applied to the filter to the
time it left as effluent.[38] By examining the growths on the filters, Sedgwick identified the bacteria responsible for these profound changes in the sewage. "The presence of abundant bacteria in the uppermost layers of the sand of the several tanks,
coinciding as it does with the disintegration of these solid, and even living, elements
of the sewage, suggests that the destruction of the latter at this level is really due to
an active fermentation or decomposition, effected by the ordinary bacteria."[39] These
studies led to a preliminary understanding of sewage filter ecology. Sedgwick, like
Frankland and others before him, rejected the idea that filters were solely or even
predominantly mechanical devices. "A field of sandy soil may, it is true, be a very
effective strainer," he wrote, "but under the best conditions, there speedily begins a
change of the profoundest significance," as proteins are converted first to ammonia

and then to nitrate, or as organic matter is released as carbonic acid.[40] But Sedgwick also rejected the theory of Frankland, that the activity of filters was chemical in nature, the direct effect of oxygen in the soil on ammonia. Sedgwick concluded that the action of filters was fundamentally biological rather than chemical or mechanical. "It has now been definitely established . . . that micro-organisms are an indispensable element in the constitution of a successful intermittent filter," he stated, " so that the essentially chemical theory has given place to one essentially vital, or biological."[41]

"Nature's Own Methods"

The Massachusetts researchers demonstrated the ability of sewage filters to purify sewage to such an extent that the effluent could be discharged directly into rivers without causing nuisance. These studies, with hundreds of pages of data, were highly influential in Great Britain, and scientists and engineers working in a number of different cities began developing explicitly biological treatment schemes.[42] In each case, the idea of the "natural" was critical in the descriptions of the process. Due to the Massachusetts research connecting naturally occurring bacteria to the purification process, "biological" was coming to mean "natural," and "natural," "biological."

In 1892, the British journal *The Engineer* publicized the recent work of W. D. Scott-Moncrieff, an engineer in Ashtead, England. Scott-Moncrieff had developed what he called a "cultivation filter" for purifying sewage, in which microorganisms were cultivated in the tanks to accomplish the purification. By emphasizing the links between the cultivation of a field and the cultivation of the helpful bacteria in the name of the process, Scott-Moncrieff invoked the correspondence in British culture between the rural and the natural. But like Sedgwick, Scott-Moncrieff also relied on the idea of natural selection to naturalize his process, referring to "naturally-selected groups" of microorganisms, each "dealing with a differentiated food supply as the work proceeded."[43] This created a zonation in the bacterial processes, which not only mimicked "what occurs in the operations of nature," he claimed, but did so specifically through the process of natural selection. "The presence of these organisms in groups, differentiated by the changing character of their environments, represent the survival of the fittest in successive stages."[44] *The Engineer* explicitly declared the cultivation filter to be "natural." To the two "natural processes, [irrigation and intermittent filtration]," they wrote, "a third has now been added."[45]

At about the same time, the city surveyor for Exeter, Donald Cameron, also began developing a bacterial treatment scheme based on biological principles. His "septic tank" relied on the action of anaerobic bacteria to digest, or liquefy, sewage solids. Cameron too, sought to characterize his bacterial treatment scheme as natural. "No system of treating sewage can be satisfactory which does not follow as closely as possible the lines laid down by nature," he wrote. "I therefore set myself to devise a system in which the work of natural agents should be forwarded to the fullest extent. . . . In this system no chemicals are employed and there is no 'treatment' of the sewage in the ordinary sense of the term; its purification being accomplished entirely by natural agencies. . . The septic tank itself is merely a receptacle designed to favour the multiplication of micro-organisms, and bring the whole of the sewage under their influence."[46]

A third treatment system that emphasized natural action was the trickling filter, an outgrowth of intermittent filtration and the experiments at Lawrence by Salford engineer J. Corbett, who explicitly acknowledged the Lawrence experiments in the development of his trickling filter. Further, he had reportedly developed his distribution apparatus after observing rain falling on porous soil.[47] Similarly, William Dibdin characterized his "contact bed" as natural, in that it relied on the action of aerobic bacteria to treat sewage. Dibdin, a chemist for London's Metropolitan Board of Works, emphasized that cultivating and utilizing bacteria was essential to purification, and that "the antiseptic treatment of sewage," because it killed the active agents of purification, "is the very reverse of Nature's method."[48]

With the development of septic tanks, contact beds, and trickling filters, bacterial treatment methods had come to the fore, and many communities looked to abandon sewage irrigation for bacteria. But by doing so, they were not abandoning their commitment to natural purification. By 1900, summarizing the advances of the previous decade on the various bacterial treatment methods, sanitary and municipal engineer H. C. H. Shenton proclaimed that "bacterial treatment for the purification of sewage is identical with the natural process by which the whole surface of the earth is purified."[49] And in 1905, answering his own question about "whether or not the bacterial system of sewage purification has been proved to be satisfactory," Professor of Chemistry A. Bostock Hill replied, "The answer is simple and definite. It must be suitable, because it is simply a development of Nature's own methods."[50]

As in Lawrence, British sanitary scientists were beginning to understand these bacterial treatment systems in the same way they understood natural systems. The scientists and engineers believed themselves to be creating a kind of natural environment in the bacteria beds and tanks—an ecosystem—that harbored the very same organisms that occurred in similar environments in nature. For years, scientists had recognized the presence of certain species as indicators of sewage pollution in rivers. Scientists noted, for instance, that the sewage "fungus" *Sphaerotilus natans* was found exclusively in areas fouled by sewage. Sir Rubert William Boyce, a pathologist who founded the School of Hygiene at University College, Liverpool, was appointed bacteriologist to the 1898 Royal Commission on Sewage Disposal. Investigating the pollution of the River Severn, he drew the connection explicitly between the habitat of *Sphaerotilus* in rivers and in sewage disposal works.[51] He investigated the conditions required for its growth in nature—"circulation, aeration, presence of H$_2$S and warmth." Given the correspondence between organism and environment, he was thus not surprised to find *Sphaerotilus* occurring in the man-made habitat of the sewage treatment plant that provided those same conditions: "With the introduction of the bacterial bed method of treatment, this organism has also made its appearance" in the sewage treatment works themselves, reported Boyce.[52] He connected the action of the "lower forms of animal and vegetable life" in rivers to similar action in both land treatment, as well as the new bacterial methods: "The process of self-purification . . . has begun to attract great attention because in the various biological processes of sewage purification, now largely made use of, a similar purification is brought about," and brought about by the very same organisms—"Sphaerotilus, green flagellata, protozoa, and bacteria of all kinds."[53] By connecting the biology of sewage treatment plants to natural habitats, studies like these helped further naturalize the biological sewage treatment process.

Contesting the Natural

Biological treatment processes like Dibdin's, Cameron's, and Scott-Moncrieff's were introduced into a climate of intense competition and conflict between competing schemes for sewage treatment. Cities were under political and legal pressure to clean up their sewage and prevent the contamination of rivers and water supplies. At the same time, the proper technical response was by no means clear. Land treatment, broad irrigation, intermittant filtration, and chemical precipitation all had their proponents, and sanitarians debated the merits in court proceedings, parliamentary committees, and scientific meetings.[54] Biology appeared as a threat to each of these camps, but particularly the irrigationists. The various royal commissions on sewage had repeatedly declared that land treatment was the sole effective means of treating sewage, and irrigationists were loath to give up their primacy.

During the 1890s, as the growth of cities and sewerage systems produced ever greater volumes of sewage to be treated, the problems inherent in treatment on land began to be manifest. As cities grew, land on their edges that might be suitable for a sewage farm increased drastically in price. Sewage farms required an acre of land for every five hundred to one thousand people contributing sewage. A city like London would require farms of 40,000 acres.[55] Berlin's 23,000 acres of sewage farms, held up as the most efficient in Europe, were larger than city itself in 1896. Further, the soils surrounding Berlin were generally sandy, and thus ideal for filtering sewage, whereas many British cities were surrounded by clays or peats that filtered water slowly. Even where suitable land could be found at adequate prices, sewage irrigation farms often succumbed to "sewage sickness" in which land became clogged with particles from the sewage and excrement-filled water ponded on the surface.

"I have seen land so sick," quipped Dr. Sidney Barwise, "that when I [handed] up to one of her Majesty's judges a photograph of the farm in question, he asked me if it was the lake district."[56]

The new bacteriological methods of purification provided a potential way out for cities. Yet, for decades, the sanitary establishment in Britain had declared that land treatment was the only "right" way to purify sewage, and the Local Government Board would loan no money for the establishment of sewage treatment unless it incorporated land treatment as the final stage of purification. For British towns to receive loans for the construction of sewage disposal works required by new anti-pollution laws, they were forced to purchase large areas of land for sewage farms or intermittent filtration. As cities increased in size during the second half of the nineteenth century, and as promising, explicitly bacteriological treatment methods were developed, cities and towns chafed under the continued insistence on land treatment.[57] One district councillor wrote to the *Times* complaining that his council had "determined to treat the sewage of a town of a thousand inhabitants, which at present pollutes an important stream, by what is known as 'Dibdin's Biological Process,'" but were prevented by the Local Government Board's refusal to provide loans for any scheme that did not include land treatment as the final purification process.[58] Larger cities were even more critically constrained.

Hearings before the Local Government Board in 1897 on Cameron's septic tank in Exeter were described as "the opening shot fired on behalf of the new system of bacterial treatment."[59] Indeed, cities like Manchester and London forced the government to revisit the question of land treatment and establish a third Royal Commission on Sewage Disposal in 1898. This was the first commission to convene since the introduction of explicitly biological treatment processes. The commission's charge was to answer three questions: "Are some sorts of land unsuitable for the purification of sewage?" as cities like Manchester argued. If that was the case, the commission further asked, "Is it practicable uniformly to produce by artificial processes alone an effluent which shall not putrefy, and so create a nuisance in the stream into which it is discharged?" Here, by the term "artificial," the commission meant both biological treatment and precipitation. Finally, given its evaluation of the treatment methods, "What means should be adopted for securing the better protection of our rivers?" S. Rideal interpreted the charge to the commission in the manner most threatening to irrigationists. The question was "whether it was possible to replace the old sewage farms" with "modern bacterial methods."[60]

Even asking the question whether "artificial" processes were satisfactory was a threat to sewage farming and its proponents. The battle between irrigationists and proponents of biological treatment quickly centered on ideas of the "natural" and "artificial." In the very charge to the commission from Parliament, methods for sewage treatment were divided into two categories: "artificial processes," and "land" treatment, which was in contrast implicitly natural. Because of this classification of methods, the commission became the battleground in which the debate over natural and artificial processes was played out. "Nature" held great power to these scientists and engineers as a guide, and opposing camps argued strenuously for the imprimatur of nature to defend their views. The idea of the natural was both an important guide for the sanitarians—they really did believe that there was a virtue in copying nature in their treatment schemes—as well as a rhetorical device to elevate their preferred treatment schemes, or denigrate competitors. Partisans of the various sewage schemes all testified before the Royal Commission, but while the commission deliberated, they also took their arguments to the public, publishing a series of books and essays arguing for their perspective. In all of these polemics, the naturalness or artificiality of treatment processes was the central argument between proponents of land application and the more recent bacterial methods.

Because of the cultural and scientific work on biological treatment, by the last years of the nineteenth century, many sanitary scientists understood bacteria beds, filters, and tanks as thoroughly naturalized. For instance, in 1898, two sanitary scientists titled their paper on bacterial treatment of sewage in filters and septic tanks simply "Some Observations on the Natural Purification of Sewage,"[61] a title that many might have previously considered appropriate only for land treatment. But the supporters of sewage irrigation fought the bacterial methods at every opportunity, primarily by contesting the idea of bacterial treatment as natural. Irrigationists reacted to the perceived "attack . . . against the application of sewage to land" by fortifying the "natural" quality of irrigation.[62] Responding to a series of papers on bacterial treatment methods in 1897, for instance, H. Alfred Roechling, a prominent German-trained sanitary engineer commented: "Nature must be left alone; in the purification of sewage, she must be allowed to do her work in her own appointed manner by her own appointed agents and her own appointed time." Rejecting the filters, tanks, and beds of the proponents of biological methods, he declared that "the only natural sewage treatment . . . was the land treatment."[63]

But as research repeatedly demonstrated the role of bacteria in sewage treatment and promoters began to equate natural treatment with bacterial treatment, irrigationists were forced to claim bacteria as well. Although the role of bacteria in land treatment was recognized by Robinson and others as early at 1879, it was never incorporated into the thinking of promoters of intermittent filtration. John Bailey Denton, the chief advocate for intermittent filtration, failed to mention bacteria at all in his reports and, as late as 1885, continued to discuss purification as a process of direct oxidation of organic matter by air itself.[64] It was only when the bacterial methods began to take hold that supporters of land treatment claimed the role of bacteria as well. As the new bacterial methods gained in favor, one of the most vociferous defenders of sewage irrigation was Alfred Stowell Jones (figure 1.4). Jones

Figure 1.4
Lieut.-Col. Alfred S. Jones. Jones was an ardent supporter of land treatment of sewage throughout his career. He managed well-regarded sewage farms at Wrexham, Wales, and for the British Army's Aldershot Barracks. Despite his severe criticism of "artificial" biological treatment, he was respected enough to be elected president of the Association of Managers of Sewage Disposal Works in 1908. *Source*: *AMSDW*, 1908

was an engineer and respected sewage farm manager, known for his careful management of sewage farms at Wrexham and later Aldershot. The son of Liverpool's Archdeacon John Jones, he came to engineering later, after first joining the army and serving in India where he was severely wounded and decorated with the Victoria's Cross for bravery. After retiring from the military in 1872, he began working in sewage treatment. His Wrexham farm had won the 1879 award for sewage farm management presented by the Royal Agricultural Society and he was later elected president of the British Association of Managers of Sewage Disposal Works.[65] Jones insisted on primacy for soils. "There was no need," he argued, to give the newer bacterial methods "a new name and delude people into believing they were introducing something new."[66] "Bacteria have worked for millions of years in land," he wrote in the *Times*.[67] The new processes were mere "gimmicks" compared to the traditional practice of sewage irrigation. Supporters of artificial methods may claim the role of bacteria, but "we" had them first, he seemed to say.

Despite his criticism of bacterial action, however, Jones recognized the new scientific landscape of bacteriology, and needed to reclaim bacteria from exclusive use by proponents of the newer "artificial" methods. "As was well known," he stated, "for the last twenty-five or thirty years he had been an irrigationist, and a false idea had gone about that he believed in nothing else." To counter that impression, "he wished to make himself right with bacteriologists of the present day."[68] Making himself "right," however, consisted only in claiming bacteria for purification in land, and insisting that there was nothing new in the bacterial filters. He criticized bacteriologists for failing "to see that the previously unrecognised microbes can do their work, as they have always done it, to most advantage in the upper layers of any porous land."[69]

The division between natural and artificial methods was clearly delineated by the commission. It had lumped together as artificial both chemical treatment and biological processes, reserving, by omission, the term "natural" for land treatment. Roechling reported that "this division seems to have given a great deal of offence" to proponents of the new bacterial methods.[70] And indeed it did. "The classification of these processes as 'artificial' would be of less importance," wrote sanitary engineer Arthur Martin, "were it not for the use which has been made of the term in the attempt to decry the new processes in favour of land treatment, which is held up by contrast as 'natural.'"[71] Martin, who was involved in the invention of the septic tank, had prepared a summary of the over eight hundred pages of evidence presented to the commission as a précis for an argument to overcome the prejudice in favor of land treatment. Selectively quoting from the record, he marshaled the evidence against land treatment and in favor of the bacterial methods. Responding to the division of methods as natural or artificial, he succinctly titled one section "Bacterial Processes not Artificial." He rejected the dichotomy between natural and artificial, writing, "All the processes considered are essentially natural, the object in each being to give natural agencies the freest possible play."[72]

Both Jones and Roechling made the contest between natural and artificial most explicit in the title of their joint 1902 book, *Natural and Artificial Sewage Treatment*, but Dibdin and other prominent chemists and sanitary engineers all weighed in on the question of natural vs. artificial treatment. In the face of the documented failures of many sewage farms, the presumptive naturalness of land treatment remained its most compelling characteristic. In contrasting land and filters, Jones praised the Royal Commission for classifying "all modes of sewage treatment under two heads, 'natural and artificial," and insisted that the distinction be maintained. "That distinction is important," he insisted, "because bacteria had long been known to take part in natural treatment on the land, and, therefore, the use of the terms 'bacterial' and 'biological,' as popular designations for certain artificial systems, was unscientific and delusive."[73] Jones thought that bacterial science, while shedding some light on the underlying causes of purification, did little to advance sewage treatment. Land treatment, was "the natural and best agent" of treating sewage, and remained so even in the face of the "preachers of microbe agency."[74]

Roechling also stated categorically that "sewage irrigation was the only natural method of sewage purification and that all the other methods were artificial." He too defended the classification of sewage treatment techniques into "natural" and "artificial." Roechling was more willing to concede the importance of bacteria, in both land and filters, and recognized all *processes* as natural, including chemical precipitation. "There is no such thing as an artificial agency," he declared, because even chemistry relied on natural laws. However, processes could be divided on the basis of how the "natural" agencies were performed, whether in works of man or in the natural soil. On this basis, he reserved the term "natural" for land treatment. "It is no longer open to argument whether a chemical process or the contact bed system . . . is artificial, or whether the land treatment is natural! For who would deny that masonry or concrete tanks and the materials contained in the same are artificial products—i.e. products formed by man—and that land is a natural product—i.e. formed by nature—and that further the soil is the natural home of bacteria." Roechling further resisted the attempts of proponents of artificial filters to appropriate the terms "bacterial" or "biological" for their processes, and insisted that the best term for these processes was "artificial self-purification in contact beds," reserving the term "natural self-purification" for land treatment."[75]

George Vivian Poore, meanwhile, decried all of the modern methods of sewage treatment, including irrigation, as unnatural. His fundamental critique was of the whole water carriage system: "Most of the shortcomings of modern sanitary methods are due to the fact that in our dealing with organic refuse we commit a scientific error—i.e. we pursue a course which is in opposition to natural law."

"This error," he explained, "consists in mixing organic refuse with water. When organic refuse is mixed with water it undergoes changes which differ widely from the changes which it undergoes when mixed with earth." He concluded that "the proper destiny of organic refuse is immediate burial just below the surface of the soil."[76]

Arthur Martin conceded Poore's fundamental point about the difference between feces and sewage, but rejected the distinction the irrigationists were making between land as natural and bacteria as artificial. They were either both natural or both artificial. Invoking Poore, he wrote, "The only mode of dealing with sewage which can strictly be described as 'natural' is to do away with it altogether, and apply its constituents separately to the soil." What was natural, he asked, about "proceeding to flood land with dirty water to the extent of twenty or thirty times the amount of the rainfall"?[77] But rather than return to Poore's vision of collecting dry excrement from each household and immediately burying it in the top layers of soil in a cottage garden, Martin advocated the modern biological treatment methods. "The agencies employed in the modern processes of sewage purification . . . are, in principle, the same as those used by Dr. Poore, . . . and are also responsible for the purification which is effected by a sewage farm."[78] Proponents of biological treatment thought that all of the bacterial processes were equally natural because both land treatment and artificial filters relied on similar bacterial action in purification.

In his presidential address to the newly formed Association of Managers of Sewage Disposal Works, Samuel Rideal, an early advocate of bacterial treatment and author of an important book on sewage,[79] also criticized the division of sewage treatment processes: "The inclusion of modern biological systems under the term 'artificial processes' was misleading, inasmuch as such systems 'can only be successful as far as they adhere to natural methods.'" "The Commissioners' classification may be to a certain extent convenient," he continued, "but cannot be said to be either scientific or complete. . . . Assisted bacterial processes, as by heat or forced aeration, are not specified; even with these, it may be observed, the term 'artificial' is inappropriate, as the first is merely what goes on in the tropics, and the second similar to the action of winds, cascades, and flowing rivers." Rideal countered Roechling's classification with one of his own: "A short expression is much wanted to distinguish between natural methods of purification on relatively large areas of land, and on relatively small ones of tanks and beds. I would suggest 'controlled natural processes' as opposed to 'uncontrolled land treatment.'"[80] These arguments over what to call bacterial treatment of sewage, whether natural or artificial, assisted or controlled, make clear the enormous rhetorical weight that sanitarians placed on the term "natural" for sewage treatment processes.

The Royal Commission's findings were long-delayed, to the consternation of irrigationists and advocates of bacterial treatment alike. As they awaited the findings, city leaders had to decide whether to move forward with sewage farms or try bacterial treatment works that in the end might not meet with the approval of the commission. "Of the anxiety with which the decision is awaited there can be no doubt," reported London chemist George Thudichum. "Many towns are postponing alterations or re-construction of existing works, or delaying to arrive at a decision regarding a plan of action until the issue of what is hopefully regarded as an authoritative statement which will indicate the real solution of their several problems. Some of these towns, although actually under injunction to remedy an existing nuisance, make no attempt at progress pending the publication of the report. It is to be hoped that it will not be too long delayed."[81]

Responding to the calls for some kind of conclusion, the commission released an interim report in 1901. If it had hoped to settle the dispute between advocates of sewage farms or bacterial processes, it was unsuccessful. Couched with numerous qualifiers, the report allowed each side to claim victory. In qualified support for sewage farms, the commission concluded that indeed, no land was "entirely useless" for treating sewage, although it found clays and peats were nearly so. Conversely, it gave guarded approval to artificial processes provided "proper safeguards" were met. The committee declined to specify those safeguards however, and the Local Government Board interpreted it to mean that cities must still purchase land for sewage farms in case the bacterial processes failed.[82] Given the commission's comprehensive examination of the artificial methods, Jones interpreted the conclusion that no land was useless as a repudiation of bacterial methods. Martin, on the other hand, cited the "strong body of opinion" collected in the commission's report and declared that "it is agreed on all hands that other processes [than land] should now be accepted."[83]

Rejecting Bacteria

Without conclusive findings, the battle over "natural" and "artificial" treatment processes continued to rage. In response to the rise of biological sewage treatment, Jones had earlier claimed bacteria for land treatment as well. He now sought to diminish biology's role in both land and filters. "These bacteria . . . were not immaculate," he argued, "and did not do all they were said to do."[84] In effect, Jones sought to claim the natural characteristics of soil without referring to bacteria at all. This would maintain the natural status of irrigation without also conferring it on bacterial methods. To discount microbiology, he argued that the cleansing action of land, as well as filters, was a result of "natural" *physical* forces, rather than bacteria. Jones collaborated with William Owen Travis, a physician and the inventor of a sewage treatment tank called the "hydrolytic tank," which he installed at Hampton. Jones and Travis asserted that sewage was clarified not by bacteria, but through a physical separation of liquid and solid components. By "bringing the liquid into intimate contact with surfaces," this "natural phenomenon" could be accelerated, they argued.[85] Both land treatment as well as artificial filtration areas could supply those surfaces, but only in land treatment could clogging be prevented. Clogging was so prevalent in bacterial filters, they claimed, that microorganisms were essentially irrelevant. To make this point, they quoted a sewage works manager who had to have his filter beds cleared of their accumulated sludge by manual labor: "The best organism I have at the sewage-works," he quipped, was not a *micro*organism, but "the man with a barrow."[86] Further, unlike the "inoffensive and innocuous" "real 'humus'" created in land treatment, Jones and Travis argued, the sludge in bacterial filters remained exactly that: "sludge . . . to all intents and purposes unaltered" from the original sewage, and teeming with dangerous, rather than helpful microorganisms.[87]

Jones and Travis's paper met with severe criticism from proponents of bacterial filters. Dibdin noted sarcastically that if the authors could demonstrate that the changes in sewage took place in the absence of bacteria, it "would be a fundamental and very important point . . . and rather startling." Rideal called the paper "very interesting—he might also say quaint," before subjecting it to detailed and withering criticism. To attribute the clarification solely to physical action, Scott-Moncrieff argued, "was flying in the face of all experience obtained since Pasteur laid the foundations of bacterial researches. . . . The resolution of organic matter into the mineral forms from which it had been evolved by the life-processes of plants and animals was a condition necessary to the continuance of life itself; and that this work was carried on by microorganisms was not an unproved theory but a fact."[88]

Despite these criticisms, Travis went even further and proposed what he called the "Hampton Doctrine," named for the city where he had installed his tank. The Hampton Doctrine "teaches that the purification process is essentially a physical operation," explained Travis. He categorically denied the role of bacteria. The doctrine, he wrote, "denies that the purification process is, in any sense of the word or under any circumstances, the result of bacterial action. These microorganisms are merely incidental." Purification, he claimed, was a "desolution" process that literally reversed the processes by which feces became initially suspended in sewage. This "desolution effect," another proponent of the doctrine wrote, "entirely overshadowed any reduction or other operation due to the action of organisms."[89] The Hampton Doctrine rejected the role of bacteria but nevertheless claimed the support of nature. Travis declared that "the Hampton doctrine has been demonstrated to be a . . . faithful representation of 'Nature's method.'"[90] Travis had decoupled bacterial action from natural action to delegitimize biological treatment. Although the Hampton Doctrine was taken seriously in some quarters,[91] the bacterial researchers subjected it to fierce ridicule, both for its portentous title as well as its rejection of decades of research supporting biological agency in sewage treatment. Critics kept waiting for Travis to retract his "Hampton heresy," but Travis insisted that "purification . . . was the result of physical actions, and therein bacteria played no part."[92]

Despite the Hampton Doctrine, the Royal Commission's long-awaited final report of 1908 categorically declared the unity of all sewage treatment processes whether they were performed in soil or in artificial filters: "We are satisfied that it is practicable to purify the sewage of towns to any degree required, either by land treatment or by artificial filters, and that there is no essential difference between the two processes, for in each case the purification, so far as it is not mechanical, is chiefly effected by means of micro-organisms."[93] The Royal Commission had officially united these two terms, declaring that both methods were natural precisely because they were both biological. In their reliance on the action of bacteria, the artificial processes had been naturalized.

Champions of bacterial methods regarded the 1908 report as "the sanitary engineer's magna carta," freeing them from the rigid rules that required irrigation as the final treatment in any sewage disposal scheme.[94] Irrigation died hard, however, with Jones and other proponents insisting long after the matter was mostly settled in favor of bacterial treatment that "the advantages of land over artificial treatment had been proved."[95] Taking qualification and reservation in the reports of the commission as proof of the failure of artificial methods, Jones argued that "all who study the Blue Books will find natural treatment the best, and the artificial ones only more or less better than no treatment at all."[96] They were fighting a losing battle, however. Rejecting the Hampton Doctrine's claim that "the microbe is little better than a myth," sanitary scientists were optimistic that sewage treatment based on bacterial action had reached some form of finality. This confidence was based on the seemingly natural characteristics of the bacterial methods. "The fact that the biological process is Nature's own method," concluded Arthur Martin, "was generally regarded as an effective guarantee of its finality."[97]

Industrializing Biology

As biology was naturalized, however, so was it simultaneously denatured. Despite the rhetoric that biological treatment was "Nature's own method," sanitary scientists and engineers considered nature's methods inadequate to the task of treating the sewage of the industrialized city. Nature must be improved upon. Engineers looked not just to mimic the natural processes found in rivers and soil, but to improve on them, to industrialize them. In short, biological treatment itself, according to these proponents, must be industrialized. The same trends that characterized the process of industrialization would have to be applied to bacterial treatment. For the natural ecosystem to become the industrial ecosystem, it would have to be denatured.

In a biological context, industrialization meant the removal of natural variability—modification of the microbial environment to achieve constant, optimal conditions. Critics of irrigation identified the chief drawbacks of the natural treatment of sewage in soil—it was capricious and unreliable. Only some soils were suitable and they varied across the landscape or even within a single field. Chemist C. Mott Tidy discussed the problems with land treatment: "The purifying power of a soil, however, is peculiar to itself. You cannot completely control aeration. . . . In fact, the soil, as a purifying agent, is, to say the least, capricious."[98] Tidy's solution to the capriciousness of nature was the industrial process of chemical precipitation, which could be potentially carefully controlled. For proponents of biological filters, however, the solution was the control and intensification of bacterial action.

Ideas of concentrating, accelerating, maximizing, regulating and controlling the bacteria suffuse the statements of early promoters of bacterial action. Dibdin, for instance, criticized land treatment as then practiced as haphazard and uncertain. In contrast, his bacteria beds were land treatment too, "but, concentrated, accelerated, and controlled."[99] Dr. D. Sims Woodhead, of the Royal College of Physicians, testified that you could "induce" decomposition, and could "control the changes better than you can in land filtration." Unlike the capriciousness of soil and farming, "you can put down a definite area, you can give a definite depth, you can keep it working under special and well-controlled conditions."[100] "By placing the action of bacteria under regulated conditions," wrote London's Frank Clowes, "their purifying action could be exercised with regularity, certainty and precision."[101] "Bacterial action in specially prepared areas," wrote Thudichum, "may be called the purely natural method," but "worked under conditions which are readily controllable."[102] "Organisms can be utilised to higher advantage," argued G. A. Hart, under the "artificial and more or less controllable conditions," of the biological filter than "in soil of variable constitution."[103]

A. C. Houston, bacteriologist to the London County Council, contrasted the natural processes of bacterial beds with the potential improvements of the chemist and bacteriologist. "All the bacterial processes in practical operation," he stated, "aim at allowing certain bacteria or groups of bacteria to gain the ascendency by a *natural process of selection*." Bacteria beds functioned well when the "special bacteria concerned in the work of purification" were able, because of environmental conditions, to outgrow "the wrong kind of bacteria." While natural selection could therefore produce a good effluent, Houston thought that bacterial treatment could be improved by "interference with nature's methods." Instead of relying on natural selection, the scientist could isolate and select "special sewage microbes" to inoculate the bacterial beds. In 1899 Houston isolated a nonpathogenic strain of "sewage proteus," which was quite common in London Sewage, and introduced "billions" of cells into the contact bed. Even though there was no improvement in the effluent,

Houston was not discouraged. Perhaps, he concluded, he had not added a sufficient number to "upset the biological equilibrium" of the bed.[104]

The intensification and control of bacterial treatment was a process of denaturing, of taking apart the natural properties of the process, removing some, and keeping only those that best accomplished the goal of treatment. W. Santo Crimp, engineer to the London County Council, made this denaturing clear when he suggested a thought experiment to explain the advantage of artificial filters over land. Take a cubic yard of a natural farm soil, he suggested, and "rearrange the particles, taking out the large ones and taking out the very small ones." Literally denatured by selective removal of certain particle sizes, "that cubic yard will do more work after the re-arrangement than it would in its natural state," he assured the commission. Having converted the soil from its natural state to one more useful by adjusting the size of the particles, engineers could denature it even further, relying on a wholly artificial material that would accomplish the same purpose. If you further "substitute . . . a cubic yard of other material which will effect the same purpose, you get this concentrated treatment," continued Santo Crimp. "You can do more work," he concluded, "by an artificial filter than you can on a natural soil."[105]

"Work" was a central metaphor that connected bacterial work to broader issues of industrialization. Industrial metaphors were common in discussions of the role of bacteria in treating sewage. Percy Frankland boasted that scientists now knew how to derive "the maximum amount of work" from the bacteria and "how to make the work most useful."[106] Cameron referred to his creation in the septic tank of separate "workshops" for the aerobic and anaerobic bacteria.[107] Another writer noted that bacteria "worked for free, took no breaks, and never struck."[108] These metaphors were critical to an industrial approach to treatment, and became the driving force behind the further development of the treatment ecosystems. There was of course an implicit comparison of the nature of bacterial work and the nature of human labor. Bacteria could be more easily induced to work than humans could, and more work could be extracted from them, presumably because they did not strike, "soldier," or get tired. However, as sewage scientists and engineers were to discover, to derive the maximum amount of work from their bacterial laborers they would need to institute control. As we will see in chapter 3, the unruliness of the shop floor extended to the petri dish.

Part of the opposition of irrigationists to the bacterial methods lay in precisely this industrial, denatured character of the tanks and filters. Irrigationists were responding not only to sewage but to the long history of urbanization and industrialization in England. Remarking on the industrial character of the purification processes and works, but also on industrial labor, irrigationists sought to maintain a connection to the land and the ideology of rural labor. Proponents of biological treatment might extol bacteria as ideal industrial workers, performing their work regardless of the speed of the process, but irrigationists criticized biological treatment for exactly these same attributes. Jones also referred to microorganisms using metaphors of industrial labor, but in terms that resonated with contemporary critiques of the effect of industrialization on laboring conditions. "Those microscopic bacteria," wrote Jones, "have been expected by their enthusiastic admirers to perform ten times the work that they always do in the land." But these bacterial laborers could only meet those expectations, he continued, "if confined by artificial means to crowded workshops."[109] H. Alfred Roechling decried the septic tank as "simply a machine for the production of sewer gas" and objected to it as "the forcing frames or forcing beds of nature in which nature was asked to do its work at express speed." Perhaps invoking in the reader images of Frankenstein's monster, he advised caution toward this runaway technology, fearing that the champions of bacterial treatment might have produced "something over which they had no control."[110] Another sanitarian likened bacterial filters to the industrialization of meatpacking: "It was like putting the pig in at one end of the machine and expecting sausage at the other. This sort of thing might be done in Chicago, but it would not succeed in our sewage works."[111] At this time, Chicago was, along with Manchester, one of the great examples of the impact of industrialization, and its famous pork processing factories, with their pig disassembly lines exemplified the industrialization of nature.[112]

In contrast to their critique of these artificial, industrialized conditions, irrigationists extolled the virtues of the natural process of applying sewage to land and sought to preserve a particular ideological vision of rural England and agricultural labor. While they did not go so far as Poore in rejecting urbanization and the resultant water carriage system, they did reject the industrial nature of bacterial treatment and insisted on maintaining a role for land and crops as symbols of England's rural heritage. John Manley, for instance, invoked "the cry 'Back to the Land'" to favor sewage irrigation over "the much advertised artificial schemes."[113] Similarly for Jones, using sewage to fertilize crops was essential to the maintenance of England's agrarian heritage. In the face of the declining number of agricultural laborers, the sewage farm represented a means to keep people on the land; by offering high wages, housing, and gardens, the sewage farm could compete with industrial wages more successfully than other farm jobs.[114] "If we are to attract men 'back to the land,'" wrote Jones, referring to legislation like the Small Holdings Act designed to attract city dwellers back to farming, "we should surely give them a chance of raising crops which may afford some return for their labour."[115] For Jones, it was a question of no less than the nature of England's democracy. "A more democratic Government would consider how much more meat, corn, milk and other foodstuffs might be raised in the country if the natural system of sewage treatment prevailed."[116]

Jones's conservative reactions to the role of bacteria in sewage treatment mirror in many respects the reaction to germ theory in public health by some prominent sanitarians. Within the sanitarian community, there was a marked strain of thought that resisted the germ theory and its implications for how the public health would be protected. Lloyd Stevenson remarked that this strain represented a conservative and religious reaction, and he linked opposition to the germ theory of disease in general with some of its practices, including immunization and animal experimentation. The language of these opponents of the germ theory is remarkably similar to Jones's and other irrigationists' responses to bacterial treatment. Anti-vivisectionists characterized the animal experimenters' methods as "always 'artificial,'" in contrast to the "'natural' methods of healing."[117] These critics of the germ theory and bacterial sewage treatment both criticized the religious terms used by their opponents, the "gospel of germs." As Jones spoke sarcastically of the "preachers of microbe agency" and "immaculate" bacteria, Collins spoke of "that blessed word," "the bacillus."[118] Perhaps most fundamentally, these critics suggested that the germ theory, as applied to either medicine or sewage, provided very little in the way of practical advances. Both groups thought that sanitarians had already solved the fundamental problems of urban health without the new theories of Pasteur and his followers. Jones asked "without detracting from the credit due to the great French savant . . . the practical man may well ask how much forwarder have we got in the main and pressing business of purifying our rivers—as a consequence of clearer knowledge of minute forms of life?"[119] Similarly, Collins wrote, "Let us see if bacteriology alone affords reliable guidance for practical purposes . . . Our modern bacteriologists tell us that the organisms found in sewer air do not constitute any source of danger, are in fact derived from the outside air by ventilation of the sewer, and are fewer in number in sewer air than in the air of our houses. What shall we say then? Shall we admit sewer air to our houses? . . . Shall we not rather conclude that bacteriology alone is an unsafe guide in our hygienic arrangements?"[120]

These critics of both the germ theory and biological sewage treatment were seeing the ground shift beneath them. They had been successfully working in public health and sanitation, and they feared the changes implied by a germ theory of disease or sewage purification. Opponents feared that public health in the tradition of Chadwick was falling before the "advance of the microbic hosts."[121]

The opposition was idiosyncratic, however, and opposition to the germ theory in a medical context did not always mean opposition to a bacterial theory of sewage treatment. Sanitarian George Wilson, for instance, author of a leading textbook in sanitation, was a prominent critic of the germ theory as expressed in immunization practice. He was also an early supporter of irrigation over chemical methods of treatment. "Irrigation," he wrote in his 1873 textbook on sanitation, "is the only process which fully meets all the requirements ataching [sic] to the disposal of sewage; in other words it is the only one which while it purifies the sewage efficiently, realises the highest profits, and may be carried on without creating any nuisance or detriment to the health of the neighbouring inhabitants."[122] By 1898, however, when his text reached its eighth edition, he fully supported the new bacterial methods—septic tanks and trickling filters—as the best means for treating sewage, and his support for irrigation became more qualified.[123] Yet this 1898 edition voiced his most explicit criticisms of germ theory, a view so anachronistic that his text became useless to teachers of modern courses in sanitation and was abandoned. This edition was his last.[124] While the general conservatism toward modern biology is evident in both Jones's critiques of bacterial treatment and holdouts against germ theory, these two groups were not synonymous.

The industrial ecosystem in general appeared as a threat to more traditional methods and structures of work. Identical issues of labor and industry, natural and artificial, arose in other industrial ecosystems as well. Delbrück's critique of Hansen's pure culture methods in the brewing industry, like that of the irrigationists, was rooted in a preference for the traditional brewer's craft in the face of industrialization.[125] For Delbrück, it was the "hand of the artisan" based on "the time-honored, practical art of our brewers, distillers, and vintners" that could "transform a impure yeast into a pure one."[126] Delbrück equated his natural techniques based on "climate" and Darwin's theory with the work of the artisan. In contrast, Hansen's system, based on laboratory work and brute-force artificial selection of the yeast, was unnatural. Hansen defended his system against the charge of artificiality, responding to Delbrück's critique in language that could have been taken from the advocates for biological sewage treatment. He argued that the distinction between natural and artificial was "wholly arbitrary and misleading"[127] and countered Delbrück by stressing the naturalness of the object of study of biological science. His pure culture system was "a biological-botanical system and not—as Delbrück asserted—a mechanical contrivance."[128]

As feared by its opponents, though, the twin projects of industrialization and modernity *were* an integral element of bacterial sewage treatment. From the first suggestion of using vital processes for purification, proponents envisioned its ultimate industrialization. Dibdin claimed his 1887 paper on the "Sewage-Sludge and Its Disposal" as the first proposal for treating sewage using biological methods.[129] In that paper, Dibdin had laid out a vision of a modern industrial ecosystem for purifying sewage. "Modern experience shows," wrote Dibdin, "that, when this subject is better understood and thoroughly worked out, in all probability the true way of purifying sewage . . . will be first to separate the sludge, and then to turn into the neutral effluent a charge of the proper organism, whatever that may be, specially cultivated for the purpose; retain it for a sufficient period, during which time it should be fully aerated, and finally discharge it into the stream in a really purified condition."[130] At the time, Dibdin had little understanding of what the details of that vision might be. He did not know which the "proper organisms" were, or how to select or cultivate them. Yet he was confident that with the advance of biological understanding, this was the future of sewage treatment.

On hearing Dibdin's vision, attendees erupted in laughter. Many of the sanitarians, both irrigationists like Jones as well as supporters of chemical precipitation, mocked Dibdin as ridiculous. Many attendees were, for one, skeptical of Dibdin's motives, seeing his plan as a way to avoid treating London's sewage in an acceptable manner. Jones, perhaps thinking that Dibdin's "organism" might be an insect or even a fish and not microscopic bacteria, remarked that "Mr. Dibdin had carried that idea a little too far in proposing the cultivation of such animals to feed upon the sewage. The question of the space and time, in which the animals were to eat up the food accumulating at the rate of 150,000,000 gallons a day, was a rather serious consideration . . . It would be a very serious thing to have to find tank feeding-room at Barking."[131] Others, however, concurred with Dibdin. "It was hardly safe to laugh at such statements," commented Dr. A. Angell, "considering the rapid strides that science was making."[132]

What Dibdin was advocating was the program of industrial microbiology more generally, the creation of an industrial ecosystem. Charles Avery, in a lactic acid factory in Massachusetts, developed the first large-scale commercial chemical production process using fermentation in 1883. Hansen introduced his system of brewing using pure yeast culture in the same year. Each involved adding specially cultured microorganisms to production vessels and adjusting the environment to maximize the organisms' growth.[133] This vision of the simplification and intensification—the industrialization—of the sewage treatment process that Dibdin predicted reached fruition in the development of the activated sludge process, introduced in 1914.

Activated Sludge

The activated sludge process developed from several research threads in England and the United States on the industrial use of microorganisms. Gilbert Fowler, a chemist and bacteriologist at Manchester University, was working with a young bacteriologist Ernest Moore Mumford who was investigating the clogging of a biological sewage filter in Scotland by an "iron organism." In seeking biological material to study, Mumford examined some canals draining a colliery near Manchester where there were iron deposits that had been laid down by a bacterium. Mumford collected these bacteria and began to establish the conditions in which they grew and deposited iron. Naming it M.7, he found that in the presence of iron, air, and a source of organic nitrogen like peptone, the digested proteins of animal milk and meat, the bacteria would precipitate iron compounds.[134]

In the meantime, in 1911, Harry Clark, chief chemist at the Lawrence experiment station, was investigating the impact of domestic sewage and trade wastes on the life of rivers. They placed fish in aquaria and began adding sewage to determine "how much sewage the fish would stand and live." They found that in order to keep the fish alive in more and more polluted waters, they had to bubble increasing amounts of air into the bottom of the aquaria. As they kept adding sewage, they noted that growths began developing on the sides of the aquaria. When they stopped blowing air, they discovered that all of the suspended and colloidal material from the sewage dropped out, turning the turbid water clear. Thinking "there might be something in the aeration studies," they quickly abandoned the studies on fish and began bubbling air into containers of sewage. After blowing air for several weeks, they were able to purify newly added sewage in a matter of just twenty-four hours.[135]

Fowler's research on the M.7 bacteria and Clark's research on aeration came together when Fowler visited the Massachusetts experiment station in 1912. Fowler was consulting with the New York Sewerage Commission on its sewage disposal plans, and, like many sanitary scientists, made it a point to travel to Lawrence and tour the experiment station. There, Clark showed Fowler the bottle experiments and explained their success in purifying sewage. Fowler had previously written that "the subject of sewage disposal is mainly a question of the separation of solids from liquids" and, in conversations with New York City public health official George Soper in New York, began thinking about how to "clot" out sewage particles from the water. When he returned to Manchester, he spoke with Mumford about the M.7 bacteria. "What about the M.7? If you can get it to precipitate with peptone, surely sewage ought to serve as well."[136] They began experimenting with cultures of M.7, bubbling air and adding iron salts and sewage. They found that sewage could be clarified with this organism, producing a "limpid sparkling and non-putrefactive effluent" that could either be rapidly oxidized on a sewage filter or even discharged directly into a stream. This procedure was similar in principle to chemical precipitation, except that the precipitating agent was biological. Rather than purchasing chemicals for treatment, the bacteria could be grown and maintained in a tank, fed on barley broth, and used to inoculate the treatment tank. Although iron salts needed to be added, they were required in far lower concentrations than with chemical treatment methods. Fowler summarized the advance in sewage treatment represented by this work as "the use of a specific organism found in Nature" to clarify sewage. The M.7 process seemed to confirm Dibdin's prediction of the future of sewage treatment in which "a charge of the proper organism . . . specially cultivated" would be aerated with sewage. The sanitary community thought Fowler's proposal to be "most interesting," "original," and "suggestive," but was nevertheless skeptical about the ability to translate laboratory experiments like Fowler's into practical treatment schemes.[137]

This M.7 research was part of a broader milieu at Manchester University, where a number of researchers were investigating the use of microorganisms in industrial processes (figure 1.5). Aside from his work with Fowler, Mumford was also an assistant to Chaim Weizmann, professor of chemistry at Manchester and future president of Israel. Weizmann was investigating the bacterial fermentation of starch to acetone and butyl alcohol, a process that would become important for the English war effort and later for the production of automobile varnishes. All of these efforts were rooted in similar methods and ideology. Both the M.7 and Weizmann's process relied on a single species of bacteria that was isolated from nature and cultivated. Scientists then determined the organism's environmental requirements and used that knowledge to grow them in large cultures that could yield a product in industrial quantities.

Fowler's M.7 research proved to be impracticable, however, and it was never adopted. Part of the problem lay in just its highly industrial character, and it suffered some of the same problems as its progenitors in chemical precipitation. The bacteria had to be cultivated separately from the sewage and were consumed with each treatment. Both the bacteria and the iron had to be added anew during each cycle.[138] The process thus failed to take full advantage of the living character of the precipitating agent, the iron bacterium, and its capacity for self-regeneration.

At the same time Fowler and Mumford were working on the M.7 process, Fowler had instructed his assistants at the Manchester Rivers Committee, Edward Ardern and William Lockett, to repeat the experiments of Clark that he had seen in Massachusetts. Ardern and Lockett placed sewage in a small quart bottle, and then started to bubble air into it until all of the nitrogen had been converted to nitrate, which was the mark of an effluent that would not putrefy if discharged into a stream. Complete nitrification took six weeks—too long to be useful in treating sewage directly. Their next step was crucial, however. Instead of discarding the sludge that accumulated in the bottle, they saved it and decanted the liquid. They then added fresh sewage and began to aerate anew. This time, nitrification took only three weeks. They kept decanting the liquid, reserving the accumulated sludge and adding fresh sewage, and with each repetition, the time to nitrification was steadily reduced, until they could completely oxidize sewage in just twenty-four hours.[139] They began to refer to this sludge as "active" to distinguish it from "ordinary" sewage sludge such as might be collected in a sedimentation tank.[140]

With the active sludge process, now renamed "activated" sludge, Ardern and Lockett felt they were on the verge of a revolution in sewage treatment. "Results so far obtained indicate that it may radically affect the whole problem of the purification of sewage," wrote Lockett in a draft of their paper.[141] On April 3, 1914, Edward Ardern reported the work to the Manchester section of the Society of Chemical Industry.[142] Ardern had invited sewage workers from around the country, and attendees were expecting to hear of a major breakthrough in sewage treatment. It was the best-attended meeting of the section ever, with perhaps two hundred

Figure 1.5
Manchester University Chemistry Department, 1911. Manchester University was a center for development of the industrial ecosystem. Chaim Weizmann, biochemist and future president of Israel who developed the industrial fermentation of cornstarch to produce acetone and butyl alcohol is fifth from the left seated in the first row. In the second-to-last row, second from the right, is E. Moore Mumford, who was a research assistant for both Weizmann and Gilbert Fowler (not pictured). With Fowler, Mumford isolated the M.7 bacterium and developed a sewage treatment process based on this organism. The M.7 process was a precursor to activated sludge. William Lockett (not pictured) was also a student of Weizmann's. To Weizmann's right is Gladys Cliffe (later Mumford), who worked as a biochemist at Activated Sludge Ltd. and later became its managing director. Reproduced by courtesy of the University Librarian and Director, The John Rylands University Library, The University of Manchester. *Source*: Manchester University Archives (DCH/3/3/1)

scientists and engineers in attendance.[143] They were not disappointed. The paper, titled "Experiments on the Oxidation of Sewage without the Aid of Filters," introduced the "activated sludge" process to the sanitary community.

There was little of the skepticism that attended the M.7 process. Attendees considered activated sludge to be potentially "epoch-making," and the reading of the paper sparked an explosion of research and evaluation of the process.[144] Within the next year alone, fifteen cities around the world began testing the process and building sewage works to utilize it. It quickly became the dominant process for treating the sewage of larger cities. By 1940, sewage from over half of the population served by secondary treatment in the United States was treated using this process, and it remains the most important method in use today.[145]

Creating an Ecosystem

As scientists saw the bacteria bed or percolating filter as the intensification of the natural process of purification taking place in soil, so they viewed activated sludge as a further intensification of these natural processes. Lockett emphasized this in titling the first draft of the initial activated sludge paper "Oxidation of Sewage by Intensified Bacterial Action with Simple Aeration."[146] The denaturing inherent in the activated sludge process is also evident in the metaphors used by sanitary scientists to describe the process. The biological metaphors used to describe the bacteria bed by Dibdin, Clark and others gave way to an industrial-chemical view of activated sludge. Scientists had thought of the filter and bacteria bed in terms of an organic whole, giving it properties of individual organisms. In 1892, Dibdin compared the bacterial community of the biological filter to an animal: "By reason of the multitude of organisms to be dealt with, a filter may be fittingly compared to a great animal." His colleague Thudichum referred to "the life of the filter" and "the mass of the filter."[147] Similarly, Clark, of the Lawrence, Massachusetts, experiment station, referred to the bacteria of the trickling filter as the "nitrifying body."[148]

Metaphors for activated sludge, however, were much more industrial. The shift in thinking about the system can be seen in Fowler's explicit comparison of biological filters and the activated sludge process. Fowler described the bacterial "deposits" of the biological filter as an "active and living agent of purification," an "active and living skin." In contrast to these organic metaphors, he discussed activated sludge in terms of theoretical optima. In the activated sludge process, he observed, "You have a continually moving surface, infinite in extent and constantly renewable, that being done always in the presence of air. That gives you the theoretical conditions for perfect aeration, and that is the new process of activated sludge known as the activated sludge process."[149]

A hallmark of industrial thinking was removing as much variability as possible from a process to achieve predictable conditions and maximal efficiency. The "active and living skin" of the biological filter suffered from heterogeneity and thus inefficiency, like the capriciousness of the soil as described by Tidy decades earlier. If you examined "the film, the living agent, on a sewage filter," explained Fowler, "the outside will be oxidized, the inside will be less well oxidized, and sometimes absolutely black and putrefactive, because it is not possible under those circumstances for thorough oxidation to take place." This variability was absent in activated sludge. "Activated sludge is essentially a homogeneous material," wrote Fowler, and thus more predictable and efficient.[150] Like the contrast between smelting copper in a reverberatory furnace and open heaps, activated sludge removed variability and speeded up the treatment process, a kind of biological furnace for sewage.

This industrialization troubled many sanitary scientists, however, who saw it as unnatural. Roechling, as might be expected, was as opposed to activated sludge as he was to bacterial filters. Focusing on Fowler's comments about activated sludge as the "concentration of the percolating filter," Roechling warned that "the more intense the treatment was . . . the smaller was the factor of safety . . . and the greater the chances of failure."[151] But even some supporters of bacterial methods were disturbed by the intensity of bacterial action in activated sludge. Arthur Martin described the difference between the sewage filter and activated sludge: "Take a filter, fling it into a whirlpool, let loose a whirlwind through it, and you have an activated sludge tank."[152] While Martin considered this advance "heroic," other engineers were not so convinced. Bacteria beds had undergone a process of naturalization in the previous two decades, and the contrast between the now seeming natural characteristics of filters and the industrial nature of activated sludge troubled engineers. In language similar to Roechling's objections to the septic tank, John D. Watson, engineer for the Birmingham sewage plant, expressed unease with the activated sludge process. Unlike Roechling, Watson was no irrigationist committed only to "natural" purification on land. As chief engineer of the Birmingham treatment plant, he had installed acres of bacteria beds and septic tanks. Nevertheless, Watson found activated sludge unsettling. Activated sludge "was indeed a natural process," he conceded, but he "was bound to conclude that Nature had been forced, and he doubted whether the best results could be obtained by thus forcing Nature." According to Watson, in bacteria beds sewage trickled "by gravitation over the surface of the bed," where "it would percolate through the . . . material employed as purifying medium." "The air," he argued, "was thus obtained naturally." In contrast to filtration, in the activated sludge process "the organisms were first cultivated in sludge, and then by air under pressure they were thrown up against the impurities in the sewage. In that way the impurities came in contact with the nitrifying and other bacteria. This was done many times. He did not know how many, probably thousands of times." Watson described the process in the bacteria bed as a gentle percolation of the sewage through the spaces in the bed, and he contrasted it with the violence of the activated sludge process. In creating this process, Watson complained, the developers of activated sludge were "forcing Nature into methods in which she would rather not enter."[153] Engineers like Watson wondered if they had gone too far in the drive to industrialize the biological processes of purification.

Activated sludge appeared so thoroughly intense, so industrial, that many researchers were no longer convinced the process was even biological.[154] Despite the ever-widening adoption of activated sludge around the world, sanitary scientists had only a vague, and often contradictory understanding of the mechanism of the process itself. "The process," wrote Fowler, "involves biochemical processes of great intensity and complexity which we none of us very completely understand."[155] Its originators, Fowler, Ardern, and Lockett, viewed it as a fundamentally vital process, but other scientists thought that physical or chemical activity was responsible for its ability to purify sewage. Others, like Glasgow's chief engineer F. W. Harris, developed a vaguely vitalistic but mysterious theory: "Activated sludge is the physical embodiment of an intangible form of energy."[156] Over the next twenty years, researchers continued to debate the underlying mechanism, as activated sludge went through the simultaneous process of naturalization and denaturing that characterized the invention of the septic tank and biological filter twenty years earlier.

Ardern recognized that the sludge was "alive" and needed to be kept so by blowing air through it. But he and Lockett understood neither the nature of the biological process nor the specific organisms responsible for the purification.[157] While Ardern suspected that physical and chemical "agencies" were likely also at play, he attributed the purifying action of activated sludge to its vital nature. He sterilized activated sludge with steam to directly test the role of organisms in purification. Modeling his own experiments after those of Schloesing and Muntz, which proved that living organisms were necessary for nitrification, Ardern showed that sterile activated sludge did not have the clarifying ability of untreated sludge. Ardern concluded that "whilst physical actions were undoubtedly involved, the purification process was essentially a vital one."[158]

Although these and other experiments showed activated sludge to be biological, researchers did not know which organisms were involved. On first looking at activated sludge under the microscope, researchers saw a variety of organisms. Not understanding the nature of activated sludge, they quickly attributed the activity of the floc to whichever creatures seemed most abundant. Ardern and Lockett found tens of millions of bacteria per cubic centimeter in the sludge, as well as a variety of protozoa, and attributed the activity to these organisms. In contrast, Illinois scientists Edward Bartow and F. W. Mohlman, in some of the first microscopical studies, found "many Vorticella and Rotifera." But the predominant organism, they found, was an annelid worm, *Aeolosoma hemprichi*. Noting that the worm was known to feed in nature "greedily and almost continuously on any small organic particles," they attributed the purifying ability of activated sludge to this worm's activity. "The activated sludge has evidently been developed by the multiplication of these worms originally present in the sewage," they declared. Soon after, though, Robbins Russell, also at the University of Illinois, had rejected the importance of the worms. Instead, he showed that purification depended on the complex interactions of different groups of bacteria, acting in succession on the sewage. He had isolated the nitrifying organisms *Nitrosomonas* and *Nitrobacter* from activated sludge as well as over a dozen other varieties of aerobic bacteria. Purification, he concluded, was dependent on a group of aerobic bacteria to break proteins down into molecules that the nitrifying bacteria could then attack.[159]

By using the same theories ecologists applied to natural systems, these researchers helped naturalize activated sludge. J. W. Haigh Johnson used an explicitly ecological approach to help understand the mechanism of purification. Studying biological filters, he had identified several ecological associations—groups of species that tended to occur together because of similar requirements for environmental conditions. In examining activated sludge, he found a ciliate protozoa *Operculia* in many samples. He noted that in natural habitats like fast flowing streams, the ciliate commonly occurred in association with the bacteria *Zoogloea ramigera*. Applying the concept of ecological associations, he reasoned that *Zoogloea* likely occurred in activated sludge as well, and attributed to this organism its purifying power.[160]

Illinois chemists Arthur M. Buswell and H. L. Long used ecological theories of succession to understand the changes in the activated sludge floc as it matured. During the startup of a new activated sludge plant, they enumerated changes in the biological community and documented the "definite succession or addition of forms, as the sludge develops."[161] Here, they used the ecological term succession, recently made popular by the research of Clements and Gleason. As with Sedgwick's research on biological filters, this ecological approach helped naturalize activated sludge, despite its highly intensified action.

Researchers also emphasized the naturalness of the activated sludge process by connecting activated sludge to the continuum of sewage treatment processes that had been previously naturalized. Fowler considered activated sludge to be part of a continuum from soil to bacteria beds to the new activated sludge plants. "The changes which go on in an activated sludge tank are essentially the same as those taking place in arable soil," he wrote. "Activated sludge contains forms of life almost identical with those found in fertile soil."[162] American consulting engineer W. B. Fuller also placed the continuum from land to activated sludge in an ecological context. "It is necessary . . . to provide an environment for the growth and multiplication of nitrifying bacteria." "Such an environment," he continued, "was provided by the soil in broad irrigation, and by the broken stone of the sprinkling filter." Activated sludge differed in that the environment "is provided by the sludge itself . . . in the form of a sponge with millions of pores, providing lodgment and environment."[163] Scientists understood activated sludge, then, to be a living community. Buswell and Long's proposed mechanism for activated sludge was widely adopted: "purification is accomplished by ingestion and assimilation by organisms of the organic matter in the sewage and its resynthesis into the living material of the flocs."[164]

Industrializing the Ecosystem

Chemists, engineers, and bacteriologists, however, sought to further intensify and simplify this already intensified biological system. "The Activated Sludge process is a still further intensification of the purification process," wrote W. E. Speight, manager of the Bolton sewage works, but perhaps, he thought, "even this process can be still further intensified."[165] "I can see a time coming," he later wrote, when "we shall have Messrs. Boots of Nottingham supplying house-holders with compressed tablets of Activated Sludge which will be dropped in the pan, flushed away, and the sewage will arrive at the outfall already purified."[166] Although he was half-joking, this comment reveals the ideological commitment among many engineers and sanitary scientists to intensification, biological simplification, and control—to the industrialization process itself.

This industrialization was again a literal denaturing, as researchers tried to remove any organisms thought to be superfluous to the purification process. The bacteria and other organisms of the activated sludge were presumed to originate in the soil, air, and sewage itself and survive in the sludge via a process of natural selection like that described by Sedgwick for the biological filter. While the biological filter had dozens of organisms, including bacteria and protozoa, as well as higher organisms like algae, insects, and snails, activated sludge was much simpler, dominated by only bacteria with a few incidental protozoa. Scientists and inventors, working on the assumption that further biological simplification would improve the process, looked for ways to encourage what they thought were the beneficial organisms and killing the others. The ideology of constant simplification was already apparent in John Thresh's 1909 prediction concerning trickling filters: "probably some sewage works manager of the future would show them how they could cultivate the useful varieties and exterminate those which were useless."[167] This ideology was now applied to activated sludge. The British Dyestuffs Corporation, for instance, advocated using the newly discovered, selective antiseptic characteristic of organic dyes (the magic bullet of Erlich) to kill the protozoa in activated sludge while leaving the bacteria unharmed.[168] The goal was a simplification even of the already simple activated sludge system. "Little is known concerning the organisms essential to the activated sludge process, but when this is known it may be possible to cultivate the necessary organisms and to keep out the others,"[169] wrote the editors of the *Surveyor* in 1923.

In this vein Polish scientist Fritz Simmer patented a modification of the activated sludge process in 1930, which he assigned to the Danish industrial fermentation company Dansk Gaerings Industri. The activated sludge process relied on the "varying" bacteria that were already in the sewage stream, the "accidentally developed associations of cultures of bacteria," as Simmer called them. Simmer thus considered the process "irrational." "The decomposition . . . takes place in an uncertain, irregular and nonreliable manner," he stated, and it was "not far superior to the natural putrefying and decomposing processes." Rather, he sought to rationalize the process by taking the techniques from "industries based on technical mycology," matching the biochemical activities of specific strains or species to the particular characteristics of the waste.

"According to the nature of the impurities . . . pure cultures of the microorganisms" would be employed at specific stages of decomposition, in a stepwise fashion, one group of microorganisms utilizing the decomposition products of the previous group. *Bacillus subtilis* would be used to degrade starch into sugars, and then removed by "filtration, decantation or precipitation." Next, yeasts would be used to degrade the resulting sugars and in turn removed before the addition of *Bacillus putrificus*, which would degrade the proteins.

Simmer, in essence, proposed to industrialize the process of ecological succession proposed by Buswell and Long. At each step, pure cultures of the appropriate microorganism would be added and then removed prior to the addition of the next species. Each successive culture used the waste products created by its predecessor. In its reliance on individual species performing only specific tasks in series, the model for Simmer's process was as much the assembly line (or more accurately, the disassembly line) as it was brewing or other industries. Further, in order to insure the rational treatment of the sewage, the waste itself must first be denatured, by removing all the naturally occurring bacteria: "It may be advisable to sterilize the waste water prior to introducing it into the decomposing plant," he suggested. His method thus required first sterilizing the sewage stream, and then growing several pure cultures of a variety of microorganisms that would be added and removed in turn. His modification of the activated sludge process was "for the purpose of simplifying the process," Simmer claimed. Yet, by simplifying the number of species involved, his process made the waste treatment inordinately complicated, requiring the segregation of the waste stream according to its main biochemical components—starches, proteins, or fats, and the separate culture of pure strains of microbes. Simmer's scheme, at least for municipal waste, was entirely unrealistic and was probably never applied in a practical plant. However, it perfectly illustrates the widespread ideology of the industrial ecosystem: to be more effective, the natural bacterial processes must be rationalized, controlled, denatured.[170]

To many researchers, the activated sludge process, with the bacteria freed from the constraints of a physical surface to grow on, began to resemble more and more a chemical manufacturing process. Thus, innovations from the chemical process industries needed to be applied to biological sewage treatment. At the time activated sludge was developed, manufacturers were just beginning to shift from producing metals, chemicals, and other products in batches to producing them using continuous processes.[171] Continuous manufacture allowed both labor and capital equipment to be used more efficiently, with no downtime. Sanitary engineers were among the vanguard in moving toward continuous processes. The trickling filter had begun replacing Dibdin's contact bed in part because it could be run continuously—sewage could be sprinkled constantly over the surface of the bed and purified as it filtered downward through the material. In contrast, Dibdin's bed had to sit full for hours with the sewage in contact with the material making up the tank. Similarly, engineers moved the activated sludge process "from batch to flow."[172] The original activated sludge process was worked on the fill-and-draw method. Sewage was added to a tank along with a portion of activated sludge. The tank was aerated for up to six hours, at which point the air was halted for several hours, allowing the solids to settle out, after which the clear effluent on the surface was withdrawn, and the process would then be repeated. Engineers quickly recognized that the process could be much more efficient if it could be conducted continuously. A continuous process required less manual labor to turn valves and utilized the capacity of the tanks to greater advantage, since they were never out of commission, awaiting filling or emptying.[173]

Seeing activated sludge as a chemical or physical process also suggested further ways to industrialize and denature the process. To many, the physical separation of solids from the liquid was the key step in purification. Many of these scientists looked at activated sludge as essentially a physical rather than a biological process. "Whether the method in use were one of filtration or the more recent method of activated sludge, the primary action—the separation of the colloidal impurity from the liquid—was a physical one," according to J. A. Reddie. Contrary to the results of Ardern and Lockett, he reported that filters composed of inert substances, like finally divided quartz, could remove the colloids from sewage, even if the material as well as the sewage were sterilized. Activated sludge that had been boiled to kill the bacteria also was able to remove colloids from solution, Reddie concluded.[74]

Proponents of physical factors returned to the Hampton Doctrine and the work of Jones and Travis on sewage colloids. Jones and Travis had first presented the Hampton Doctrine to counter the equating of "bacterial" with "natural" in describing sewage treatment. By emphasizing the physical processes in settling out impurities, Jones had sought to diminish the role of bacteria and thus the naturalness of bacterial methods. Proponents of physical factors now marshaled this argument to produce a denaturing of the activated sludge process. Unlike the proponents of the Hampton Doctrine, however, these scientists recognized the role of bacteria in purifying sewage. But they divided the activated sludge process into two key steps, which they labeled clarification and purification. Clarification was the removal of particles and colloids from suspension, clearing the water. They attributed the clarification to various chemical and physical properties of the activated sludge floc. Purification, in contrast, was the oxidation of this organic matter to its inorganic constituents, carbon dioxide and nitrate primarily. To these scientists, purification was the only step that seemed to require biological activity.

Activated sludge, as an industrial process, could be rationalized, broken down into its component steps, so that each could be accomplished more efficiently. Like Simmer, these engineers were creating an assembly line of sorts for sewage treatment. Watson and Francis O'Shaughnessy, of the Birmingham treatment plant, developed a scheme they named the "flocculated sludge process." Different parts of the plant would perform each function separately—clarification in activated sludge tanks and purification on trickling filters. The sewage could thus be passed through the activated sludge tanks more quickly, streamlining the process and increasing the capacity of the plant.[175]

Others though, taking this denaturing further, saw the separation of activated sludge into component stages as simply one step toward the replacement of a living floc with an artificial one. By emphasizing the physical process of clarification, engineers began to decouple the treatment process from the biological activity of the floc. From this perspective, there was no fundamental difference between clarification by activated sludge and clarification by the older techniques of chemical precipitation. "The coagulum which is formed by the chemicals used in the ordinary process of precipitation is nothing more or less than an artificial sludge," wrote Arthur Martin.[176] Sanitary scientists began searching for such artificial sludges that would mimic the biological sludge's ability to clarify sewage without the difficulty of growing the activated sludge organisms.

In chemical precipitation, chemical precipitants were spent in the process and had to be continually added anew. In the activated sludge process, biological reproduction replenished the sludge, and it could be used repeatedly. "If we regard the activated sludge as a biological precipitant, probably the living portion of it plays a double function, firstly in acting like any other non-living material possessed of large surface area as an adsorbent for the organic matter in the sewage, and secondly in being able with the aid of oxygen to regenerate the precipitant or reactivate the sludge," wrote H. T. Calvert.[177] Chemists searched for artificial substitutes for each of these two functions. The more straightforward substitution, as implied by Calvert, was to find an inorganic sludge that would provide the large surface area of the biological floc. Just about any small particle was proposed. Fowler himself suggested using river silt, which he called "activated silt," and others suggested using clay.[178] Calvert proposed using coal dust, allowing the extraction of both energy and ammonium sulfate from the sludge. Swiss scientist Paul Zigerli patented what he called the "Z" process, which relied on using asbestos fibers as an artificial activated sludge.[179]

Finding an artificial sludge that could be reactivated was the more difficult project, but essential if the advantages of activated sludge were to be duplicated. Ralph Stevenson, a researcher at the Great Western Electro-Chemical Company, a chlorine manufacturer, developed a scheme at the Palo Alto, California, sewage treatment plant analogous to activated sludge, but using a chemical precipitant rather than bacteria. "Sewage purification now accomplished by biological action could just as well be obtained chemically," he argued, and it would possess many advantages, in particular that it did not "hinge on the action of temperamental bacteria which are very apt to 'turn up their toes' at just the wrong moment."[180] Like the activated sludge process, they first needed to build up a required volume of sludge, but they relied on chemicals rather than biological growth. This sludge was then added to a mixing tank with fresh sewage and a small amount of additional iron salts. The sludge would settle out, precipitating the sewage with it. By treating the settled sludge with chlorine, they could "regenerate" the spent coagulant and use it repeatedly for approximately fifty cycles of regeneration.

Chemist Emery J. Theriault, of the U.S. Public Health Service, undertook the most explicit research program for developing an artificial sludge. Theriault argued that activated sludge worked too quickly for biological processes to be responsible and concluded that "the participation of biological elements . . . in the primary clarification of sewage is definitely ruled out." Rather, he argued, clarification was due to the physical and chemical characteristics of the "the floc itself . . . apart from embedded bacteria, secreted enzymes, or attending protozoa."[181] Theriault conducted tests of the mineral composition of the activated sludge floc, and reported that the composition was "unmistakably" that of a zeolite, a naturally occurring mineral commonly used to clarify drinking water in municipal water purification plants.[182] Theriault proposed a "biozeolitic theory of sewage purification," which emphasized physical and chemical action and freed the clarification process from the need for bacteria. The source of the zeolite, he argued, was the feces themselves, rather than the attached bacteria, and the clarification of the sewage was "accomplished by the adsorption of the organic matters on the sludge zeolite, without . . . bacterial intervention." Theriault suggested that dried, or sterilized, sludge could be used to clarify sewage as easily as live, activated sludge. Whereas activated sludge was reactivated by bacterial oxidation, dried sludge, he argued, could be "regenerated" by treatment with sodium chloride, making it possible to develop a treatment scheme that no longer relied on bacteria at all.[183]

These attempts to develop artificial sludges, however, were all unsuccessful. Birmingham's "bioflocculation" process never spread widely, and when the city built a new treatment plant in 1930 it used the activated sludge process without modification. By 1938, it had abandoned bioflocculation altogether.[184] When Palo Alto built its treatment plant, it abandoned the artificial sludge as well as advanced treatment in favor of a cheaper gravity clarification system and reliance on dilution of the effluent in San Francisco Bay.[185] Despite Theriault's confidence that he had determined the mechanism of activated sludge purification, his research inspired no innovation in the sewage treatment field. More than ten years after Theriault's papers on biozeolites, an engineer with the city of Baltimore reopened Theriault's investigations but determined that "activated sludge as a zeolite is of negligible value."[186]

These efforts to replace activated sludge were ideological as well as technical projects, quests for control over the capriciousness of biology and nature. There was an assumption that the more industrial the process, the better its performance. Chemistry could be more easily rationalized than biology and made to more closely resemble a well-controlled manufacturing process. Despite the abundant evidence to the contrary, however, and the failure of scheme after scheme, engineers and sanitary scientists were loath to abandon their quest for complete control over biology.

Yet it was the very nature of biological processes that made them valuable, and artificial replacements were not easily found. Artificial sludges, it turned out, had the same drawbacks as their intellectual forbear, chemical precipitation. Chemicals added to sewage to precipitate the solids were lost and had to be continually added anew. In contrast, the biological nature of activated sludge meant that the precipitating agent—the biological floc—could be renewed by the addition of air and the regenerative power of the organisms themselves. Another perceived drawback of natural biological processes, Simmer's "accidentally developed associations" of bacteria, was also in fact one of their great strengths. The bacteria themselves would adjust their populations to the nature of the incoming waste stream. As the composition of the sewage varied, so would the bacteria. As it turned out, the vital character of activated sludge could not be replaced, and biological sewage treatment could not be arbitrarily denatured. The industrialization of the process seemed to have reached its limits. Buswell, arguing for the importance of the vital nature of activated sludge, wrote: "The term 'cultivated sludge' used by one author, and 'animate energy' used by another contrast as strongly as any with the term 'coagulated' or 'agglomerated' sludge, used by those favoring the colloidal theory."[187] Buswell and Long's initial biological theory of the process became dominant, and his metaphors of "cultivation" and "animate energy" prevailed over the chemical metaphors of "coagulation" and "agglomeration."

The industrial ecosystem was created through a simultaneous process of naturalization and denaturing. Theories and practices of evolution and ecological science were applied to microorganisms that produced useful products, making the production systems into ecosystems. These ecosystems were then systematically simplified and regularized, making the ecosystems industrial. But there were limits to the extent an ecosystem could be denatured and still function. In 1919, British microbiologist A. Chaston Chapman commented on the rapid progress that chemists had recently made in understanding enzymes and biochemical changes. Chemists were now able to replicate biochemical changes in the laboratory, using "various purely chemical processes." "Many of the changes," he noted, "can be readily brought about by what may be termed, in this connection, 'artificial' laboratory methods." Nevertheless, he warned, "there are still a great many important natural processes . . . which cannot be successfully imitated." "For these," he continued, "we are still dependent on the living organism."[188] The industrial ecosystem itself resisted its ultimate industrialization and maintained its character as a complex, living system.

2

Public vs. Private: "Nature Must Be Circumvented"[1]

In 1914, sanitary engineers in the United States formed an organization with the unlikely name "National Septic Process Protective League." Eventually representing over two hundred municipalities, the league was established to battle the Cameron Septic Tank Company of Chicago, which had sued the cities for infringing its patented sewage treatment process. After a "long and stubborn contest in which many engineers and municipalities . . . cooperated to resist" the patents, Cameron and the league reached an agreement in 1919. Relieved that the controversy might finally be over, the *Engineering News-Record* praised "this remarkable league of states, cities, companies and individual engineers who have stood united in the long fight."[2] The Cameron Septic Tank case was just one of several sewage patent cases in the early part of the century in which public health officials, sanitary workers, and municipal and state engineers organized against the commercialization of science and engineering. They formed groups opposed to the patents, published detailed critiques, testified in court, and encouraged municipalities to resist the demands of patent syndicates.

The conflict over patents centered on issues of public over private interests in sewage treatment, municipal infrastructure, and science and engineering more broadly. By "public," I mean public in multiple senses, all of which contributed to the opposition to patents. Biological sewage treatment was established on research performed by public bodies in the public interest. The research was funded by the public itself, typically through municipal governments and local taxation. The research was also made public—disclosed to the broader scientific community—in journal articles, in scientific meetings, and through the movement of scientists and engineers nationally and internationally. Finally, and perhaps most fundamentally, scientists and engineers considered the bacterial action responsible for treating the sewage to be essentially a natural biological process. These engineers believed that as "Nature's means or methods,"[3] sewage treatment processes should belong to the public at large. Because the controversy implicated the idea of natural processes as common property, the very question of what was natural and what was artificial came to the fore in this context as well. Whether a sewage process could be privatized by the grant of a patent turned on the question of whether the process was considered natural or artificial. As a result, many engineers charged, inventors and courts distorted nature itself.

Much of this early twentieth century opposition to patents resonates with current controversies over the patenting of software, business practices, pharmaceuticals, or research performed by public institutions. Scholars have typically dated these concerns to the late twentieth century, when developments in biotechnology and related public policy led to an increased emphasis on patents. A constellation of factors in the 1980s led to the establishment of the so-called triple helix of industry, government and university involvement in technological innovation and intellectual property. During this period the key techniques for producing recombinant DNA were developed and patented, the Supreme Court case, *Diamond v. Chakrabarty* permitted the patenting of life forms, and the U.S. government passed the Bayh/Dole act and other legislation that encouraged researchers to file for patents on work funded by the government. As a result, governments and academia made patenting an important part of their research, and licensing and joint undertakings with industry increased substantially.[4]

As Angela Creager points out, though, much of the recent history of biotechnology has focused on the manipulation of DNA as the novel characteristic in the structure of the new industry rather than on the altered relations between laboratory and market exemplified by patenting.[5] The sewage patent controversies highlighted those relations. These patent conflicts called into question the proper role of the scientist and engineer in society. Sewage treatment patents generated so much opposition because of the important relation between sanitary science and engineering and the public interest. A significant group of engineers worked in local and state government, on sanitation, roads, and public utilities. These municipal engineers, working for public agencies rather than corporations, were instrumental in the development of modern sewage treatment. The contrast between the public character of sanitary engineering practice and the privatization inherent in patenting produced among them a profound unease. They considered research on sewage treatment and its benefits to belong to the public and reserved particular opprobrium for municipal engineers who, despite the public nature of their work, patented the processes they developed while working for cities. This history complicates histories of engineering by David Noble and others, who emphasize the instrumental role of engineering in the rise of corporate capitalism as engineers mobilized the powers of science and technology.[6] In the sewage treatment patent controversies, however, engineers were central in critiquing patenting and privatization of research and the reach of corporate interests.

Patent law and policy has not been static. In Britain, patents were originally a grant from the British Crown for a monopoly to produce a given product, but were eventually restricted to only new inventions. They were still granted on an ad hoc basis and no formal administrative structure for patenting existed. In the United States, the 1789 Constitution specifically empowered Congress to provide for patents and copyright. The first U.S. patent law was enacted in 1790 and specified that a patent could be granted to "any useful art, manufacture, engine, machine, or device, or any improvement therein not before known or used." Through a variety of legislation and legal interpretation, a modern, bureaucratized patent system began to emerge in the mid-1800s out of the "pre-modern" assemblage of common law and traditional practice that had existed previously.[7]

Despite the legal systematization of patenting and modernization and bureaucratization of the process, sanitary engineers remained deeply opposed to certain classes of intellectual property. In the nineteenth century, courts in the United States and England began expanding on what class of invention could qualify for the term "manufacture," and began to treat a process, or a method of making a product or performing a task, as a manufacture itself. The courts expanded what was considered a patentable invention to include these less tangible manufactures. The sewage patent controversy called these longstanding decisions in the U.S. and English courts into question.

Despite the opposition, however, U.S. courts ruled the sewage patents valid. For the courts, the key determination was not whether sewage processes should be public or private, but rather whether they were natural or artificial; the term "useful Arts" enshrined in the U.S. Constitution as the subject matter for patents, after all, is related to "artifice" and "artificial." In ruling sewage processes to be fundamentally artificial, they became valid subject matter for patents and were privatized. With the privatization of sewage processes over the first half of the twentieth century, research on sewage treatment shifted increasingly from public, municipal agencies to industrial research laboratories. With this shift, patenting of research became the norm rather than a contested exception. The sewage patent cases were instrumental in establishing patents in industrial microbiology in general. In these cases, scientists, engineers, patent officials, and the courts fought out fundamental questions on the patentability of bacterial processes as well as the microorganisms themselves. The failure of this opposition movement to curtail the patenting of sewage treatment processes had direct implications for the modern expansion of patenting to the classification of living organisms themselves as intellectual property.

The Septic Tank Syndicate

In 1896, Donald Cameron, city surveyor for the city of Exeter, England, received a British patent on his design for a tank that he had built for the city to treat its sewage. Many previous sewage treatment techniques had prevented the putrefaction of sewage and its resultant odors by killing the resident microorganisms. In contrast, Cameron's tank relied on the action of anaerobic bacteria to digest, or liquify, sewage solids. Preliminary treatment by the septic tank allowed the sewage to be more easily treated by other chemical or bacteriological means. Cameron, coining the term "septic tank" to contrast his tank with the *anti*septic nature of previous processes, promoted his invention in the engineering press and to the British Royal Commission on Sewage Disposal. Cameron received a British patent for the septic tank in 1896 and was issued a U.S. patent in 1899. The Cameron Septic Tank Co. was established soon after to receive royalties for the use of the invention in the United States.[8]

The Exeter treatment plant attracted the great interest of sanitary engineers around the world. Cameron even erected an observation chamber in the tank so visiting engineers could climb down through a manhole and see the septic process in action from behind a viewing window. Engineers in the United States saw great promise in the system and began designing their own treatment plants. But instead of modeling their tanks on the Exeter design, engineers used the idea of septic action in general in tanks of a wide variety of shapes and configurations. The Cameron Septic Tank Co. notified cities that had installed any tank taking advantage of anaerobic action that they were in violation of the patent, for the American patent differed from the British in one key respect. When he filed for a U.S. patent, Cameron claimed to have invented not only a tank of a specific design in which the septic process took place. Rather, he claimed the septic process itself. The company demanded 5 percent of the construction cost of the plant as a license fee, as well as an additional 3 percent for each year the plant had been operating without a license.[9]

Leonard Metcalf, a Boston consulting engineer, was surprised to receive a letter from the Cameron Co. claiming infringement from a tank he had designed and installed. He initiated the broader opposition to the patent by presenting a paper to the American Society of Civil Engineers that criticized Cameron and encouraged his "brother engineers" to resist the patents. Cameron's tank, he argued, was anticipated by the long history of sewage treatment tanks going back to 1852.[10] Other engineers soon joined Metcalf, and on the basis of their arguments, many cities refused to pay royalties. Few engineers expected that the Cameron patents would be upheld. F. Herbert Snow, a consulting engineer, city engineer for Brockton, Massachusetts, and later chief engineer of the Pennsylvania State Department of Health, did "not think for a moment that proprietary rights in the septic system need be a subject for serious apprehension."[11] Another engineer, busy designing a sewage plant for an Illinois town, assured its mayor, "I am paying no attention to the claims of the Cameron Company."[12]

Some towns and cities succumbed to Cameron's demand for royalties. Plainfield, New Jersey, confronted by the Cameron company, sought the assistance of New

Figure 2.1
The septic tank at Saratoga Springs, New York, before operation. For a sense of scale, the ceiling is 9 ft. 6 in. high. Sewage would later fill this tank to a depth of about 8 ft. After treatment in the septic tank, the effluent would be purified in separate bacteria beds. The Cameron Septic Tank Company sued Saratoga Springs and other municipalities for infringing its patent. *Source*: F. A. Barbour, "The Sewage Disposal Works at Saratoga, N.Y.," *Journal of the Association of Engineering Societies* 34 (1905): 47

Jersey's Sewage Commission in defending themselves, but the attorney general determined that the state did not have funds to hire special counsel to defend the city. The Sewage Commission encouraged Plainfield to contest the suit, but the city settled with the Company as the cheaper alternative to going to court.[13] In contrast, Saratoga Springs, New York, faced with an infringement suit, joined with other cities to defend in court their use of the septic process (figure 2.1). The progress of the trial and subsequent appeals were followed closely by the engineering profession. The most prominent sanitary engineers of the period testified for the cities, and each development in the case was reported in the chief newspaper for the civil engineering profession. Engineers could follow the course of the legal proceedings, and the texts of the trial and appeals court decisions were reprinted in full.

In 1907, the trial judge agreed with the municipal engineers on the validity of the septic patents, stating: "There is no new . . . function performed by the tank. It receives and holds sewage as it always has." Nor had Cameron invented a new process, the opinion declared; rather, he had simply relied on a "process of nature." "As a closed tank, pure and simple," the judge ruled, "it holds the sewage placed therein . . . and enables one of the processes of nature to go on therein. Such always has been and such always will be the function of such a tank. That the process described goes on therein may have been a new discovery, but the process of nature is not caused by the tank."[14] Cameron appealed. While the appeals court agreed with the trial court on the apparatus claims—that is, there was nothing novel about the design of the tank itself—it sustained all of the process claims. The septic process itself, the liquifaction of sewage solids using anaerobic bacteria, was found to be patentable.[15]

The engineering community reacted with shock. "This is the reverse of all former decisions in lower courts, both state and United States and of the opinions of engineers generally," reported *Municipal Engineering*.[16] In response, engineers organized more broadly. "Without some organization, all municipalities having sewage tanks would practically be at the mercy of a strong company," wrote Anson Marston, an engineering professor at Iowa State University who had designed a number of septic tanks. In response to the threat from Cameron, Marston became chair of an organization representing mostly small Iowa towns.[17] Two months after the appeals court decision, engineers representing over twenty municipalities and state boards of health formed the larger "Association for the Defense of Septic Process Suits"[18] to carry an appeal to the U.S. Supreme Court. Although the appeal failed, the engineers were not deterred. On studying the appeals court decision, they argued there could be many ways of constructing and operating a septic tank that would not infringe the patent—for instance, by simply cleaning the tanks every few months. The association thus advised its members not to pay any damages to Cameron until their specific plant had been determined to infringe.[19] The League of California Municipalities also organized a group to defend its member cities against the Cameron patent claims, with the state and the Association for Defense of Septic Process Suits both providing legal assistance.[20] As a result of this nationwide organizing, the appeals court decision was widely ignored. The *Engineering News* later reported: "Numerous cities, counties, states, the United States itself, have ignored the patents. So have public institutions, private corporations and individuals. And so have unnumbered engineers."[21]

When Cameron initiated additional suits even after the presumed expiration of their patent, the League of Iowa Municipalities called for another organization to defend infringement suits, and formed the "National Septic Process Protective League." Within five months, over seventy municipalities had joined this organization.[22] Finally, in 1919, their resources exhausted by the endless litigation forced on them by the organized resistance to the patents, Cameron offered to settle all remaining claims with the league for a nominal sum of $5,000. The *Engineering News-Record* noted this final disposition of the case with its testimony to the "remarkable league of states, cities, companies and individual engineers."[23]

Activated Sludge Inc.

In the midst of this fight with the "septic tank syndicate," the *Journal of the Society for Chemical Industry* published Ardern and Lockett's 1914 report on activated sludge. With the septic tank patent battles fresh in their minds, however, the response of the engineering community to the activated sludge process was conditioned to a great extent by concerns over its patent status. On a visit to European sewage treatment plants late in 1914, Edward Bartow, professor of sanitary engineering at the University of Illinois, went to Manchester to meet with Gilbert Fowler, where he saw the experiments with activated sludge (figure 2.2). Bartow was tremendously excited by the results, but he expressed great concern to Fowler over the patent status of the process. "Have you made any arrangements to have the process handled by any firm in the United States?" Bartow asked. "I hope that no firm will get hold of patents on the process and cause trouble," he continued, "such as was caused by the Cameron Septic Tank Company."[24]

Fowler replied that he was "anxious to avoid anything like the experience of the Septic Tank Syndicate" and described his plans for distribution of the activated sludge process that he hoped would seem unexceptional to American engineers: "In any work that is done in the States for which I can be held in any way responsible the idea would be that I should be paid ordinary consulting fees and that if Messrs. Jones & Attwood's appliances are used they should be paid an equitable trade price. . . . No question of royalty would in general arise if reasonable fairness be maintained on both sides." However, despite these assurances from Fowler, Bartow's wish that "no firm will . . . cause trouble" was not realized.[25] For from a very early stage in the research on activated sludge, Fowler was quietly involved in negotiations with Walter Jones of Jones & Attwood Ltd., a sewage and hydraulic engineering firm, to establish a syndicate to develop and market the activated sludge process. American cities would ultimately be tied in litigation for decades and be forced to pay millions of dollars in royalties.[26]

Initially, Fowler, with "very clear recollections of the troubles of the Septic tank Syndicate"[27] and concerned over how an interest in a patented process would affect his reputation, "did not care to be connected with a patent."[28] As the commercial potential of the process became clear, however, Fowler began to abandon his prior convictions and applied himself wholeheartedly to the business aspects of the

"Harnessing the forces of Biotic Energy in the service of Man"

Figure 2.2
"Harnessing the Forces of Biotic Energy in the Service of Man." This image of Gilbert J. Fowler is from a volume of papers honoring Fowler by his students at the Indian Institute of Science. Fowler moved to India right after the controversy with the city of Manchester over the activated sludge patents, where he pursued research on the nitrogen cycle, sewage manure, and other industrial uses of microorganisms. *Source*: Frontispiece, *Some Studies in Biochemistry by Some Students of Dr. Gilbert J. Fowler* (Bangalore: Phoenix Printing House, 1924)

scheme. Fowler's business associates had advised him not to take out the patent himself, as this might be a blemish on his reputation and affect the appearance of scientific objectivity. Rather, he should allow the syndicate to take out the patents.

A potential impediment still existed, though, in the legal question of whether Fowler or Manchester's Rivers Committee, for whom he had directed the research, had rights to the patent. Having checked with the town clerk, he was told that legally "the Committee can have no claim whatever upon me." He then argued that the committee had no *ethical* claim on the patent either, for "while it may be difficult to disentangle the precise origin of any detail what is clear is that no member of the Rivers Committee has invented anything."[29] Fowler allowed Walter Jones to patent the invention in Jones & Attwood's name and, in turn, Fowler received a £1000 insurance policy.[30]

Fowler was technically truthful when he wrote to the Manchester Rivers Committee, "I have therefore no patent or share interest in any Company,"[31] since he only later converted his insurance policy into shares. But many on the city council nevertheless began to suspect that Fowler, who was being quietly paid by Jones & Attwood as well as Manchester, was "serving two masters."[32] The council became concerned both for the city's interests in the sewage process developed with its staff, facilities, and funds and for the reputation of the city. Throughout these early negotiations, it is clear that patents were viewed by both Fowler and Manchester as somewhat sordid; something that neither wanted to be publicly associated with.

The Rivers Committee, although unhappy with Fowler's actions, was legally constrained, because Fowler, as a consultant to the committee and not an actual employee, was beyond their control. Faced with a split on the council and the inability to prevent Jones & Attwood from patenting the activated sludge process, Manchester made an accommodation with the firm that the city should have access to the process free of any royalties. Many on the council, while acceding to the agreement, wanted to formally express the view "that they do not approve of the connection of Dr. Fowler with the syndicate formed for the purpose of exploiting certain patents taken out by Messrs Jones & Attwood Ltd." This attempted censure narrowly lost, however.[33]

Activated sludge promised to be an extremely lucrative business, and sewage patents were a key component of Jones & Attwood's business strategy. "The Sewage Work of the World is a big thing," wrote J. A. Coombs, Jones & Attwood's chief sewage engineer, "and the firm are by no means selfish in trying to corner it."[34] In attempting to corner the market, Jones & Attwood pursued a patent strategy similar to the more well known examples of the telephone and electric utility sectors, in which AT&T and General Electric sought to control whole industrial sectors by controlling the patents for key innovations.[35] Having already purchased the initial activated sludge process from Fowler, Jones & Attwood sought to purchase

Figure 2.3
Bioaeration plant at Sheffield, 1919. City engineer John Haworth reportedly developed this process after observing the action of turbulence in rivers. In these plants, the activated sludge followed a serpentine path through the plant, and was aerated using mechanical paddles. Jones & Attwood tried to entice Haworth to patent the process in the firm's name. Instead, he intentionally left it in the public domain. The bioaeration process competed successfully with Activated Sludge Ltd. in Great Britain but was never widely adopted in the United States. The mechanical agitators are at the far ends of the channels. *Source*: Edward Bartow Papers. Courtesy of University Archives and the William A. Rice family

and file patents on any variations of the process that were being developed by other municipal researchers. John Haworth, city engineer of Sheffield, was working on a system using mechanical agitation, rather than diffused air, to aerate and activate the sludge. Inspired by the aeration in a natural river, Haworth designed a winding concrete channel in which the water was propelled by huge paddles (figure 2.3). Because municipalities might get around the activated sludge patents by using this process instead of Jones & Attwood's diffusers, Coombs spent a great deal of effort to entice Haworth into a deal similar to their arrangement with Fowler. With a promise of royalty-free use of the patent for the city and a substantial payment for the researcher, the firm hoped to tempt municipal workers like Haworth to give up their inventions. Arguing that the researchers, as municipal employees, "had no legal claim to any financial recognition" and, because cities paid for the research, "stood no financial risk in the development of their ideas," Coombs described their arrangement as "generous" and "unprecedented," if so sensitive as to require being kept "private and confidential."[36] In the end, Haworth refused. Even if Sheffield would be exempt from paying royalties for his own invention under Coombs's scheme, he did not want to "place other municipalities in the position of paying the Activated Sludge Limited fees."[37] Indeed, Haworth's convictions were broader, and included the public interest in general. Occurring during the First World War, many of the remarks by these engineers were suffused with statements of patriotism, peace, and higher cause. "I have not been working for myself," he stated, but rather the municipality of Sheffield, and "the cause."[38]

Soon after Ardern and Lockett's pathbreaking paper, activated sludge was taken up by many cities in both England and the United States, where engineers established experimental plants to test and develop the process. Twelve cities established experimental plants in 1915 alone, the most prominent being Milwaukee, Wisconsin.[39] Consulting engineers for Chicago as well recommended the activated sludge process. When engineers in the United States first heard of activated sludge, they assumed that its use would be "free to all" since there had been "no public mention of patents on either the tank or the process."[40] Rather, engineers looked to the published papers of the Manchester researchers for guidance. "The papers of Ardern and Lockett describe the process so well that each engineer or chemist feels that he is capable of following along the lines described in their paper without additional assistance," Bartow reported to Fowler concerning Fowler's desire to consult for American cities that might build activated sludge plants. Regarding the use of Jones & Attwood's appliances for diffused air, Bartow noted that "American practice has dealt so largely with mechanical water filtration that the supplying of air does not seem to them to be a difficult problem and they are going ahead with various devices of their own."[41]

As news reached America that the original process had indeed been patented in England and that applications were pending for U.S. patents, engineers immediately thought of the septic tank, and began to organize against the activated sludge patents in a similar way. At the 1916 meeting of the American Society of Civil Engineers, engineers called for the establishment of a special committee.[42] This committee gave several prominent engineers, including Earle Phelps, Edward Bartow, and George Hammond, the task of establishing the "priority of knowledge of the activated sludge process" essentially reprising the role of Metcalf, fifteen years earlier on the septic tank. These engineers all submitted affidavits to the commissioner of patents "to prevent, if possible, the patenting of this process by English engineers to whom Fowler has signed over his interests and rights."[43]

The acting secretary of the committee was T. Chalkley Hatton, recently hired as chief engineer for the Milwaukee Sewerage Commission. He was a veteran of the septic tank battles as well. As a municipal and then consulting engineer in Wilmington, Delaware, he had been a founding officer of the Association for the Defense of Septic Process Suits in 1908.[44] Hatton was particularly incensed over the activated sludge patents, as Milwaukee had hired Fowler to consult on their activated sludge experiments, and only after months of correspondence and the approval of a consulting contract, did Fowler intimate that the process was patented by Jones & Attwood. Because Hatton assumed these patents would be on the specific apparatus that Jones & Attwood had designed, he appeared little concerned since he was moving forward with plans and equipment of his own, rather than Jones & Attwood's apparatus.[45]

After further correspondence, however, Hatton began to worry. "In your last letter just received you suggest to my mind inquiring about patents," wrote Hatton. "I should like to know from you, what if any patent rights have been obtained which might govern the use of the process with which we are experimenting."[46] So began years of tense correspondence and decades of litigation. Milwaukee had designed their Jones Island treatment plant from scratch, using data from their own experimental plant, yet Jones & Attwood claimed infringement. "The design of the plant is entirely original with us," wrote Hatton. "In fact, the only comparison . . . is in the process of using 'activated' sludge," which he considered to be, if not public, at least included in Fowler's consulting fee.[47] Again, Hatton wrote Fowler, "I cannot help but feel that much of the information so far secured here in Milwaukee . . . is original."

The Milwaukee Sewerage Commission negotiated with Jones & Attwood for several years over the issue of patents and licenses, finally reaching an impasse in 1922.[48] In 1925, Activated Sludge's American representatives began contacting cities across the country, pressing for royalties. As with the septic tank, many cities refused to pay, arguing that their activated sludge plants were designed on the basis of experimentation performed by the cities themselves. Again, similar to the septic tank case, cities banded together to resist the claims of the sewage treatment syndicate.[49] Activated Sludge sued the Sanitary District of Chicago in 1925 and Milwaukee in 1929. In 1933, the Milwaukee case was decided first, in favor of Activated Sludge.[50] Armed with this initial victory in the Milwaukee case, Activated Sludge filed suit against over one hundred cities in 1933, ranging from tiny West Concord, Minnesota, population 613, to Los Angeles and Houston.[51]

After over twenty years in the courts and numerous appeals, the patents were finally held to be valid and infringed. Even though Ardern, Lockett, and Fowler worked for Manchester, the court chose to accept the fiction that they all, in fact, were working under Walter Jones's direction, and that Jones was the true inventor of activated sludge.[52] Unlike the septic tank case, the monetary damages were substantial. Milwaukee was assessed damages of almost $5 million and almost had to shut down its treatment plant altogether. They were later able to settle for $818,000, while Chicago was assessed $950,000. The other hundred-plus cities and towns in the United States were forced to pay an additional $600,000.[53]

Ultimately, neither the opposition to the patents for the septic tank nor activated sludge won in the legal arena. After controversy in the patent examiner's office, the patents were all issued.[54] And after mixed results in the lower courts, the appeals courts all ruled in favor of the patentees and against the municipalities and their supporting scientists and engineers. While some legal historians might argue that the decisions in these cases were appropriate and based on sound legal arguments reflecting the state of patent law at time,[55] the patent struggles were about more than the specific claims of the various patents. Rather, they centered on a much deeper critique of patents, widespread in sanitary and municipal engineering.

Patents and Professionals

When Metcalf presented his paper on the antecedents of the septic tank, he hoped to alert the civil engineering community to the tactics of the Cameron Septic Tank Company. But he also hoped to elicit a discussion on the broader question of patenting any "methods and devices used in any engineering works." In the ensuing discussion within the engineering societies, it became clear that the status of patenting within the engineering community was complex and was bound up with many issues concerning the professional position of the engineer. Whether an engineer was self-employed as a consultant, worked for a city or county, or worked for a large industry or a small engineering firm all affected his position on patenting.

When Metcalf wrote his paper, the ethics of patenting in particular, and role of commercial interests more generally, had been debated on and off among the recently established professions of science and engineering. As Paul Lucier describes, in the 1840s and 1850s, conflicts in the United States and England over patents on kerosene exposed many of the divisions among scientists and inventors over the propriety and ethics of patents. Some disparaged patenting as beneath the dignity of their calling. "No true man of science will ever disgrace himself by asking for a patent. . . . He cannot and will not leave his scientific pursuits to turn showman, mechanic, or merchant," said geologist Charles T. Jackson. Other scientists, however, thought there was no impropriety in taking out patents. For Lucier, the distinction between pure and applied science became one of the relation of scientific work to commerce. Engineering supposedly embraced a relationship with commerce. The distinction was nowhere near as clear, however, and a debate over the ethics and propriety of patenting continued within the engineering profession itself.[56]

Respondents to Metcalf's paper took up the discussion of patenting in general, and the paper spurred further discussion in engineering journals over the ethics of patenting. Engineers aligned on the different sides chiefly by their professional position; municipal and consulting engineers opposed patents; engineers working for firms supported them. James Owen, County Engineer for Essex Co., New Jersey, after relating his experience with the demands for royalties from Cameron, tried to "lay down the law" on the patent question in general: "It is entirely unprofessional for a civil engineer to get any patent on any work which he has constructed."[57]

Incited by James Owen's comments, Archibald Eldridge, a railroad engineer for the Burlington Road,[58] made a presentation the following year to the American Society of Civil Engineers entitled "Is It Unprofessional for an Engineer to Be a Patentee?" Eldridge was incensed over Owen's critical attitude toward patenting and defended the practice against charges of money-grubbing on the part of engineers. "Engineers are simply business men," he stated, "and, as such, are entitled to full compensation for their wares, the products of their brains."[59]

Engineers were in the midst of trying to establish themselves as a respected profession.[60] To many professional engineers, the practice of patenting materials and methods, and then charging clients for their use, reduced their status in the eyes of the public. If engineers hoped to be among the "learned professions" of medicine and law, they would need to adopt a code of ethics similar to those professions. Physicians had adopted a code that prohibited the patenting of methods and devices used in their practice; engineers would have to do the same. In a professional practice, "thoughts and ideas must and should be untrammeled by any question of individual interest of the engineer himself." This was less a moral question than one of the interests of the profession. If an engineer "acquires a patent and uses it there is certainly no moral delinquency, but it lowers the dignity of the profession, and lowers its appreciation by the public at large."[61]

To many sanitary scientists and engineers, the assumption that they might be favoring a particular plan only because of a monetary interest threatened their integrity. William Dibdin, for instance, who developed the contact bed for sewage treatment, wanted to clearly separate himself from the idea that he had a monetary interest in the design. He testified to the Royal Commission on Sewage Disposal immediately after Donald Cameron's appearance in which Cameron was pushing his recently patented septic tank. After Cameron stepped down, the chair welcomed Dibdin, "I think you are Dr. Dibdin, and the inventor of what is known as the Dibdin process?" In a seeming rebuke to Cameron, Dibdin responded somewhat testily, "In the sense of being an inventor of any particular system, or having proprietary rights, or anything of that kind—I have nothing whatever to do with that at all. I do not appear before you in that character."[62] Fowler, too, was certainly aware of the opprobrium held for scientists and engineers involved in the commercial development of municipal technologies, and it undoubtedly played a role in his initial reluctance to patent activated sludge, and the later secrecy of his involvement.[63]

Some engineers ascribed the debate over patenting to structural differences in the conditions of employment for engineers. To J. P. A. Maignen, the engineering profession was divided into three general classes of employment: "the consulting engineer, the salaried engineer, and the independent engineer."[64] For the consulting engineer, "his reputation is his patent" and he advanced according to the structure of scientific rewards. For many engineers, lower on the hierarchy of status, the monetary rewards of patents were of much greater importance than reputation. For Jones & Attwood engineers J. A. Coombs and Harry Partridge, for instance, royalties and bonuses from patents developed while employed were a substantial part of their income. At Jones & Attwood, the bonus from a patent might represent as much as 25 percent of an engineer's annual salary. There was thus a much greater incentive for rank and file engineers to patent their work.[65] The financial incentive for salaried engineers to file patents may have been instrumental in Jones & Attwood's pursuing the activated sludge patents to begin with. J. A. Coombs, who was responsible for first connecting Fowler to Walter Jones was under great financial pressure when he first started at Jones & Attwood. He had "considerable debt" when he left his previous employer, and was forced to "sacrifice twelve good patents" and the income they represented on changing employers. This financial pressure may have been behind his pursuit of new patents, particularly for the activated sludge process.[66]

To the defenders of patents like Eldridge, the only ethical consideration required was for engineers to sell their patented products "openly and above board" rather than "'sneaked' into a piece of work or a contract." If an engineer disclosed his interest, that was sufficient. The client could gauge whether the engineer's financial interest influenced his judgment. Eldridge concluded his paper by claiming that "the question is now settled, and . . . an engineer can patent any article which he may invent." Samuel Whinery, a New York consulting engineer who wrote frequently on the question of patents, considered the profession to be much more divided. "There exists a radical difference of opinion in the engineering profession, particularly among civil engineers." Another engineer noted that "the prejudice against patented inventions . . . is more widespread than persons of intelligence and good sense would suppose."[67] Indeed, the question was not settled at all, as engineers across England and the United States criticized the sewage patents and organized to prevent their enforcement.

Sewage and the Public Interest

Many of the differences among the engineering profession toward patents lay in the differing relationship between engineers and their conceptions of and commitment to the public interest. Patenting violated many of the core beliefs of the scientists and engineers of the period over questions of public vs. private interests. Sanitary researchers considered their work to be in the public interest and were highly critical of attempts to privatize that research. In the discussion following one of Cameron's first papers reporting the results of his septic tank, British municipal engineer J. Lemon remarked, "I should like to go as far as this, and make it illegal for a man to take out a patent for anything affecting the public health."[68] Indeed, the prominent English sanitary engineer Arthur Martin, one of the co-patentees of the septic tank in the United States, later regretted his decision to be involved in the patent and declared, "There is a widespread feeling that it is not desirable that any individual or company should possess a monopoly in a process which is essential to the effective discharge of a public service."[69] A 1911 editorial for the New York publication *Municipal Journal and Engineer* decried the recent "troubles" over patents in sewage treatment and wished that state and local entities, like Lawrence, Massachusetts, or Essex, England, had patented the inventions in the name of the public and "given the same freely to the world."[70] Milwaukee's T. Chalkley Hatton, writing to Fowler, expressed his commitment to the public interest in the strongest terms: "Whatever information we obtained here is public property and no patents will be applied for. Speaking from my point of view I am, and always have been, utterly opposed to patenting any process with which I have been connected, and I may say right here that I am giving every aid in my power to help others in this country interested in this process to carry on their experiments to the end that by this means the most good can be given to the greatest number."[71]

The relation of the engineer to the public and public interest was highly contested during this period and was reflected in the various professional organizations.[72] It was the municipal engineers, in general, who were more closely aligned with the public interest than were other branches of engineering, putting the municipal engineer in "frequent antagonism" with pecuniary and commercial interests. The British Association of Municipal and Sanitary Engineers and Surveyors was formed in 1873 with the express aim of insulating municipal engineers from the influence of private interests. The new office of town surveyor was charged with constructing works "of the utmost importance in connection with the life, health, and comfort of the people," explained Lewis Angell, the association's first president. Yet "there are many conflicting commercial interests in a town—a gas interest, a water interest, a railway or a tramway interest, &c., various such directors have seats in the local authority. It is, therefore, almost impossible for the local Surveyor to discharge his duties faithfully and impartially without giving offence to interested parties"[73] To protect the municipal engineer, Angell and the association sought rules to prevent the summary firing of town surveyors.

Professional organizations of municipal engineers in the United States also emphasized the public interest.[74] The American Society of Municipal Improvements was started in 1894 as a way for city officials to share information, but it quickly morphed into an organization almost exclusively of municipal engineers, eventually changing its name to the American Society of Municipal Engineers in 1930. James Owen, the early critic of patents, served as president of the society in 1908. John Alvord, another critic of the septic tank patents, was on the society's Sewerage and Sanitation Committee, along with other prominent consulting engineers in sanitation.[75]

The conflict between public and private interests in municipal engineering was perhaps best exemplified by Morris Llewellyn Cooke, at the time city engineer for Philadelphia, and his struggle for control of the American Society of Mechanical Engineers during the 1910s. Cooke hoped to see engineers become more and more influential in government, but he argued that would happen only if "the engineer will consider in every decision and act there shall be the clearest possible recognition of the public interest."[76] Cooke accused engineers of being too often compromised by the potential for lucrative consulting arrangements with public utilities and other commercial interests that conflicted with that of the city or its citizens. It was impossible, he charged, to find qualified engineers to work in municipal government who had not been tainted by their "service of private interests" in the electrical, concrete, or asphalt industries.[77] Cooke campaigned for a new code of ethics for engineers that would recognize "public interest as the bedrock of professional practice" and would declare that "it is unprofessional for an engineer to safeguard any private or special interest at the sacrifice of public welfare."[78] Patents, it was clear to Cooke, were one of the many ways private interests worked against the public interest.[79]

The objection to sewage patents was a part of the larger opposition to the use of patented materials in municipal civil works of all kinds. Cities, with their increasing reliance on complex technological networks for transportation, gas, and electricity, became more and more vulnerable to monopoly. Public fear of monopoly in providing city services was well grounded. One group of capitalists, for instance, formed a trust to control "all the asphalt in the world." By securing the natural supplies of the material and absorbing competing paving companies, the "Asphalt Trust" hoped to set its own price for all municipal paving in the United States. City governments like New York feared that there would be "no way of delivering the city from the tightening grasp of the monopoly."[80] Other cities were plagued by transport, electricity, or water supply monopolies.[81]

Patents were a key aspect of monopoly in urban networks. The electric utility monopolies were based to a great degree on controlling patents.[82] In the transportation system, many paving materials and construction methods were patented. Engineer Daniel Luten's National Bridge Company sought to monopolize concrete bridge building by patenting most of the commonly used methods of bridge construction. National's unethical use of patents and the disdain over its business practices were emblematic of why twentieth-century municipalities viewed patenting with suspicion. Even though the bridge patents were of questionable validity, the company could force or persuade cities to pay royalties through legal intimidation. Luten used the (false) claim that its patents had been upheld in numerous court cases to force cities to pay royalties on its patents for bridge design. Although the patents were ridiculed in the press (and ultimately found invalid by the courts), many city governments found it cheaper to pay royalties than fight in court. Following the adoption of a Luten design by the city of Topeka, the Topeka newspaper complained: "Topeka will pay tribute to Daniel B. Luten of Luten bridge-patents fame. . . . 'It is cheaper to pay the royalty than to try to build a bridge not infringing on the Luten patents'. . . . Luten holds forty-one patents on the construction of bridges. If a bridge is desired with an arch of a certain degree of curve, Luten has a monopoly on that kind of an arch. There are 360 deg. in a circle, but Luten was satisfied to secure only forty-one patents."[83]

Because of these and other unethical business practices, city officials were distrustful of patents in general. Indeed, following the formation of the National Septic Process Protective League, the National Municipal League considered forming a permanent association to fight all "fraudulent or unjust patent claims," not just the septic patents. "Unless the cities of this country show a disposition to make a united defense of their rights," it warned, firm after firm would come forward with dubious patent claims. By vigorously defending against the Septic Tank Syndicate and forming a new organization, it hoped to show that "the municipalities of the country have 'millions for defense but not one cent for tribute.'"[84]

Much of the skepticism toward patents was also rooted in the commonly held belief, reinforced by the actions of dishonest inventors like Luten, that patented processes were frequently worthless—schemes for extorting money from cities and towns rather than valuable advances in technology. A 1876 report of the Local Government Board in England listed 417 patents for sewage or manure processes between 1856 and 1875,[85] while J. W. Slater listed 434 patents between 1846 and 1886 for chemical treatment processes alone. While some of these patents were indeed useful, the bulk of the precipitation patents were widely derided by much of the engineering community, described by Slater as "bogus patents." "Not uncommonly," he wrote, "to furnish the basis for a patent, some well-known and often-used agent was proposed, and with it, to give an air of novelty, some useless or even mischievous matter."[86] Lime, for instance, was recognized as an efficient precipitation agent. But for sake of novelty (and thus patentability) inventors combined lime with a wide variety of other substances, most of which added nothing to its efficacy. Thus, there were patents for lime in conjunction with soil, sea weed, charcoal, zinc, ashes, soot, potash, carbolic acid, or magnesia.[87] To the active ingredient of alum in the ABC process, the inventor had added a small amount (0.04 percent) of blood that could have no effect on the sewage at all but supposedly corresponded to the teachings of Moses. Common salt was added to some mixtures, sugar to others, and alcohol to still others. Soap and ammonia—ingredients cities tried to remove from the sewage—were added in yet another. Arcane ingredients such as "the water from electric batteries," were ingredients in other patented processes.

Similarly, engineer George E. Waring had patented in the United States a system of sewering cities that relied on small-diameter sewers that would separate stormwater from sewage. The system was built in Memphis, and Waring's company, the Drainage Construction Company, sought to build the system in other cities. Many sanitarians, like Rudolph Hering of Philadelphia and C. W. Chancellor of Maryland, criticized Waring and the patents for claiming principles of sewer construction that were long known and used. Further, Hering concluded that the system itself was ineffective. "Colonel Waring has taken out patents for a *combination* of certain features relating to sewerage, and collects a royalty of ten cents per running foot of sewer," wrote Chancellor. "This 'combination' happens never to have been made before—singly, however, the features have been used, partly as long as forty years." He continued, "Why should the people of Baltimore pay so large a royalty for *practically nothing*?"[88]

Aside from the distrust caused by blatantly "bogus patents," many sanitary engineers felt that sewage was too inherently variable to submit to a patented treatment scheme. Patents, specifying a particular set of steps or procedures, would be too rigid to deal with the extraordinarily variable composition of sewage. Depending on the local climate, water use, industrial activity, population, time of day, or season, the sewage would change, and so would the process needed to treat it. With biological sewage treatment, this variability in sewage was compounded by the natural variability of microbial populations derived from local soil, water, and air. The very natural character of the treatment process and its variable qualities prevented the specification of a universal procedure. "One thing is certain in my mind," wrote Edward S. Philbrick, an engineer for Boston. Sewage is "a local question very largely, and that *patent systems* are no more likely to be widely applicable in our profession."[89] Municipal engineer J. Lemon placed the septic tank into this category of suspect patent systems that ignored local variation: "The moment we get a scheme prepared and suitable for the locality—because you must suit your scheme to the local conditions—down comes one of these sewage disposal inventors . . . and unsettles their minds as to the utility of the scheme they have adopted. . . . The practical effect of many of these inventions has been to retard the proper disposal of sewage. . . . This scheme [the septic tank] requires a good deal of consideration before we are going to say it will be suitable for all parts of England and under all circumstances."[90] One of Hatton's early criticisms of the septic tank was also related to this variability. "The strength and quality of the sewage from any community varies hourly," wrote Hatton, "and yet a septic tank is constructed to purify automatically every kind of sewage in the same way and during the same period of time." Despite this fundamental drawback, the septic tank had been installed widely in the United States, a situation Hatton blamed on the deceptive salesmen of the "patentee."[91]

Research articles evaluating new systems of sewage treatment invariably included a section titled something like "local conditions," because without an understanding of local conditions, the reader would not be able to evaluate the system in its proper context. Without understanding the particularities of place and sewage, one could not treat it. Cities that considered the activated sludge system established testing stations, at great expense, to try the procedure out on their particular sewage under their particular climatic conditions, with the particular bacteria that developed there. The Milwaukee station was established "to try out those processes of sewage treatment applicable to large installations" and "design a sewage disposal plant which would, without any reasonable doubt, best fulfill the conditions of the city of Milwaukee."[92] The process developed in Manchester had never been tried in a cold winter climate like Wisconsin's, nor with the particular mix of industries present in Milwaukee. The very site for the experimental plant in the main river valley near the harbor was specifically chosen so that it could receive a representative sample of the city's industries and neighborhoods. Thus, the activated sludge process, as developed and used in Milwaukee, was different from the activated sludge process of any other city, and was different from the activated sludge process invented by Fowler. The septic tank designed by Snow for Saratoga Springs differed from other cities'. Each was designed with the peculiarities of place in mind. Engineers for Milwaukee and Saratoga Springs considered their systems unique and thus not covered by patents.

But even in those cases where city officials regarded patented materials or processes as useful, there was still a prejudice against them. As part of the progressive reform of city governments, most states had passed laws requiring competitive bidding for municipal projects.[93] Because only the holder of a patent could submit a bid for its use, patented materials or processes precluded competition among bidders. While some states interpreted the requirements for competitive bidding to disqualify patented articles in almost all cases, other states interpreted similar laws more liberally, allowing the specification for patented materials if all bidders could use the "patented article on reasonable terms, which shall be the same to all bidders."[94] Actually implementing procedures for fair competition proved more difficult, however, as there was frequently collusion between patentees and contractors. Thus, despite these safeguards, Samuel Whinery noted in 1915, there remained a "pronounced prejudice against the use of patented articles or processes" on the part of "the public and public officials."[95]

As cities feared the Asphalt Trust and other monopolies, they similarly feared a trust like the one envisioned by Jones & Attwood for sewage treatment. "The great field of sewage purification offers inducements to capital," noted F. Herbert Snow, the designer of the Saratoga Springs septic tank. "Any plausible scheme to create a monopoly of the business can find financial backing." He then connected monopoly control to the control of patents. "New companies are being organized for this purpose. In every instance patents are the basis for the enterprise."[96] English observers also feared the power of the sewage syndicates in very similar terms. One correspondent, perhaps thinking specifically of the septic tank syndicate, wrote to the *Times*, concerned that the 1898 Royal Commission might recommend (and require) only certain processes for municipal sewage plants. "There is money, a great deal of money in this sewage question. . . . It further opens the door to the creation of patent rights controlling indispensable details of the machinery to be subsequently forced upon all local authorities by the action of the Local Government Board. Has the Government so much as made the stipulation, loudly called for in the interests of the public, that no such private rights shall be, directly or indirectly, founded upon any results arrived at in the progress of this inquiry?"[97]

The ire of the sanitary profession, however, did not fall on engineers like Coombs, working in a private firm. Rather, it was reserved for those who most clearly crossed the boundaries between public and private—those who worked for cities and should have put the public interest highest. It was by patenting that the municipal engineer or chemist transgressed this boundary. The opprobrium held for Cameron and Fowler was related to this perceived transgression of the public interest. They were the most notable (and notorious), but others were also subject to professional censure. In contrast, John Haworth was lauded for resisting the allure of patents and standing up for the public interest.[98] In patenting activated sludge, Fowler had transgressed the norms of the municipal engineering profession, in which the public interest was held key, and crossed into the social world of business. Fowler may have thought that his pursuit of what he called "a little petti-fogging profit"[99] was inconsequential, but for many engineers of the period he had clearly violated the norms of his profession.[100]

Despite the importance of the public interest to municipal engineers, the U.S. courts barely referenced the public interest in sewage treatment in their validation of the sewage patents. Yet this was precisely the overriding concern of the sanitary engineers and the municipalities they served.[101] The courts were enforcing a presumed public interest as incorporated into patent law, "to promote the progress of useful arts."[102] But by doing so, they reinforced private, monopoly control over processes critical to the public health, to the objection of municipal engineers.

Public Research

Almost all of the research conducted on sewage treatment in England and the United States was performed by municipal governments or other local agencies, and many of the sewage patents had their origins in publicly funded research. Sewage testing laboratories built on a tradition of municipal science that began with water supply, sanitation, and pavement testing in the 1890s.[103] George Whipple, the sanitary engineer, noted in 1900 that "the laboratory idea is fast taking hold of our municipalities."[104] By 1906, the existence of municipal laboratories working on the sewage question was so commonplace that when Baltimore constructed an experimental sewage station, reports noted that it was of interest primarily for the particular arrangement of distributers and tanks "rather than for any novelty attaching to such a municipal laboratory."[105] Funded through charges to rate payers, cities spent great sums researching alternatives for treating sewage. Chicago spent $500,000 on investigations into sewage disposal, and Milwaukee almost $800,000. Many other cities had spent upward of $50,000 or $100,000. Langdon Pearse of the Chicago Sanitary District estimated that U.S. municipalities and states had provided over $1.5 million for research on sewage disposal from 1905 to 1925.[106] This was undoubtedly an underestimate.

Because the public was funding the research, city engineers felt that the public was entitled to its benefits. John D. Watson, borough engineer for the city of Birmingham, was highly critical of the activated sludge patents at the time they were first filed, on just these grounds. Describing his and his many colleagues' involvement with sewage research, he stated, "We were paid directly by the public, either a municipality or other public body, and we gave all our time and attention to this kind of work . . . and my feeling was and is that work of that kind should not be capable of being transferred to any company for the purpose of making money."[107] Watson was thus very disappointed when his colleague F. R. O'Shaughnessy patented the flocculated sludge process that the entire municipal sewage district was developing. The patent application was opposed, and O'Shaughnessy had to surrender his right to the patent.[108]

T. Chalkley Hatton's opposition to patenting sewage processes extended to his own engineering staff as well. Carl H. Nordell, design engineer for the Milwaukee Sewerage Commission, worked on the air supply problem for the activated sludge tanks and developed diffusers made of basswood, testing them in one of the first experimental tanks built in Milwaukee. But without Hatton's knowledge, Nordell patented these diffuser designs. When he refused to sign a license allowing Milwaukee free use of the design, he was forced to resign. Hatton criticized him for filing patents on inventions "developed and tested at the testing station at the expense of the Commission." Hatton appeared more upset at Nordell's patenting than concerned that Milwaukee might have to pay royalties to use the diffusers, as commission staff was highly skeptical of their utility.[109]

One federal government researcher, Leslie Frank, who had begun developing a continuously operated version of the activated sludge process on the basis of Ardern and Lockett's paper, submitted his own patent application on the process but dedicated the patent to the public. Rather than try to take commercial advantage, his stated goal was to prevent the monopolization of the process by private interests. The attempt initially succeeded, as the patent examiner denied Jones's activated sludge applications on the basis of the previously filed Frank patent. But Jones was able to argue that one of his patent applications, on diffusers for the M.7 process, predated Frank.[110]

As municipal institutions, these experimental laboratories were dedicated to developing research that would benefit other cities as well. Many engineers considered the patents unethical because much of the progress in sanitation had been made possible by the free exchange of information and scientists between cities and countries. The research was widely published in the original sense of the word—of being made public. Papers were presented to professional meetings and published in journals. Notable papers were often republished in other journals, or abstracted or excerpted in the general engineering press. Cities and states considered it their duty to share the results of their research. As Ardern described the publication policy of the Manchester Rivers Committee, "It had always been the policy . . . to publish any research work which they thought might be of assistance to other local authorities or to those engaged in the problem of sewage purification."[111] Langdon Pearse, sanitary engineer for the Sanitary District of Chicago, described the policy of their sewage experiment station in 1924: research would be "broadcast and used by anyone, without fear or favour."[112] Ten years later, following the court decisions in the Activated Sludge case, he feared that the sewage patents were changing the behavior of his profession: "In the past," he wrote, "the practicing engineer in the United States has been free handed in disclosing to the public the discoveries and improvements made in the art, particularly when done at the expense of the public. Whether such a policy is now to be reversed remains to be seen."[113]

In addition to the circulation of scientific publications, the engineers themselves circulated widely, to professional meetings as well as to sewage experiment stations. Activated sludge had its origins in the visit by Fowler to the Lawrence, Massachusetts, experiment station. Once British scientists and engineers began developing the activated sludge process, scientists from around the world visited their facilities. Over one hundred engineers from Russia, the United States, and Britain visited the Salford, England, works in 1914 alone.[114] Scientists visiting Manchester, like Bartow, brought ideas for activated sludge back to the United States, and U.S. researchers working on activated sludge, like Hatton, returned to England to present the results of their work in developing the process further.

Because scientific results were so widely shared and engineers built on the accomplishments of others, many researchers found it difficult to assign exclusive credit to any particular discovery. However, the patent system required that the "inventor" be identified and given exclusive rights to an invention. Leonard Metcalf argued that while an inventor might be *legally* entitled to patent an improvement in sewage, if it were merely an adaptation of "the knowledge and discovery of others," he would not be *morally* entitled to the patent.[115] Fowler seemingly held similar beliefs about activated sludge, proclaiming that "for large advances of this kind no one person can claim credit." Yet by assigning the rights to activated sludge to Walter Jones, he acted to give Jones the sole credit for the invention.[116] I don't want to describe some golden era of scientific openness—these researchers could be petty, concerned about whether they were getting proper credit for advancing the field, and worried that other researchers might scoop their results—but the impression in the literature is one of great excitement, as scientists and engineers made rapid progress on a critical problem facing society and shared their results promptly and widely. In the three years following the publication of Ardern and Lockett's first paper, for instance, over three hundred articles on activated sludge appeared in the engineering journals.[117]

In contrast, the allure of patents led to secrecy and less open publishing. There was a culture of secrecy around the development of patents at Manchester University where both Fowler and Chaim Weizmann were pursuing patents for bacterial processes. On Weizmann's insistence, Fowler signed statements recognizing Weizmann's claims to secrecy surrounding work on his acetone fermenting bacteria.[118] Fowler, on his part, also kept many aspects of the activated sludge process secret until patents could be filed. While soliciting information from other municipal engineers, Fowler was secretive about his own research on activated sludge.[119] At the same time, though, he quietly provided the results of the experiments performed by Manchester employees to members of his syndicate prior to their being published.[120] This secrecy extended to Fowler's participation in professional meetings. In discussing a paper read by Ardern at a meeting of sewage managers, many respondents who were taking up work on the activated sludge idea shared the specifics of their apparatus with the other engineers and managers present. For managers planning new sewage plants, the details on tank capacity, sewage flow, or placement of air pipes that early experimenters shared were critical. Fowler, however, would only say that there were "points of engineering detail which had yet to be settled." He added that "he could not speak publicly about them at the present moment," undoubtedly because patents had not yet been filed, and Fowler did not want to disclose Jones & Attwood's apparatus.[121]

The publishing of research results also had legal implications. Once a process or apparatus was publicly disclosed, it was typically no longer eligible for patent protection. Ardern and Lockett disclosed their initial research on activated sludge by reading their paper before the Society of Chemical Industry and publishing it in the society's widely circulated journal.[122] Because of this open disclosure to the scientific and engineering community, researchers considered the process to belong to the public. The U.S. patent examiner initially denied Jones's patent application for the process on this basis: Ardern and Locket had publicly disclosed the process—thus, it could not be patented. In high irony, even Jones at first considered the *process* to be public. When the Manchester Rivers Committee expressed nervousness about Fowler's "giving away valuable information to private persons," Jones used the publication of Ardern and Lockett's paper to justify the company's actions. "The effect of that first paper read on this subject was to render the process public property. It was then open for anyone to design, invent, or patent appliances calculated to perform the mechanical side of the question."[123]

Monopolizing Natural Processes

In making this statement, Jones differentiated between "process" and "appliance." He considered the process public, but appliances, the mechanical means of implementing the process, could be patented. As it turned out, however, it was not just the devices that were patented by Cameron, Jones, or others. It was the processes themselves. The most fundamental critique of the sewage patents centered on the contested position of the bacterial processes involved in treating sewage.

Proponents of bacterial processes considered biological sewage treatment to rest on natural processes. Thus both critics of patents and inventors themselves thought of the septic process and activated sludge as "natural." Metcalf, for instance, described the septic process as "essentially a natural one, a process of Nature."[124] The Merchants Association of New York declared Saratoga Springs' septic plant an illustration of "the employment of Nature's forces."[125] But it was not only opponents of the patents who viewed the process as natural. The patent examiner for the septic tank had initially denied the patent on just such grounds, arguing that the process occurred by itself in ordinary sewers and stagnant ponds, and was not patentable.[126] Even Cameron, before he patented the septic tank, considered the process to be natural in a fundamental sense. "In this system . . . there is no 'treatment' of the sewage in the ordinary sense of the term; its purification being accomplished entirely by natural agencies."[127] Similarly, activated sludge was considered by its inventors to be natural as well. Fowler wrote, "The changes which go on in an activated sludge tank are essentially the same as those taking place in arable soil."[128]

But as a natural process, biological treatment had another fundamental property. It could not be patented, privatized, or monopolized—it belonged to the public at large. As the trial judge in the septic tank case expressed it, "This development of bacteria is a fundamental truth in nature. In patent law it is a principle. It is a process of nature, but a principle, and not patentable." As "nature's means and methods," he continued, these bacterial processes are "common property, and cannot be appropriated and monopolized by any one."[129]

Processes had occupied a complicated position in patent law since the eighteenth-century steam engine patents of Watt. In early patent law, patents were to be granted for a "manufacture," which was considered to be a tangible thing. It was clear that an apparatus, like a steam engine, might fall in that category, but what about a "method of lessening the consumption of steam in steam engines" to make them more efficient? When Watt developed such an invention, his patent was challenged on the grounds that it was not a manufacture and thus not a valid subject matter for a patent, setting up a landmark case in *Boulton v. Bull*. The court debated whether Watt's invention was a method, a principle, or a machine. Was a method a manufacture if it did not produce some new substance? Or could a method "detached from all physical existence whatever" be subject for a patent? The judges of the court initially split on this question, and Watt's patent was not found to be valid.[130]

Over the next half century, however, courts began defining processes in such a way that they increasingly came under patent protection, perhaps paying heed to Chief Justice Eyre's warning in *Boulton v. Bull*. Eyre noted that "two-thirds, I believe I might say three-fourths, of all patents granted . . . are for methods of operating and of manufacturing" and cautioned that "we should well consider what we do in this case, that we may not shake the foundation upon which these patents stand."[131] At first, patents on processes were allowed because the courts judged that the patents were in fact not on a method, but on an altered machine that would implement that method. Thus, Watt's patent on a "method for lessening steam consumption" was considered to be, in fact, a patent for the critical modification he made to engine design, that of a separate condenser. Patents on chemical processes were not on the processes themselves, but rather on the vendible product of the chemical process—the medicine or dye or other chemical produced. By the mid-nineteenth century, however, processes themselves, independent of their products or implementing machinery, began to be fixed in law as patentable in both the United States and England. Eyre's opinion in the Watt patent case had begun to establish this distinction. "The word 'manufacture,' in the statute . . . applied not only to things made, but to the *practice of making*, to principles carried into practice in a new manner . . . Under the *practice of making*, we may class all new artificial manners of operating with the hand, or with instruments in common use, new processes in any art producing effects useful to the public . . . I think these *methods* may be said to be *new manufactures*."[132] It was not until 1842, however, that the courts firmly established methods as proper subject matter for patents in the United Kingdom.[133] These decisions in England influenced patent law in the United States. In 1877, a process was defined by the U.S. Supreme Court as "a mode of treatment of certain materials to produce a given result. It is an act or a series of acts, performed upon the subject matter, to be transformed and reduced to a different state or thing. If new and useful, it is just as patentable as a piece of machinery." This decision severed the connection of a process with any particular device: "The process requires that certain things should be done with certain substances, and in a certain order; but the tools to be used in doing this may be of secondary consequence."[134] Processes were thus legally distinct from apparatus, and each category of invention was patentable.

Distinguishing either a process or machine from a principle was perhaps more difficult for the courts. Patents were to be granted for invention, not discovery. The discovery of a natural principle could not come under protection of a patent. Around the middle of the nineteenth century, courts in England and the United States, in a series of related cases, began to grapple with these distinctions. "A principle," the U.S. Supreme Court ruled, in 1853, "is a fundamental truth." "A principle is not patentable," the court declared.[135] A treatise on patents from this period reiterated one of the justice's remarks from *Boulton v. Bull*: while a "mere principle" could not be considered a "manufacture," "a principle so far embodied and connected with corporeal substances, as to be in a condition to act and to produce effects in any art. . . becomes the practical manner of doing a particular thing. It is no longer a principle but a process."[136] As the Supreme Court ruled, invention lay, not in the discovery of the principle, but in the development of "the processes used to extract, modify, and concentrate natural agencies."[137] Altering nature thus became a prerequisite for finding an invention. As intellectual property scholars Brad Sherman and Lionel Bently concluded, the mid-nineteenth century marked the period when "it was clear that in law it was the artificial or created nature of the final product, its distance from Nature, which ensured that an object became an invention rather than a mere discovery"[138]

Processes, then, were differentiated from machines on the one hand, and principles, on the other. They were a distinct kind of invention, and became valid material for patents. By the time of the sewage battles, legal scholars considered the law settled on this matter.[139] Yet, while the courts and legal scholars may have delineated the philosophical and legal outlines of processes, principles, and machines, the distinctions were still highly problematic in the scientific and engineering community, and there was no consensus. Patentees, patent examiners, and the scientific community at large were commonly confused by these distinctions, and despite the rulings of the courts, engineers resisted the notion that processes themselves could be patented and monopolized, with one engineering commentator noting that there persisted a marked "prejudice against process patents."[140]

Sanitary engineers drew a fundamental distinction between the ethics of patenting apparatus and patenting processes in sewage treatment. Much of the apparatus used in a sewage plant—pumps, distributers, diffusers—was patented, and engineers organized no protests over these patents. Rather it was patents on processes themselves that created the most controversy. "The validity of patents issued for appliances to facilitate the [septic] action is not questioned," wrote F. Herbert Snow.

However, "the exclusive right to the natural process may well be," he argued.[141] Hatton, on learning of the activated sludge patents responded to Fowler, "I imagine such patent rights, if any, must have been issued on appliance rather than process."[142] Birmingham engineer John Watson likewise thought that apparatus and process patents were fundamentally different. In discussing the activated sludge process at its inception, Watson thought it fine to patent "mechanical devices." But, "so far as he could see, to patent a process itself was calculated to put back the hands of the clock" by "strangling further experiments."[143]

This distinction between apparatus and process on the part of the engineers can be seen in the different reception of Cameron's septic tank patent and another patent for a sewage digestion tank. Because of Cameron's royalty demands, many cities turned to other means of treating sewage, most notably the so-called Imhoff Tank, designed by German sanitary scientist Karl Imhoff. Even though this tank too was patented, it did not produce the outcry of Cameron's patent. Imhoff's patent was on the design of the tank itself—on apparatus—rather than on the process. In contrast to the many tanks that were used to digest sewage, including the septic tank, the Imhoff Tank was a two-story tank that separated the physical settling process from anaerobic digestion. Engineers recognized the novelty of the design, and were willing to accept its patenting. Further, Imhoff's royalty demands were seen as modest, recognizing a commitment to the public interest in sewage treatment rather than a desire for profit.[144] As a result, many cities converted their septic tanks to Imhoff tanks, and Imhoff tanks became the standard treatment technique prior to activated sludge.[145]

In the *Cameron* case, the distinction between process and apparatus led to nonsensical results that helped reinforce the widespread prejudice against process patents among sanitary engineers. The appeals court denied the apparatus claims for the septic tank, ruling that the design of the tank was anticipated and thus not novel. The process claim, however, was left intact, leaving builders of septic tanks in a very confused situation. "The decision of the Court of Appeals permits the public to build without infringement tanks in strict accordance with Cameron's apparatus claims," explained the Association for Defense of Septic Process Suits in disparaging the opinion. "While the public may do this much without infringement the one thing the public must *not* do is to permit natural agencies to become operative, for then, the Court decides, Cameron's process patent claims are infringed. Nature must be circumvented in some way (the Court does not say how), or legal difficulties will follow."[146] By patenting a process of nature, Cameron had twisted reason, logic, and nature itself.

Scientists and engineers likewise struggled with the distinctions among principle, process and apparatus in the activated sludge patents. Earle Phelps, sanitary scientist at the U.S. Public Health Service, decried the "important fact too often overlooked by engineers, namely, the distinction between a process of sewage disposal and an apparatus or device for carrying out that process." For Phelps, there were "but two essential or basic processes, namely anaerobic decomposition and aerobic oxidation . . . Upon these two real processes of organic decomposition all modern biochemical methods of sewage treatment are based."[147] Devices in which either of these processes occurs for treating sewage might be patentable, but the fundamental process of bacterial action was not.[148]

The distinction between principle and process was made more complex by the actions of the sanitary profession itself. As discussed in chapter 1, bacterial treatment processes had not always been viewed as natural. When first introduced, many sanitary scientists considered the septic tank, bacteria bed, and biological filter to be artificial processes, compared to the natural process of treating sewage by distributing it on soil. Researchers on bacterial treatment undertook an explicit program to naturalize these "artificial" processes by emphasizing their reliance on natural bacteria. Thus, Cameron emphasized how his septic tank was based on "natural agencies." At the same time, however, engineers also denatured these biological processes through a continual concentration, intensification and control of the bacterial activity. These simultaneous processes of naturalization and denaturing were fundamental in creating the industrial ecosystem of the sewage treatment plant. But it also allowed researchers, lawyers, and judges to emphasize distinct aspects of the system. Opponents of patenting saw the ecosystems as fundamentally natural and thus not open to appropriation. The courts, however, negated that view by declaring the denatured biological processes artificial, one of the "useful arts" that might be patented under the U.S. Constitution.[149]

The Cameron Septic Tank Company denied that it was trying to patent the natural process of anaerobic action itself. "The patent . . . makes no pretense that Cameron was the inventor of putrefaction, and of course, such a pretense would be absolutely absurd," argued one spokesman.[150] Rather, Cameron claimed his invention was to separate the anaerobic action from the aerobic, in a way that was not present in nature, making the septic process an "artificial method." The trial judge had concluded that the septic process was natural. Does "nature's method and means of doing a thing become became an 'artificial method' of doing the same thing," he asked, simply by providing "an artificial place . . . like a tank, constructed by human hands, in which nature by its own instrumentalities is to perform its own process"? No, the judge concluded. The bacteria "develop, as they always had done, and always will do, under certain conditions, and when developed they liquefy certain materials, as they always had done, and as they always will do if not prevented."[151]

Unlike the trial judge, however, the appeals court accepted the argument that the septic process was indeed an artificial method. To Saratoga Springs' argument that septic action "is a process of nature and one which cannot be covered by any one," the appeals court ruled that the septic process was not natural, but rather artificial. They emphasized the industrial nature of the process as the feature that distinguished the microbial activity in the septic tank from a natural one, ruling that "the distinctively novel feature is the . . . separate workshop for the microbes."[152] Cameron had separated the processes of aerobic oxidation and anaerobic putrefaction, making his septic tank an artificial, rather than natural process.

Similarly, the courts ruled that the activated sludge process patents were not on "the discovery of the bacteria, nor their characteristic activities" but rather on the means for providing conditions for the bacteria "to function to the best advantage."[153] The appeals court went into great detail describing "the processes of nature" involved in sewage purification. Starting with nitrification and the nitrogen cycle, the opinion described aerobic and anaerobic decomposition in soils and "running streams," and how in sewage irrigation and trickling filters "nature's process was followed or approximated," because the bacteria were attached to soil particles or stones. "In nature's processes and in all artificial filters . . . the aerobic bacteria were fixed, and the polluted water or sewage was brought to the bacteria," they asserted. In contrast, with activated sludge, "the situation is reversed, so that the bacteria instead of being fixed are put into circulation and brought to the sewage." According to the appeals court, activated sludge thus differed from "nature's processes" of purification and could be patented.[154] While Fowler and others, in order to naturalize the novel process, emphasized the continuity between the natural process of land treatment and activated sludge, the appeals court placed a clear demarcation along that same continuum between natural and artificial. Sewage irrigation and filtration were "natural," the court implied, because the bacteria grew on fixed surfaces. Activated sludge was artificial (and thus patentable) because the bacteria were freed from surfaces.

"The Chaos of Pure Science and Natural Processes"

Controversy over the patentability of natural processes permeated other industrial ecosystems as well. "Microorganisms appear to have gone into business," declared patent lawyer J. Howard Flint in 1930. "Imperceptibly their myriads have crept into industry, where they are nourished and pampered for the work they do." While extolling the potential of the innovations coming from industrial microbiology, he wondered, "Will patent law protect investments in this new factor in industry?" Flint saw the critical impediment to patent protection to be how the courts interpreted the kinds of problems that the engineering and scientific communities had been exposing in the sewage patents. "Can any good patent at all come out of the chaos of pure science and natural processes involved?"[155] The decisions in the septic tank and activated sludge cases helped firmly establish the patentability of bacterial processes in general.

Two key cases in industrial bacteriology addressed these issues, and both cases relied heavily on the findings of the court in the Cameron septic tank case. One of these cases arose from the same milieu of industry, chemistry and bacteriology that produced the activated sludge patents, the Victoria University of Manchester where Fowler studied and taught. Chaim Weizmann, future president of Israel, was appointed in the chemistry department there in 1904 and worked on industrial fermentations. Weizmann and the activated sludge researchers were closely connected. One of Weizmann's assistants was E. M. Mumford, who with Fowler had developed the M.7 process of sewage treatment using iron bacteria. William Lockett was also a student of Weizmann's. Weizmann and Fowler collaborated on studies of industrial bacteria. Another student at the time was Gladys Cliffe who later worked for Activated Sludge Ltd.[156]

At Manchester, Weizmann isolated a bacterium that fermented starch to produce acetone and butyl alcohol. By the beginning of World War I, acetone had become essential to the British war effort in the production of smokeless cordite. With its supplies cut off from the continent, England was desperate to find replacements. Weizmann's bacteria were able to accomplish the efficient conversion of cornstarch to acetone, with butyl alcohol as a by-product. Weizmann was at first reluctant to file a patent, but at the urging of the head of research for Nobel's Explosives Company, he applied for a patent for the process in 1915.[157] The process was implemented on an industrial scale, first in Canada and then in Peoria, Illinois, and Terre Haute, Indiana, closer to the sources of cornstarch (figure 2.4).[158] Following the war, it was found that the butyl alcohol could be used in the manufacture of quick-drying lacquers used in automobile finishes. With this potentially large commercial market, the production of the alcohol became the primary goal, with acetone as a by-product. Weizmann licensed his process to the Commercial Solvents Corporation, which began commercial production. A competitor, the Union Solvents Corporation, soon began producing butyl alcohol using the same bacterial process, and perhaps the same bacteria.

In 1931, Commercial Solvents sued Union Solvents for patent infringement. In its defense, Union Solvents claimed the invention unpatentable, as a patent on "the life process of a living organism."[159] The court case centered on the question of whether the production of butyl alcohol and acetone was simply a "natural function" of the bacteria, and thus not a patentable invention.[160] To demonstrate that the process was natural, Union Solvents presented the court with evidence of the natural soil environment of the bacteria and attempted to show that the same processes that occurred in the fermentation vessels also occurred in the natural soil habitat of the bacteria. "The bacteria . . . by their life processes during their growth

Figure 2.4
Farmers unloading corn at the Commercial Solvents Corporation plant in Terre Haute, Indiana, 1923. Following the war, Commercial Solvents licensed Chaim Weizmann's patent for making butyl alcohol and acetone from cornstarch using the bacteria Clostridium acetobutyricum. The plant had originally been set up in Terre Haute during World War I to take advantage of nearby sources of cornstarch. *Source*: Community Archives, Vigo County Public Library

and multiplication, produce, as a kind of by-product butyl alcohol, ethyl alcohol, acetone and gases. . . . This is the natural function of these bacteria. There is not the slightest proof in the record that they can go through their life processes . . . without necessarily producing the products named above."[161] Union Solvents' expert was the by then renowned bacteriologist Edwin O. Jordan, who had first studied sewage bacteria at the Lawrence Experiment Station. Jordan applied the ecological perspective that he had shown in his studies at Lawrence, and argued that the butyl alcohol bacteria were producing these products "in the soil feeding upon the substances found there."[162] Commercial Solvents' response to this argument reveals the contradictory nature of the industrial organism. Commercial Solvents engaged the services of Dr. Elizabeth McCoy, a bacteriologist at the University of Wisconsin and a leading expert on the particular class of bacteria used in the process. Asked if she considered the industrial use of the bacteria as "carrying out the natural functions of those organisms," McCoy replied, "They are, of course, natural functions in the sense that the organisms are doing this process." But, she continued, "they are not natural functions in the sense that they occur in nature necessarily."[163] People, by providing the right conditions in an artificial environment turned a natural process into an artificial one.

The courts of the time were willing to consider patents on processes *using* bacteria, but they were unwilling to grant patents on living things themselves. Union Solvents charged that Weizmann was trying to patent the bacteria themselves. This was the same argument that had been used in the Cameron case, but the court had ruled that the septic patent was valid as it was on a process, not the bacteria. Union Solvents tried to use the Cameron decision to claim that no such process was claimed by Weizmann, only the bacteria themselves: "If the Weizmann patent is a patent on anything at all, it is a patent upon this alleged Weizmann bacteria as such, and that only."[164] To Commercial Solvents' reply that "we have not attempted in the patent to patent the organism itself," the court replied, "you hardly could." "Of course, you could not do it," agreed Commercial Solvents' lawyer. Instead, Commercial Solvents argued that bacteria were simply an "agency of nature, the use of which for practical purposes can be patented," in the same way that "chemicals, minerals, or anything else" could be patented.[165] The courts agreed with Commercial Solvents, citing *Cameron* as precedent for their decision.

Another important contemporary patent case in industrial microbiology also rested on the decisions in *Cameron* as well as *Guaranty Trust*. The U.S. Patent Office saw this case as an opportunity to definitively rule against microbial patents. In the end, however, with the help of the decision in *Cameron*, the Patent Office was overruled and microbial process patents continued. In the 1920s, Samuel C. Prescott and a Japanese microbiologist Kisaku Morikawa studied the bacterial flora of the ferment used to produce rice wine, in search of undescribed bacteria that "may have technical significance and find use in industry." They isolated a bacterium that fermented the sugar in rice to isopropyl and butyl alcohols when grown in pure culture. Their name for this new species emphasized the industrial character of their research: *Bacillus technicus*.[166] They sought a patent on the culture of *Bacillus technicus* as well as the process of using the culture to produce the higher alcohols. The Prescott patent came out of a similar environment of sewage and industrial bacteriology at the Massachusetts Institute of Technology as had Weizmann and Fowler at Manchester. Prescott was a student of William Sedgwick, who had conducted the first studies on the bacteria of sewage filters. His first job after graduating from MIT was as a chemist and bacteriologist at the Worcester sewage treatment plant, and he was the author of the primary text in the early 1900s on the bacteriology of water and sewage. Along with MIT graduates and professors Earle Phelps, William Sedgwick, and Lawrence Experiment Station biologist Stephen DeM. Gage, he formed a loose association of industrial and sanitary bacteriologists named the "Bug Club" that would discuss advances in sanitary and industrial microbiology. He later succeeded Sedgwick as head of the Department of Biology and Public Health at MIT, but he shifted the focus of the department from sanitation toward industrial microbiology.[167]

In their patent application, Prescott and Morikawa claimed a "ferment consisting of a culture of the hereinbefore described Bacillus technicus." The patent examiner rejected the claim of the "ferment," "as being for essentially a living organism." "Living organisms are not patentable," he declared.[168] The inventors replied by arguing that bacteria were fundamentally different from other organisms: "Of course the statement that a living organism is unpatentable sounds plausible because when we think of living organisms we think of a man or a dog or a bear or something which has a separate and known existence and is old." Rather, they were claiming "a culture of the bacillus . . . free of admixture with other organisms."[169] Following this reply, all official responses to the patent application were drafted by the Principal Examiner in the division, Vernon I. Richard. For several years, Richard had been faced with numerous claims on both bacterial processes and the bacteria themselves, and was unwilling to approve patents for what he considered to be "principles of nature."[170] Richard tried to use the Prescott and Morikawa patent to establish a standard in the patent office of denying claims on bacteria and many bacterial processes. Richard stated that his decision rejecting the Prescott and Morikawa application was to act as a precedent in the Patent Office for all similar future claims.[171]

Richard provided the inventors an analogy to support his contention that claiming the "ferment" was the same as claiming the bacteria. As Sedgwick had naturalized bacterial action by invoking natural selection and ecology, Richard's analogy served to naturalize the *Bacillus technicus* in the face of attempts to industrialize it.

A 'culture' of bacteria consists of a substance, which contains the bacteria, and maintains it in living conditions, and is analogous to a mass of water containing gold fish. The gold fish are grown, that is, 'cultured' in water . . . and almost identical words will describe the culture, handling, and maintaining of bacteria. It frequently occurs that a previously unknown or unidentified organism is found in a pond of stagnant water, but its discoverer never for a moment believes he made or invented it. . . . The 'ferment' of applicant cannot be an article of manufacture. Nor is it a composition of matter, any more than water containing a gold-fish is such a composition.[172]

For Richard, the inventors could claim a process for isolating the bacillus, a new process for culturing it, or a new process "in which it might be used to produce either a new or old substance." But the inventors could not claim the bacterium itself.

But neither, argued Richard, could they receive a patent for the process they had described, for the bacterial fermentation was simply an inherent character of the bacteria. "The function of all bacteria is fermentation," he stated. "B. Technicus fermentation produces butyl and isopropyl alcohols. That is their natural function and these claims recite nothing more . . . It is a 'principle of nature' that these bacteria produce those alcohols."[173] The inventors gave up on trying to convince the patent office to approve the claim on the bacteria itself and withdrew the claim on the ferment. But they insisted on the patentability of their "process of producing mainly butyl and isopropyl alcohols . . . with a substantially pure culture of the herein described Bacillus technicus." Once again, the examiner denied the claims: "These claims are for a law of nature." The patent office's response demonstrates that despite rulings like *Cameron*, there was no consensus on the status of process patents.

Prescott and Morikawa appealed the denial to the Patent Board of Appeals. Richard saw the opportunity to establish his precedent and provided a detailed examiner's statement that first presented the biology of single-celled organisms, their habitat, their food, their biochemical processes, and, finally, the relation between their biology and their use in industrial processes. By focusing on the biology of the bacteria, Richard sought to naturalize them. He forcefully stated his opposition to the patentability of living processes: "Bacteria grow. Man has never made one, and, so far as known, cannot make one. Their function is fixed without human agency; why then should a man receive a patent which is for no more than that function?" Prescott and Morikawa's patent "is not a claim for a process in which the power of the bacteria to ferment material is made use of," stated Richard, "but a claim for the power, the fermentation itself."[174]

In the appeal, both sides cited the decision in the Cameron septic tank case. The inventors emphasized the similarities between the industrial production of alcohol using bacteria and the septic tank. "The Cameron case was a bacteriological case. The patentability of bacteriological processes such as is here under discussion has long been recognized." The brief placed the process using *Bacillus technicus* in the context of the septic tank: "In the words of the Court in Cameron v. Saratoga Springs," Prescott and Morikawa "put the force of nature into a certain specified condition."[175] By doing so, they had invented a new process, eligible for patenting. Richard, for his part, made a point of distinguishing this case from *Cameron* by emphasizing the specific steps "each caused by an act of man" by which the sewage was "subjected to the action of anaerobic bacteria." In contrast, he argued, Prescott and Morikawa were claiming only the fermentation itself, with no separate steps caused by human action.

The inventors also emphasized the similarity of their process to other chemical processes. "It is merely confusing the issue to talk about living organisms," they stated. "The question why we call bacteria things of life, yeast a plant and a crystal growing in a solution an inorganic compound and devoid of life is interesting philosophically but not important from the point of view of the Patent Office. The organism in question is a chemical agent . . . In a sense it is comparable to a synthetic dyestuff first made in a laboratory. This reagent when placed in a certain chemical environment, that is, a sterile fermentable sugar mash, operates in some specifically unknown way to effect a chemical reaction which is new and which produces a new result of high importance in the arts."[176] Even the name, *Bacillus technicus*, seemed to imply that the bacteria were more chemical than living organisms.

During the six years that the Prescott patent was being examined and adjudicated in the Patent Office, the *Guaranty Trust* case was decided in the appeals court. The Patent Board of Appeals relied on this decision, itself decided on the basis of *Cameron*, and rejected the argument of the examiner. Quoting *Guaranty Trust*, that Weizmann "obtained his patent not for the bacteria *per se*, but for a process which consists in the employment of certain bacteria," the Board of Appeals argued that the Prescott case was even stronger. The board accepted the analogy between the bacteria and a chemical reagent. The appellant has "discovered a new reagent, the bacillus technicus," it declared.[177]

The failure of the engineers' movement to keep sewage treatment processes in the public domain was thus extended, with *Guaranty Trust* and *Prescott*, to other industrial ecosystems. Whether or not these cases were properly held according to their merits and the law, however, the scientific community of the time was not comfortable with the outcome. Patents for processes remained deeply troubling to many scientists and engineers. There was also still great confusion as to what was actually patented. The *Chicago Tribune*'s report on the activated sludge decision seemed to imply that the courts had validated patents on the bacteria themselves. "The microbes known in biochemistry as bacillus M7, which the sanitary district, the city of Milwaukee, and other municipalities employ by the billions to purify their sewage, received the austere consideration of the United States Court of Appeals yesterday. The court affirmed . . . that the city of Milwaukee has no right to use the little organisms without paying royalties."[178]

The Rise of Private Industrial Research

In the 1930s, the growth in municipal research on sewage leveled off; industrial research and its patenting culture began to increase substantially. Yet the status of sewage patents was still in flux. The example of Edward Mallory illustrates the unsettled position of sewage patents in the 1930s and 1940s. Mallory developed a procedure for operating activated sludge plants according to an "equilibrium index" and patented his system of measurements, mathematical calculations, and process interventions that operators would use to run a plant. Because of the long history of skepticism toward sewage patents, however, Mallory felt the need to defend himself by discussing the role of patents generally in encouraging innovation. These arguments, though, were the same ones that had been rehashed by engineering supporters of patents for years, and had failed to persuade critics of the sewage patents. So, lest any city intend to use Mallory's control system without paying royalties, he concluded with a warning that municipal engineers might have found more persuasive. He reminded the reader of the "very expensive error" made by the many municipalities in the activated sludge patent litigation.[179]

The loss of the activated sludge cases portended great changes in professional practice, as an industrial model for research and patenting began to take hold in sanitary engineering.[180] Mallory embodied this new direction in research and innovation in sewage treatment. Mallory was employed by an industrial research laboratory, Lancaster Research, a subsidiary of Lancaster Ironworks, and he developed his system as a consultant for the city of Tenafly, New Jersey. This kind of work would previously have been performed by an independent consulting engineer, like Snow or Metcalf, who would have no proprietary interest in any particular design. But by tying patented processes to sewage installations provided by the patent holder, firms like Lancaster Ironworks sought to extend their legally sanctioned monopoly in patented processes and materials to the much more lucrative business of supplying "complete treatment plants, purchased from manufacturers."[181]

The shift away from public, municipal research toward industry was evident in Great Britain as well. In 1929, the Research Secretary of England's Association of Managers of Sewage Disposal Works decried the "jealousy and desire for secrecy" among sewage scientists and wondered whether they were "losing their respect for the purely scientific and academic side of their professions in favour of the attractions of its commercial application."[182] This question was answered fifteen years later, during a symposium commemorating the fortieth anniversary of activated sludge: "There was "an unfortunate trend," bemoaned D. H. A. Price, of the Severn River Board. "Without exception, all the original work [on activated sludge] was done under the auspices of local authorities. Not only that, but the subsequent developments which had taken place had in the main, although not exclusively, been the work of people employed by local authorities . . . Perusal of the Institute's Journal over recent years suggested that such activities were not now so prevalent as in the past."[183]

This shift in the structure of the sanitary engineering profession became the focus of a committee appointed in 1944 by the American Society of Civil Engineers to investigate how "proprietary processes" in sewage treatment might affect the ability of the profession to serve the public. The committee identified many areas in which the prevalence of patented materials was distorting the ethical relationships between engineer and municipal client. In its 1946 report (the same year the courts finally fixed damages in the Activated Sludge case), the committee attributed the bulk of the problem to the privatization of sewage research—the rapid increase in industrial as opposed to municipal research into sewage treatment processes: "The development and promotional methods used by industry are now being brought to bear in some force upon municipal . . . sewage treatment," the report stated. The sanitary field, suggested the committee, was rapidly approaching the state of mechanical, electrical and chemical engineering, in which "a great majority of engineers . . . are employed directly by proprietary interests, and disinterested consultants are relatively very scarce." This was precisely Morris Cooke's criticism thirty years earlier of engineers involved in the critical transport, electric, and gas networks in cities. Now it applied to sanitation as well. In the face of these changes, sanitary engineers, the committee acknowledged, were "not prepared to cope," as they were "unfamiliar" or even "unsympathetic" with the "methods of patent protection."[184] But the committee offered these engineers little more advice than to get used to it. When the report concluded there was little the profession could or should do to stem the changes in practice taking place, the committee signaled the end of the engineering campaign against patenting.

3

Craft vs. Science: "Be an Operator, Not a Valve Turner"[1]

In a 1936 magazine for sewage treatment plant operators, called *The Digester*, the Illinois Department of Health provided the following advice to the men who ran the state's activated sludge sewage treatment plants: "Think of the aeration solids as your 'workmen,'" it suggested, referring to the bacteria in the activated sludge tanks. "Retain as many as is necessary to do the work as it comes," it advised, "but not too many, as they all require air to stay in condition." "Don't keep any more working organisms on the 'payroll' than necessary," the article warned. "The excess workmen are just 'shovel-leaners.'"[2]

In the narrative of this article, the bacteria were the workers; the human workers, the managers. Envisioning bacteria as workers was not unusual (figure 3.1). For decades, microbiologists had been conceiving of bacteria in terms of industrial work. "The idea of the activated sludge process is to get uncounted multitude of bacteria to work," wrote Milwaukee's T. Chalkey Hatton, for instance.[3] Similarly, the task here was to try and obtain the maximum amount of work from the bacteria through control of the biological processes. Like a shop foreman managing his workers, the sewage treatment plant operator was to manage his microbial laborers. What sets this article apart from the usual metaphors, however, was that the advice itself was part of a concerted effort by engineers and other supervisors of sewage treatment plants to exert increasing control on their *human* laborers—the operators themselves. For, as sanitary engineers were discovering, controlling the bacterial process required control of the labor process.

In this realization, managers of the industrial ecosystem were joining trends in industry more generally. Sanitary scientists often noted the relation between sewage treatment and industrial processes. "Treatment methods are just as truly chemical manufacturing processes as are the production of sugar and dyes," wrote Arthur Buswell in his treatise on sewage treatment chemistry.[4] "The Works Manager," wrote John Farmer in 1914, "should look at the disposal of sewage in the light of a manufacturing process—the sewage discharged on to the works being the raw material and the final effluent the finished product."[5] So seeing their treatment plants as

Clostridium Acetobutylicum

Stained and Magnified 1500 x

THESE bacteria are the
real workmen in the
factories of Commercial
Solvents Corporation. The
human element is only
necessary to supply them
with food and the condi-
tions required for their
most efficient functioning.

Figure 3.1
Frontispiece to industrial pamphlet describing the process for producing butyl alcohol at
Commercial Solvents Corporation, 1929. The microorganisms of the industrial ecosystem
were commonly presented as "workmen," placing the role of the human workmen in an
ambiguous position, as both workers and managers of the bacterial laborers. In process control, who was to be controlled, the bacteria or the people? *Source*: Commercial Solvents Corporation, 1929, Community Archives, Vigo County Public Library

factories, sewage treatment engineers began to transfer the management techniques
of chemical manufacturing into the sewage treatment industry. One hallmark of
industrial organization at the time was scientific operation. Sewage treatment plants
were joining a wide variety of other process industries in implementing scientific
control of their production processes.

Prior to the rise of scientific control, many industrial processes were managed
by factory workers who had developed their craft through long experience. Operators used that experience to make continuous adjustments to achieve a high quality
product. In the latter part of the nineteenth century, however, many of these industries began to institute new, scientific approaches to control, leading to profound

changes in the labor process. Where workers had previously possessed autonomy, with scientific control their actions were increasingly regulated by sensors, alarms, and distant supervisors, "guided by signal lamps and a deviation indicator . . . stationed in a central instrument-control room."[6] As Harry Braverman, David Noble, Stephen Bennet, and other scholars have argued, scientific control also meant controlling the labor process.

Process control was just one part of the innovations in industrial management that came under the umbrella of "scientific management." Best known through the advocacy and writings of F. W. Taylor, scientific management was a concerted effort to rationalize all aspects of industrial work. This management movement helped lead to a profound deskilling of work and reduced industrial workers to simple valve operators.[7] As in other industries, the establishment of process control in sewage treatment led to a prolonged struggle for shop floor control between operators and engineers. Who would determine the proper adjustments to make in a plant and when? Would sewage treatment plant operators, like other industrial operators, be reduced to "simple machines," or would they be able to exercise independent thought, analysis, and control of both their work and the sewage processes.

In industries such as brewing, sugar production, smelting, or even the production of heavy chemicals, scientific control supplanted earlier craft approaches. Workers' skill was transferred to centralized management, and their own jobs made more menial. In sewage treatment, however, this trajectory differed. First of all, biological sewage treatment arose contemporaneously with scientific control; in a brand-new industry based on a poorly understood and complex biological processes, there were no craft approaches to supplant, and operators developed their own understanding of the treatment process at the same time as the scientists and engineers. Craft approaches developed in parallel and in competition with "scientific" process control. Further, where the impetus in industry was to remove decision-making authority from the worker, the emphasis on process control in sewage treatment placed the worker at the center. Rather than concentrate decision making authority in the scientists and engineers who designed the process, scientific control kept plant operators at the center of successful sewage treatment. This exposed a contradiction in the efforts to rationalize sewage workers. Instituting process control in sewage treatment required the creation of a new kind of treatment plant operator: technically trained, capable of performing laboratory analyses, and committed to the ideology of control. Where the incorporation of process control in industry deskilled the workers, coopting the expertise of the craft worker, in sewage treatment the requirements of a technically trained labor force led to the institution of training programs and a concerted effort to "skill" the workforce—in not only the practical knowledge needed to run a plant but also more basic scientific principles of bacteriology and chemistry. Sanitary engineers and scientists focused on developing and educating a workforce that had the necessary skills to implement scientific control. Through certification and licensing, establishing schools in chemistry and bacteriology, and creating professional organizations, sanitary scientists helped create a workforce skilled in both basic and applied chemical and bacteriological skills.

As a result, where craft approaches were superseded in many industries, in sewage treatment craft management has persisted and has been remarkably persistent in the face of a century of concerted effort to rationalize the management of sewage treatment plants. This persistence is related to characteristics of complex processes like activated sludge itself as well as contested relations among operators, engineers, and science. Sewage treatment plants were complex socially, chemically, and ecologically produced systems, living ecosystems subject to all the complexity that interacting populations responding to a variable environment produced. Their management required more complex theories of control than were typically offered by scientific management. In the face of the constant variability and local specificity characterizing the sewage treatment plant, these techniques based on operators' craft were often more effective than scientific control. Scientific process control proved unable to take into account the biological complexity of the sewage treatment process. The skill of the experienced workmen to manage the biological treatment process remained necessary, and they resisted the deskilling and ultimate alienation of their work.

Process Control

The notion of sewage treatment as a "process" amenable to control derived from recent advances in chemical engineering. By the end of the nineteenth century, analytical chemists had shifted their orientation from product to process. Instead of analyzing only the chemical characteristics of feedstocks and products, chemists began to analyze intermediate characteristics that would tell them if their process was performing efficiently. They emphasized measurement and control of the conditions of pressure, temperature, or other factors that were critical to the chemical transformation effected by the industrial process.[8]

Using these new techniques, chemists sought to wrest control of industrial processes from industrial operators. In the earliest moves toward process control, scientists introduced laboratory and instrumental analysis to replace the craft approaches of skilled operators in which they used close observation to judge the state of a chemical process. In making sulfuric acid, for instance, British factory workers would judge temperature by touching the outside of reaction chambers, or regulate draft by observing moisture on the undersides of the chamber caps. As one alkali manufacturer instructed, "If the inside of the cap be quite dry, with a slight crystalline formation upon it, which turns green when spit upon, the amount of steam is altogether insufficient . . . If, on the other hand, the caps be dripping wet, it may be taken for granted that the steam is too strong."[9] In contrast, chemist and inventor James Mactear argued that these craft techniques—"indications afforded by colour (and smell!) of gas, drip samples, temperatures" —were all insufficient for managing the sulfuric acid production process. "For while all seems to be going on rightly," he noted, "the escapes vary in the most irregular way, and it is not until a considerable time after an irregularity has occurred in the working that it can be discovered by these indications."[10] In 1877, Mactear introduced a control system in which waste gas was analyzed in the laboratory for the presence of sulfur dioxide, a technique that reportedly improved the precision of the control and made the process more efficient.

But improvements in control like these occurred too haphazardly for leaders in British chemical engineering. In 1882, F. A. Able, a founder of the Society of Chemical Industry in Great Britain, complained about the state of the British chemical industry in his presidential address at the first annual meeting of the society. "The control of the processes is left in the hands of men whose only rule is that of the thumb, and whose only knowledge is that bequeathed to them by their fathers." Rather, Abel advocated for a rigidly hierarchical industrial organization based on his observations of German dyeworks. Here, the industrial works would be supervised by a scientific director, under whom would serve trained chemists as department heads, then assistant chemists, and finally the "common workmen, who have, of course, no knowledge whatever of scientific principles." Abel emphasized another important aspect of process control here. Controlling the chemical process meant controlling the labor process. No longer would the process be under the control of the worker. Rather, both the process and the worker would be under the control of the manager. The "common workmen" were to be reduced to "simple machines, acting under the will of a superior intelligence." In this structure for process control, the operations were "being watched by, and constantly being under the control of the chemists."[11] The goal of modern, scientific process control was thus to supplant the traditional knowledge of the worker with the scientific knowledge of the chemist or engineer.

Reducing the common workmen to simple machines was to alienate the worker from his work. For Marx, capitalist industrial organization estranged the worker from his or her work in two important ways. First, under capitalism, the product of the workers' labor did not belong to the worker but rather to the owner of the business. Second, the organization of industrial labor estranged the worker from performing fulfilling labor. Work became drudgery. Work, for Marx, was an essential part of the human experience, satisfying both material and emotional needs. Labor was both physical and intellectual. Alienation "estranges man from his own body, from nature as it exists outside him, from his spiritual essence, his *human* essence," he wrote.[12] In this important sense, alienation was a denaturing of the human experience. Industrialization involved a denaturing of not only natural processes but also human experience.[13] With sewage treatment and the industrial ecosystem, transforming a natural biological process into a chemical manufacturing process was similarly a process of denaturing—of both biology and humanity.

The decoupling of intellectual and physical work, the alienation of labor, was a central aspect of a broader trend in industrial management, called "scientific management."[14] While the drive for process control predated scientific management, it was incorporated as one of its many themes. Scientific management sought to rationalize almost every facet of industrial organization, from the shop floor to the management floors. In particular, proponents of scientific management sought to control the actions of workers, from the size of shovel they used, to how long it should take to unload each shovelful of ore from a railroad car. Under scientific management, as in the Illinois activated sludge plant described in the introduction, there would be no "shovel-leaners."[15] The science of shoveling, as Taylor termed it, or any other industrial task required a great amount of information on the best manner of accomplishing a task. For the manager, the rules of thumb that workers used incorporated valuable information for process control. One of the problems for scientific management then was information. How could the manager obtain this information that previously had only passed down from parent to child? For Taylor, this information was taken directly from the worker, in the form of time and motion studies of especially skilled workers.

In a summary of the principles of scientific management, Taylor identified "the first of these principles" as the "development of a science to replace the old rule of thumb knowledge of the workmen; that is, the knowledge which the workmen had, and which was, in many cases, quite as exact as that which is finally obtained by the management, but which the workmen nevertheless in nine hundred ninety-nine cases out of a thousand kept in their heads, and of which there was no permanent or complete record." The first task of the manager then was the "deliberate gathering in on the part of those on the management's side of all of the great mass of traditional knowledge, which in the past has been in the heads of the workmen, and in the physical skill and knack of the workman, which he has acquired through years of experience." Once management acquired this knowledge, it could assert better control. The traditional knowledge would be studied, distilled to mathematical formulas, converted into a science, and then "applied to the everyday work of all the workmen."[16]

Under scientific management, the worker would no longer need to think about what he was doing. He was to simply do what management told him. The goal, as described by Harry Braverman, was to separate the work of conception from that of execution, mental from manual labor.[17] "The manual worker should not have any avoidable head work to do," declared one British engineer. "Make it unneccesary for the workman to think."[18] In the modern world, mental and manual work were to be divided.[19]

During the decades surrounding the turn of the century, the tenets of both process control and scientific management were implemented in industry after industry, in chemical manufacturing, sugar production, brewing industries, smelting, and machine making. Process control techniques were imposed on the older craft techniques, "the rules of thumb . . . bequeathed to them by their fathers,"[20] resulting in substantial changes to the structure of labor and transforming skilled workmen into "simple machines."[21]

Process Control in Sewage Treatment

Engineers working on sewage treatment were familiar with process control from their experience with precipitation processes, which had many similarities to chemical manufacturing. The early notion of scientific control in sewage, and its relation to chemical processes, can be seen in J. W. Slater's proposal for control of a sewage precipitation plant. One of the key problems for sewage treatment processes was the constantly varying strength of the sewage coming in to the plant. Sewage varied "according to the locality, the season, the weather, and even the hour of the day." As Slater outlined,

In all towns there is generally a well-marked difference between the day and the night sewage. In a residential town the sewage, from midnight till five or six in the morning, is very much reduced in strength, as well as in quantity. Often it consists of little more than the surface-water from the streets and roofs, and of the ground springs, which find their way into the sewers to a considerable extent. As the day advances the flow of sewage becomes more copious and more offensive, and is at its worst from two to eight p.m. . . . On Sundays, in a residential town, the sewage differs little from its condition during the rest of the week. In manufacturing towns it is on Sundays more purely excremental in its character, the industrial waste waters being in great part absent. The same holds good with respect to public holidays. In small agricultural centres the sewage on market days is perceptibly stronger than on other days, and requires an extra share of attention.[22]

To account for this constant variability, Slater outlined a scheme for the arrangement of pipes, mixing pits, and other aspects of the treatment plant that allowed the plant foreman to monitor the character of the incoming sewage. Slater then specified a series of tests that the manager would conduct on the incoming sewage to adjust the treatment process. Depending on the results of laboratory tests, the manager could, "by simply reaching his hand to a cock or a valve, make the corresponding change in the quantities and proportion of the mixture used."[23]

Process control was first incorporated into bacterial treatment with the septic tank. Chicago engineer John Alvord cautioned that the septic process was "sensitive and delicate" compared to precipitation, but he argued it was nevertheless "ameanable [sic] to control" provided the operator "watched and governed" it. Alvord proposed a scheme for regulating the septic process in which the operator would continuously measure the strength of the sewage and its temperature and then adjust the time it took for sewage to pass through the tank based on the results of these tests and previous experiments.[24]

The ability of process control to take these variations into account underlay debates over the use of automatic machinery in the sewage disposal plant. With his septic tank, Cameron had introduced the first automatic machinery to regulate the filling of bacterial beds like Dibdin's—the aerobic portion of his "workshop"—with the effluent from the septic tank. Sanitary engineer Arthur Martin, one of the co-inventors of this automatic system, campaigned for its adoption in England using arguments taken directly from the scientific management movement. But these schemes were criticized for a variety of reasons. One criticism centered on the effect of the automatic apparatus on labor. "One frequently hears it assumed," wrote Martin, "that the object of using alternating gear is merely to save the wages of the labourers who would otherwise be required to open and close the valves."[25] Others objected that automatic working "substitutes a mere machine for the human intelligence." These critics argued that because sewage varied so much in character over time, the period of bacterial action needed to be adjusted according to the constantly shifting composition and strength of the sewage. Instead, the machine at the center of the automatic apparatus allowed for only a fixed period of contact. Gilbert Fowler had installed automatic apparatus at the Manchester treatment plant (prior to the invention of activated sludge). He discovered, however, that proper management still required "constant personal supervision." After several years of operation, he had altered the apparatus to allow for more adjustment of the rates of filling and emptying depending on the varying conditions of the beds, and considered that automatic apparatus was not justified "by any possible advantage."[26] A. S. Jones, of course, with his commitment to land treatment and interest in rural labor, criticized the apparatus as well. "I prefer good intelligent hand labour in anything that has to do with the treatment of sewage."[27] For these critics, human intelligence exerted a far better control over the process than automatic machinery.

For Martin, however, relying on the operator's intelligence was misplaced. "No one will deny the need for intelligence in the centre of the working of a bacterial installation, but most engineers will probably agree with the author in his opinion that it is more conducive to efficiency that the intelligence in question should be that by which the works were designed than that of the labourer who is often promoted to the important office of manager."[28] Martin acceded that sewage could change its character completely "from one hour to another."[29] But for him, eliminating the uncertainty of manual control, and substituting the intelligence of the engineer for that of the laborer, was more critical, even if the resulting control was less effective. The management of sewage works was "an engineer's question," he stated.[30]

Martin's support of automatic working was a means of retaining control over a plant. In a large city like Manchester or Birmingham, "the men who designed the works retained control of their operation," he argued. But where an engineer consulted for a city and then handed "the works over to the local authority," he lost control over its operation. Only through automation could he maintain that control.[31] Chemist W. J. Dibdin expressed similar frustration with the loss of control after a sewage plant had been designed. "Many who have had to do with sewage treatment have keenly felt the disadvantages they have been labouring under when they see a delicately worked out process handed over to the fumbling of a farm labourer, or superannuated foreman, or an engine driver without the slightest knowledge of the real engine which he has to drive, namely, the process entrusted to his charge." Dibdin's solution was to hire trained chemists to "superintend the collection of samples, to submit them to analysis, to give instruction to the works foreman as to varying the details of the daily routine of manipulation" (figure 3.2). Although Dibdin and Martin agreed on the problem, they differed a bit on the solution. "The remedy is obvious," wrote Dibdin. "If you want to brew—appoint a brewer. If you want engineering—appoint an engineer. If you want a chemical process properly carried out—appoint a chemist."[32]

The drive for process control exposed long-standing tensions in sewage treatment plants between scientists and engineers on the one hand and operators on the other. Sewage treatment was grounded in two competing traditions: the industrial, chemical processes of precipitation, but also the more rural, agricultural traditions of irrigation and the sewage farm. Traditional and even progressive farmers chafed under the supposed "scientific" management of the local boards that were responsible for overseeing sewage treatment in British towns. In 1876, for instance, George Ferme, a successful dairy farmer who contracted with his local town for its sewage, published a diatribe on the worthlessness of "scientific" sewage farming. Ferme complained of meddling "scientists" on local boards who had no practical experience in farming. "It can serve no good purpose, for instance, for surgeons on a local board, as is their habit, to correspond with persons all over the country on the subject of farming, asking meaningless questions, the answers to which absolutely depend upon situation and circumstances; nor to go about inspecting sewage farms, without even a farmer in their company, but arguing with one another under the idea that they are finding out some new business–'an entirely new industry.'" These enthusiasts for sewage farming, he continued, "write under the delusion that their ideas are 'real exact science,' and that their heads are stored with 'a variety of somewhat abstruse chemical and physiological reasons.'" In contrast, Ferme declared that "farming is a profession which requires not only early training, but much experience and sound judgement." Only experienced farmers, he continued, could estimate the "infinity of details . . . correctly in practice." "Persons who are not *bred* farmers should never attempt to manage a farm," he concluded. Experience was the

Figure 3.2
"By chemical analysis we thus measure the invisible." This illustration demonstrating the use of the "micro-filter" in sewage analyses is from Dibdin's chapter on analytical methods. Dibdin explained the necessity for laboratory analysis rather than sensory techniques: "The useful and necessary guides of appearance and smell are insufficient to indicate when matters are in a critical state and the process is in danger of breaking down. . . . This can only be done by methods of chemical analysis." *Source*: W. J. Dibdin, *The Purification of Sewage and Water* (London: Sanitary Publishing Co., 1897)

key. Ferme ridiculed the scientists who thought they knew how to manage a sewage farm. "Science is a word too freely used by many who have not been through the mill of experience," he declared.[33]

This conflict between practical experience and science later resurfaced in the irrigationists' critiques of biological treatment. Jones and Travis, in criticizing the bacterial theories of the new biological treatment systems, defended sewage works managers for their "practical intelligence." It was not the fault of managers, they argued, that sewage disposal works were inefficient. Rather it was the fault of insistence on incorrect scientific theory. "Without detracting from the credit you have given to all observers with the microscope and other laboratory instruments, from M. Pasteur downwards," Jones remarked, "I would ask you to encourage by your notice a large class of workers who have long been labouring in a quiet way to render sewage harmless by the most convenient practical means available in their respective districts, whether as chairmen or members of sewage committees or managers of works."[34] Similarly Travis defended sewage works managers against the criticisms of scientists. Reacting to a comment by the president of the Association of Managers of Sewage Disposal Works, that "many schemes of sewage disposal were ruined by inefficient management," he replied that "that was a statement which he thought ought never to have been made."

Anyone who had seen, as he had done, managers manfully struggling with an impossible proposition would agree that it was unfair to blame them for what occurred. It was not their fault that sludge accumulated in the tanks, and required removing therefrom, or that its effluent was highly charged with suspended impurities. It was not their fault that the beds or filters became choked, and that the material had to be taken out and cleaned. It was not their fault that a foul effluent resulted. It was due to the disabilities of the methods adopted which had been dictated by an erroneous theory. . . . It was unfair to attach blame to the sewage works manager when the operation did not coincide with the view of the theorist."[35]

Supporters of bacterial treatment responded by charging the irrigationists as being anti-scientific and "loudest in exclaiming against that which they will not or cannot understand."[36]

The conflicts over scientific management and process control built on this history of distrust. Operators viewed engineers as holding abstruse theories but no practical knowledge. Engineers, in turn, viewed operators as ignorant, needing to be schooled in the rational management of their plants. "If the work is to be trusted to men ignorant of the first principles," wrote Dibdin, then "failure will be directly invited; and no one will be to blame except those in authority who thus deliberately throw away the teachings of science."[37]

"Skilled Control of the Process Will Be Essential"

It was within this context of industrial transformation that the activated sludge process was introduced in 1914. Laboratory control had spread throughout the process industries. Taylor had published his *Principles of Scientific Management* in 1911 and had implemented his program in leading industries in the United States. British industries were following many of the same trends.[38] By 1916, according to one observer, process control was so integrated into chemical manufacture that "practically all works . . . and frequently each department of a factory maintains its own control laboratory."[39]

Although sanitarians quickly recognized the advantages of the activated sludge process, they also identified one major reservation. Activated sludge was "temperamental"; it could lose its "activity" and thus its remarkable ability to clarify sewage.[40] Because of the fragile nature of the activity, from the time of the first descriptions of the process, scientists on both sides of the Atlantic recognized that activated sludge plants, to a greater degree than older processes, would require highly trained, "scientific" supervision. Edward Ardern wrote in an early summary, "The aeration process being essentially a highly intensified bacterial action, will require closer scientific control than is demanded by the usual filtration areas."[41] In his first reports on activated sludge, Hatton wrote that it "requires constant and expert supervision."[42] A review of the new process in the American journal *Engineering Record* also stressed scientific management. "It seems probable, from the nature of the activated-sludge process, that it will require more intelligent supervision than most processes that have hitherto been in use."[43] "Skilled control of the process will be essential," remarked Sheffield's John Haworth simply.[44]

Researchers and sanitary engineers quickly focused on the two interrelated aspects of the skilled operation they deemed essential—controlling both the bacterial and human workers in the plant. The first task was defining the characteristics of "process control," how the treatment process could be scientifically managed. Researchers investigated the activated sludge process and the factors related to its success. The second, more involved, task was the development of a skilled workforce that could "intelligently" implement the scientific control developed by the researchers.

From the initial development of the process, activated sludge researchers focused on "control" as a central theme of their research. "The neglect of any one of a number of factors, such as air supply, sludge percentage, or detention period, will cause this loss of 'activation,'" according to the *Engineering Record*.[45] As early as 1915, one year after their first publications describing the activated sludge process, Ardern and Lockett described the "factors controlling the activated sludge process," including the rate of air supply and ratio of sludge to sewage.[46] The greater the quantity of incoming sewage, the greater the quantity of activated sludge required to treat it. Early investigators determined that a constant proportion of approximately 20 percent activated sludge in the aeration tanks would be sufficient to purify the influent. Operators would thus adjust the rate at which they returned the activated sludge from the settling tanks to maintain this constant 20 percent proportion. This means of plant management was distinctive and illustrates the shifts in emphasis with process control. Instead of looking at the quality of the effluent and adjusting the process on this basis, operators would measure and adjust the amount of sludge in the process tank. In a similar vein, Lockett also emphasized the measurement of the process, rather than the outcome. He concluded that maintaining a minimum level of dissolved oxygen in the activated sludge tank was the appropriate control point. By sampling the tank every six hours for dissolved oxygen and using "this test as a 'control' . . . a greatly increased output" from the plant was possible. "This test," he concluded, "is superior to all others, as an indication of the condition of the sludge and working capacity of an activated sludge plant."[47] Researchers settled on sludge volume and oxygen concentration as the most useful measures for controlling activated sludge, and suggested that difficulties of operation could be overcome by using them as control tests.

Despite these efforts to establish scientific process control, however, activated sludge proved to be unruly and difficult to control. It was "temperamental,"[48] "sensitive,"[49] and "extremely fickle and easily upset."[50] As researchers discovered, activated sludge was not a straightforward industrial chemical process. Its control required an explicit consideration of activated sludge as a *biological* process.

"A Living System"

In July 1927, police raided "house after house" in the western Chicago suburb of Melrose Park, searching for illicit home distilleries. Continuing gang warfare in Chicago had forced the Genna crime family to relocate its home alcohol still operations from the city's South Side, and Melrose Park became a center of this illegal industry. In one day, police seized ten thousand gallons of alcohol and destroyed five stills. Except for a few gallons needed for evidence, the police poured all of the seized alcohol into the town's sewers. Raids continued for the rest of the year, with thousands of gallons of mash and alcohol dumped into the streets. At the peak of alcohol production, hundreds of households were involved. The equivalent of a railroad car of sugar was shipped into the town each day, fueling production of over one million gallons of alcohol a year. In addition to the tens of thousands of gallons that were dumped in the streets by police and revenue agents, the residue from the daily production of alcohol was also sent down the drains of Melrose Park. All of it made its way into the sewers and on to the Des Plaines River Treatment Works.[51]

Completed in 1922 to relieve the pollution of the Des Plaines River, the Des Plaines sewage treatment plant in Chicago used the newly developed activated sludge process. Equipped with a laboratory for conducting control tests, it also incorporated the latest theories of scientific process control into the management of the plant. It performed well for half a dozen years before the process inexplicably began to fail. In June 1927, plant managers noted, the normally clear effluent of the activated sludge process became turbid, and the dissolved oxygen demand spiked. The activated sludge itself had changed as well. "Instead of well-formed, spongy flocs," the Sanitary District of Chicago reported, the sludge had become "slimy and gelatinous" and would not settle in the clarifiers. Sewage sludge flowed over the weirs and out of the plant. The activated sludge process was failing.

When operators examined the flocs under a microscope, they found "masses of intertwined filaments" that they identified as the bacterium *Sphaerotilus natans*. This species had been found recently in a number of activated sludge plants around the world, and sanitary scientists had labeled the resulting condition "sludge bulking."[52]

In searching for the cause of this bulking problem, Sanitary District scientists conducted a systematic survey of all the sewers contributing waste to the plant. The
Melrose Park sewers, they discovered, had the odor of "fermented grain or sugar,"
and chemical analysis of the sewage indicated an enormous amount of sugar, equivalent to almost thirteen thousand pounds a day draining into the plant. The Sanitary
District managers read the newspaper reports of the police raids and recognized that
the illegal stills were the source of the carbohydrate and the cause of the filamentous
growths in the plant and the failure of the treatment process. By changing the nature
of the food for the activated sludge microorganisms, the stills had altered the bacterial community, leading to treatment failure.

As this episode illustrates, sewage treatment plants, despite the rhetoric of sanitarians, were not simple chemical processes. Managers were quickly confronted with
the complexity of the biological processes of sewage treatment that necessitated a
more complex theory of control. The idea of process control had originated in the
chemical industry, and early efforts to control the activated sludge process were all
based on this chemical model. But, as shown by the experience in Melrose Park, the
chemical model for process control proved inadequate. Not only were their plants
industries, they were ecosystems as well. As Fowler had emphasized, "to be properly 'managed,'" activated sludge had to be understood as a "living system."[53] To
properly manage these living systems, sanitary engineers had to adopt an explicitly
ecological view of their treatment plants. Recognizing the importance of biology in
managing plants, English engineer F. C. Vokes lamented his profession's ignorance
of biology. The chemist and engineer "usually persuaded the sewage to become purified," he remarked, using a "somewhat dictatorial manner." The biologist, in contrast, "got into intimate contact with the organisms themselves, and studied their
habits and customs on the spot." Vokes was convinced that the proper management
of sewage works would require the biologists' approach.[54]

But how to manage these living systems proved elusive. These were not simple
biological systems, like Weizmann's, where a single species of bacteria was inoculated into a sterile medium. Rather they relied on the activity of a wide variety of
microorganisms and more complex animals, interacting with both each other and
a constantly varying environment. Temperature varied over the course of a day and
season; the strength and composition of the sewage depended on industrial schedules, household schedules, and the weather. All of these changed the environmental
conditions for the bacteria. Scientists quickly recognized that controlling the activated sludge process would rely on both chemical analysis as well as insights from
the emerging science of ecology.

Scientists began to recognize that the problem of sludge bulking lay in competition between two groups of bacteria: the floc-forming bacteria that created an
easily clarified effluent and the filamentous bacteria that caused bulking. "Activated

sludge is not a pure culture of a single organism, but is composed of a variety of different organisms," explained Hovhannes Heukelekian, a chemist and bacteriologist at New Jersey's Sewage Experiment Station. "Under certain environmental conditions," he continued, "the proper biological balance may be upset, giving rise to the preponderance of certain organisms and a different type of a floc. The change in the biological balance is due to the fact that there are specific differences of optimum environmental conditions even among closely related organisms." "What can the operator do to check bulking?" he asked. "The first maxim is constant vigilance, to keep ahead of the process so that it does not get out of hand." He recommended daily examination of the sludge with a microscope to watch for an increase in filamentous bacteria as well as daily measurement of dissolved oxygen.[55]

Investigating his Long Island plant that was "continually troubled with bulking," George Anderson reported, "I found that we were working with living matter existing under conditions every bit as exacting as those surrounding man or animal."[56] Pursuing his analysis of the activated sludge as a living system, Anderson concluded that "there was death or septic action going on in this living matter when dissolved oxygen disappeared." Anderson sought tests that would allow him to "control and regulate" the quantity of sludge in the system, to prevent "over-feeding" and loss of dissolved oxygen. Working from these ecological assumptions, Anderson identified tests for determining the food supply and oxygen requirements.

The necessity of an ecological approach to control became apparent in other industrial applications of microbiology as well. Bristish microbiologist A. Chaston Chapman set out a strategy for managing industrial organisms in general: "A very thorough knowledge of the nature of those organisms and of the influence of environment on their chemical activities is essential to efficient and successful factory work," he argued.[57] "For every organism," he continued, "there is a particular set of conditions, the observance of which is absolutely essential to successful working." These conditions were to be based on a knowledge of the "probable habitat" of the organisms and included factors such as temperature, aeration, and acidity. "No matter how good the plant may be, how carefully all the purely chemical operations may have been carried out, or with what success the various engineering difficulties may have been surmounted, the success or failure of the manufacture depends on the selection of the right organism and on a precise knowledge of the conditions under which that organism is capable of doing its best work."

While Chaston Chapman emphasized the identification of a single, superior species for each fermentation, industrial microbiologists recognized that they were often working with complex environments in which many species of bacteria, fungus, and mold could potentially thrive. These scientists recognized the importance of an ecological understanding of their systems—the relation between organism and environment and the struggle for existence. By adjusting the conditions of the fermentation,

producers favored the growth of beneficial microorganisms at the expense of contaminating species. This was the fundamental basis of Delbrück's "natural" process of brewing. Makers of lactic acid conducted their process at high temperatures
to favor the lactic acid bacteria. In industrial alcohol production, if the yeast could
maintain high growth rates, it would exclude other organisms. Manufacturers thus
built up yeast populations in a series of fermentations conducted in increasing volumes of medium in the laboratory before using them in the factory, and they carefully controlled the pH of the medium to preclude infections by bacteria.[58]

Despite the realization that activated sludge was a complex ecosystem, however,
scientists and engineers continued to think of the sewage plant more in terms of a
chemical process and emphasized simple methods of control based on this chemical model. As research on process control in activated sludge progressed, investigators developed additional tests that could help them fine-tune aeration and the
amount of sludge to either return to the process or waste. While ecological thinking
informed these attempts at process control, simpler notions of control adapted from
chemical manufacturing nevertheless dominated. "A noticeable improvement has
occurred during the past ten years in the development of analytical and operating
tests which furnish a more precise basis for the control of the process," noted the
Sewage Works Journal in 1932. These improvements centered on laboratory tests
for determining the optimal supply of air and the condition of the sludge. Evidenc-
ing faith in the scientific project, the journal continued, "Probably in the next ten
years, operating tests will be worked out to enable the plant operator to measure
these variable factors accurately, so that the plant may be operated at maximum efficiency."[59] Some of the control tests, such as oxygen demand or oxidation-reduction
potential, were based on explicit theoretical considerations of the biology and kinetics of the process, while others, such as sludge volume, were more empirically determined. Other researchers suggested pH, sludge volume, sludge concentration, or
oxygen demand. By the 1940s, there were many competing theories of process control, some even patented.[60]

An "Unsavoury" Occupation

All of these tests had two common but conflicting threads. They all displayed a faith
in the ideology of control—that chemists and engineers would be able to identify
the proper conditions and regulate the treatment process. And they all placed the
operator of the plant at the center of successful sewage treatment. In each of these
schemes, the operator would have to measure some characteristic of the activated
sludge process and make appropriate adjustments. The article on process control in
The Digester was typical in centering the role of the operator: "Much depends on
the operator in order to obtain best results from an activated sludge plant. A rather

delicately balanced biological process must be maintained through his adjustment and regulation of the plant."[61]

The emphasis on controlling the treatment process quickly led to an emphasis on operators and controlling the labor process within the plant. Design engineers saw a literal loss of control when operators took over a plant. "After the plant is completed and put into operation," wrote a New York engineer, "control passes out of the hands of the engineer."[62] How was the engineer to regain control? Through scientific management and control of the operator. As historian Monte Calvert concluded about scientific management in general, it "was a way for the engineer *as* engineer to retain decision making powers."[63] The sanitary science profession—engineers, bacteriologists, chemists, and public health officials—instituted a number of efforts, all designed to create a labor force for operating sewage disposal works according to the dictates of scientific control. These efforts, including certification and licensing, training, and professional societies, were all oriented toward controlling operators in order to control the process and led to a continuous struggle between engineers and operators for who would exert shop floor control and control the sewage treatment process.

In 1901, the bailiff, or farm manager, of the Cuckoo Hall sewage farm, in Enfield, England, wrote to the *Sanitary Record*: "Now that the bacterial method of sewage disposal is becoming widely taken up by Local Authorities, it is necessary . . . to have a considerable scientific knowledge, which ordinary bailiffs do not require." He looked to a newly formed association of sewage plant workers to help provide that knowledge. "I am very pleased to see that the 'Association for Managers of Sewage Disposal Works' is likely to come into existence."[64] Other sewage farm and disposal works managers also supported the idea of a new organization. "I believe there are a great number waiting for such an association, and ready to join if it is started," wrote the manager for the Walthamstow works.[65] Many managers felt their occupations were considered "somewhat unsavoury" and looked to this organization as a means of advancing "the status of sewage farm managers generally."[66] In 1901, eighty-eight managers throughout England established the Association of Managers of Sewage Disposal Works (AMSDW). Shortly after the formation of this association, J. H. Barford, manager of the Maidenhead works, professed that "Unity is Strength" and looked to the association for the "combined and individual benefit of each one of us." Barford saw professionalization as the means to improve the position of sewage works manager "both socially and financially." For Barford, this meant being less like the "ordinary labouring man . . . and more nearly related to that of professional and business men." An emphasis on mental work is what was to distinguish his profession.[67] Keeping up with the changing technologies, greater respect and pay, and job security were some of the many reasons managers began to organize. By the end of the decade, there were almost two hundred members.

The AMSDW was the first professional association for sewage treatment plant managers and operators, but managers were not the only professionals with an interest in the sewage treatment plant. Although they all worked in sewage treatment, there was a professional divide between sewage scientists and engineers on the one hand and sewage works managers and operators on the other. The impetus for professionalization of sewage plant operators was complicated and rested on the needs of the state for efficiently run sewage treatment plants, professional claims of sanitary engineers, and the desire of operators themselves for the recognition and protection that professionalization offered. It thus was generated from the conflicting demands of independence and control.

Andrew Abbott describes professions as a system in which individual professions compete for the social and cultural control of tasks in the workplace, what he terms "jurisdiction." Professions split, merge, appear and disappear, subordinate or dominate as they seek to settle conflicts over jurisdiction. When a new arena of technical expertise emerges, any previous settlement among competing professions must reach a new equilibrium.[68] As the 1848 Public Health Bill created the new field of urban sanitation in England, doctors, public health officials, engineers, chemists, plumbers, and later bacteriologists all competed for control of the physical, scientific and administrative space of sanitation. The 1848 bill established the positions of officer of health, surveyor, and inspector of nuisances.[69] But the position of officer of health, through its higher pay, greater public respect, and protection and insulation from local politics, became the dominant public health position.[70] Municipal engineers and surveyors began to organize. As the Royal Sanitary Commission met to consider revising the public health laws, Lewis Angell, surveyor for Stratford, wrote Parliament to complain that the commission was ignoring their professional expertise. "Engineers and Surveyors holding Local Board appointments are disappointed that *their* evidence has not been taken with reference to the *working* of their department," he wrote.[71] The only way for them to be recognized professionally and exert control of their workplace was to explicitly organize as a separate profession. In 1873, at a meeting of the Institute of Civil Engineers, Surveyors formed the Association of Municipal and Sanitary Engineers and Surveyors.[72]

As sewage treatment plants expanded, and biological sewage treatment, with its emphasis on bacteriology, came to the fore, a new professional identity was defined and professional organization established. This new profession included the representatives of what Martin Melosi calls the "old public health" of miasma and filth as well as the bacteriologists, those preachers of "microbe agency" and germ theory. Thus the field of sanitary engineering was notable for its interdisciplinary and comprehensive approach to the urban ecosystem and included the surveyors and engineers working for municipalities, other civil engineers, medical officers of health, chemists, bacteriologists, and others. All came under the rubric of

Sanitary Engineering.[73] The Institute of Sanitary Engineers was formed in England in 1897. The Sanitary Engineering Section of the American Society of Public Health was formed in 1912, essentially as a new professional society. Founders emphasized its interdisciplinary nature, combining as it did the "medical officer, the sanitary engineer, the laboratory worker, the statistician and the sociologist," as what distinguished the profession from that of civil engineering or medicine.[74] Sanitary engineering, then, represented the group of interests that contested control over the biological sewage plant with the managers and operators of sewage plants.

In both England and the United States, managers' associations developed within a few years of the first explicit sanitary engineering associations, and the tensions between these groups can be seen in the goals of the sanitary engineering associations and their attempts at control of the new managers' and operators' organizations. Membership in the AMSDW was limited to those having responsibility for a sewage disposal works. "Now, the Association is an association of sewage works managers," emphasized Samuel Rideal, seemingly needlessly. But, he continued, "The title of the Association, 'The Association of Managers of Sewage Disposal Works,' had not been selected without careful consideration on the part of the Council." From this carefully considered name, it was clear that the association was to exclude laborers in the plant, on the one hand, and, on the other, the municipal engineers that might supervise the works managers, the consulting engineers that designed the plants, or the academic scientists and engineers that studied microbiology and public health more generally. Some engineers, like A. S. Jones, saw the new organization as a trade union movement that would increase the pay and respect of the profession, and he supported it, as did the important journal the *Sanitary Record*. But many other engineers feared that the association was formed with the expressed goal of freeing its members "from all supervision and control of the engineer."[75] Scientists like Fowler feared the organization would be too exclusive, to the detriment of sewage disposal.[76]

The shift to biological treatment from land treatment was an important impetus for the new organization. As the 1898 Royal Commission on Sewage Disposal investigated the new bacterial methods, many expected that the number of sewage treatment plants would rapidly expand at the expense of sewage farms. Not only did William Ellis mention this in his letter to the *Sanitary Record* supporting the formation of the AMSDW, but Rideal, the first president of the association, focused on this issue in his inaugural address: "The duty of the Commission was defined as to whether it was possible to replace the old sewage farms with what are called modern bacterial methods." He noted that the organization had perhaps specifically elected him because, as an early advocate of bacterial treatment, he represented the future path of sewage treatment. In this changing world, it was "very difficult to predict what class of man would be the sewage works manager in the future." In the past," he said, the sewage works manager "was thought to be a farm labourer; a little later on he was to be a market gardener; then he was to be acquainted with engineering; and afterwards had to become a chemist."[77] What would the position of plant manager look like in this changing world of sanitation? The AMSDW hoped to have a crucial voice in determining the qualifications and character of the new manager.

In the United States, both treatment plants and associations of sewage plant managers formed later than in England. As in England, treatment plant operators varied widely in their backgrounds. But unlike those in England, associations of plant managers were from the start composed of operators and managers as well as other sanitary professionals. As a result, the professional contest over control of the sewage plant took place within these new organizations rather than between the AMSDW and sanitary engineers, like in England.

The New Jersey Sewage Works Association was the first association in the United States, formed in 1915. The organization grew out of the efforts of the New Jersey Department of Health. The purpose of the organization was the "advancement of the knowledge of designing, construction, operation and management of sewage works." But this was put in the context of the needs of the state. The organization would enable managers, engineers, and sewage equipment manufacturers "to cooperate intelligently with the State Department of Health."[78] Unlike the AMSDW, membership was open to every sector involved in sewage treatment, including managers, operators, engineers, chemists, bacteriologists, public officials, and manufacturers.[79] The New Jersey association became the model for other states. In 1927, Wisconsin, Illinois, and Indiana formed a Midwestern regional organization, with a similar membership structure, open to engineers and other sanitary professionals as well as operators. By 1940, there were twenty-two sewage works associations in the United States and Canada.[80] In the 1920s prominent members of the state and regional associations began an effort to coordinate the individual organizations, and in 1928 they formed the Federation of Sewage Works Associations (later the Water Pollution Control Federation, and now the Water Environment Federation).[81]

There were a number of tensions incipient in the development of these organizations. Were they to serve the interests of the operator? the state? the manufacturer? For the operator, professionalism involved respect, "the protection and improvement of the status of sewage works operators."[82] But, because these organizations were not exclusively operators' associations, the interests of other members often dominated, and operators' goals were ignored. Further, engineers and scientists held greater prestige and power in the profession, and this was reinforced by state departments of health and other public officials. The sewage works associations thus became vehicles for enforcing the engineers' ideology, the practice of laboratory process control, and thus the operation of the plant. "Sewage treatment is not an empirical matter which can be successfully accomplished by any set formula or 'rule of thumb' method,'" declared two of the founders of the Central States Sewage Works Association. Rather scientific practices must prevail, and the objective of the organization was to transfer these practices and their underlying ideology. "The development of efficient operating personnel . . . is one of the chief concerns for the Central States Sewage Works Association and constitutes one of the principal motivations for the formation of the organization."[83] For the engineers organizing the association, efficiency meant process control. One of their first actions, for instance, in a move clearly directed at operators, was the formation of a "Committee on Sewage Control Tests" in 1928 to devise and present simple control tests for sewage plant operation in readily understandable terms."[84] In the balance between independence and control in the profession, control began to dominate. Engineers had designed the sewage treatment plants. They now sought to expand their professional jurisdiction into their operation.

These new sewage treatment plants became the site for a contest over professional authority. Process control became the means. Laboratory process control, based on laboratory techniques and scientific understanding of the biological treatment process, was the domain of scientists and engineers, and using these abstractions, they tried to exert control over the operators, determining what adjustments would be made in the plant (figure 3.3). For Abbot, "Professions expand their cognitive dominion by using abstract knowledge to annex new areas, to define them as their own proper work. . . . Knowledge is the currency of competition." In Abbott's analysis, "Control of the occupation lies in control of the abstractions that generate the practical techniques." As Abbott also notes, "The techniques themselves may in fact be delegated to other workers." Thus scientists and engineers controlled the sewage plant by generating the rules for how and when operators would turn valves and flip switches. With the invention of the biological sewage treatment plant, operators and managers, sanitary scientists and engineers all competed for jurisdiction over its control and operation. The application of scientific theories and methods as defined by scientists—process control—became the currency of professional competition in the sewage treatment plant.[85]

The new sewage works managers organizations were forced into accommodation with the sanitary engineers. After several years, to placate those like Fowler, concerned over being excluded from the organization, the AMSDW established a category of honorary membership for sanitary engineers and scientists who were not involved in the day-to-day management of plants. In addition, while the organization was controlled by a board consisting only of full members (and thus managers), the offices of president and vice president were reserved for renowned sanitary engineers and researchers, not managers. Officers of the New Jersey association were all day-to-day operators or managers of sewage plants, but the impetus for the organization came from the New Jersey State Department of Health. The Central States

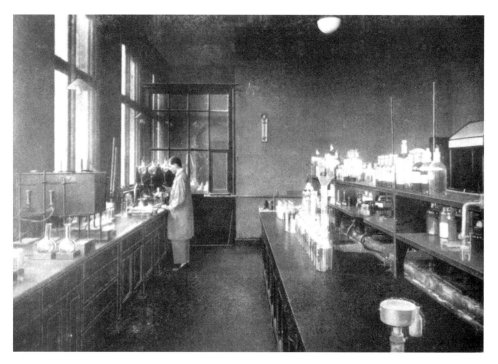

Figure 3.3
Sewage works laboratory at the Davyhulme, Manchester, activated sludge plant, 1938. Laboratories became an increasingly important part of the sewage disposal plant, not only to evaluate the quality of the effluent, but to control the treatment process. *Source*: City of Manchester Rivers Department, *Descriptive Notes of the Davyhulme and Withington Sewage Works*, 1938

Sewage Works Association came out of meetings between sanitary engineers working for large consulting firms, public health officials, and engineers working at the largest sewage plants in the region. While one organizer mentioned a goal of the new organization as "uplifting the dignity of sewage plant operators," in the same breath he noted that the organization would help state sanitary engineers "in the enforcement of proper operation of sewage plants."[86] Similarly, the Michigan Sewage Works Association arose out of a similar blend of interests and personnel. Operators initially held the offices of the association, but after several years, although operators remained a majority within the association, offices were increasingly held by consulting engineers, state officials, and academics.[87]

In the United States, the state organizations often reflected the conflicting demands for professionalization on the part of the operators, and control on the part of sanitary engineers and state public health officials. Along with the goal of improved status was the disciplining of "careless or ignorant supervision" of sewage plants. F. Herbert Snow, for instance, in one of the first papers delivered before the new Section on Sanitary Engineering, wrote, "The enforcement of compulsory skilled supervision of sewage purification works involves the exercise of not only local but also state and national police power." He encouraged the section to establish a committee to report on "standard methods of operation and supervision of sewage purification plants."[88]

A key part of controlling the labor process was to control who could become an operator, to prevent the "placing of some street sweeper in charge of plants."[89] In early sewage treatment plants, operators were typically drawn from the ranks of skilled and unskilled labor rather than from the technically trained. In England, the development of sewage treatment plants from sewage farms or precipitation works meant that many early plant managers were either farmers, farm managers, or "mechanical engineers" (closer to what we would consider today as mechanics than engineers), with little scientific training.[90] In the United States, early treatment plants in large cities might be operated by chemists or engineers, while plants serving smaller cities might be in the charge of a laborer. In 1921, the Rochester, New York, plant for instance was in the charge of a chief engineer, with another engineer and chemist below him. In addition, there were "two tank operators, relief man, mechanic, electrician, labor foreman, and 17 laborers." In contrast, the personnel at the activated sludge plant of a smaller city, Sherman, Texas, consisted of "one man who also tends the city incinerator."[91]

With the emphasis on scientific process control, the skills needed for operating a plant changed. "Management was too often supposed to be an easy business which might be left to some old roadman, too old to keep the roads in order," recalled H. Maclean Wilson, president of the AMSDW.[92] In a 1909 discussion of the sewage works manager of the future, John Thresh, AMSDW president, noted that "the time was rapidly passing when a person entirely ignorant of sewage and of its purification could be placed in charge of a works." In the near future, he advised, works managers would need "knowledge of engineering, of chemistry, of bacteriology, of biology, of meteorology, of public health."[93] "The work of a sewage works manager to-day is very different from what it was twenty years ago, when some of us were appointed," recounted Stephen Flinn in 1911. "Some of us were not even asked whether we knew anything about sewage. In those days, if we had heard of such things as bacteria or bacilli, we should probably have thought they were a new kind of disinfectant or hair restorer."[94] "Those days of ignorance are past," he concluded. "The sewage works manager of to-day must be a specialist, scientifically trained and fully equipped for the performance of his important duties."[95]

Another aspect of professionalization supported by the new organizations was the implementation of certification and licensing. State regulations requiring that operators be certified would prevent the staffing of plants with unqualified people, and protect the positions of those operators who were certified. Sewage treatment plants, almost always part of the municipal political structure, were subject to the same problems of political patronage in staffing as any municipal agency, with operators subject to the whims of changing political administrations. Stories abounded in the sanitary literature of the hiring of unqualified operators. One manager recounted the story of a qualified operator replaced following a change in city government by someone trained only in recreation and playground supervision. A sewage equipment manufacturer recalled, "I have seen several cases where men, politically selected, too old and feeble to get around, have been placed in charge of modern sewage plants."[96] Another typical complaint came from Paul Hansen, an engineer with the Illinois Water Survey. Describing a visit around 1912 to six plants in Illinois, he noted that none was operated adequately. Indeed only one plant even had an operator. "This particular operator was somewhat old and decrepit and seemed to pride himself on how little he knew regarding the plant . . . his lack of knowledge of the mode of operation of the plant resulted in the by-passing of crude sewage into the stream."[97]

In the United States, individual states began implementing more formal requirements, but ran into obstacles from both municipalities and operators. In 1918, New Jersey, spurred by the creation of the state Sewage Works Association, became the first state to require the licensing of operators. However, it was not until 1933 that Connecticut established the next licensing law, followed by New York and Ohio in 1937.[98]

Although many operators saw the benefits of certification, there were critical tensions in the process. Many operators, lacking certification and formal education, were nevertheless highly skilled at managing the treatment processes. Yet the emphasis on scientific training cut off opportunities for advancement for these operators, because higher class licenses often required specialized engineering education and a college degree. "Practical" men would be unable to advance to supervisory positions, these operators feared, regardless of their experience running a plant. "What chance has the man ordinarily educated who has worked himself up to the position of superintendent when another man comes in with a technical knowledge and can get a license," asked one man during a discussion of New Jersey's licensing law.[99] "Too much stress is being put on the desirability of College training," stated Homer Blakely, an operator at a California plant. "Too many men like myself came up the hard way without benefit of college training. But because they do not have College training, shouldn't disqualify them from top grading, and corresponding pay brackets, if they can deliver the goods."[100]

Because of resistance from operators to mandatory certification, most states were unable to pass certification programs unless they were voluntary. "Remember that there is no legal requirement that any operator apply for a certificate. Application is purely optional," assured the Illinois Department of Public Health in their newsletter to operators.[101] The department hoped to convince operators that its plan was "fair to all, regardless of education," and it established a grading system in which experience could substitute for formal education. Operators there received a "service rating score," documenting their skills through practical examinations. Operators needed to demonstrate, for instance, ability to conduct laboratory tests such as dissolved oxygen or suspended solids, keep records, and conduct process control.[102] Many states refused to establish mandatory certification, so operator associations established voluntary programs. By 1940, fifteen states had certification programs, although only four were mandatory.

In England, there was no formal certification. From its beginning, however, the AMSDW carefully limited its own membership and hoped that treatment plants would view membership as a prerequisite to employment in positions of responsibility.[103] To further control who could be hired through this process, it established an examination for membership. This exam became increasingly rigorous, requiring applicants to perform laboratory analyses on sewage samples and answer a broad range of questions on the biology and engineering of sewage treatment plants. In order to standardize the tests, the association shipped uniform samples of sewage by passenger train to the applicants throughout the country. Applicants would perform tests on the sewage and submit their results to the association.[104] As it had hoped, membership in the association was increasingly recognized as a standard qualification for employment.[105]

In England, too, however, many operators complained about the requirements for certification. "The case of the manager or foreman at the smaller works is causing some concern," concluded the Institute for Sewage Purification (successor to the AMSDW). "Such works are usually, and are likely to remain, in the hands of men who do not pretend to a high standard of theoretical and technical knowledge. Inside their admitted limitations, many of these men are conscientious and efficient, and carry out their duties in a thoroughly satisfactory manner, but they could hardly hope to satisfy the increasingly high standard of knowledge demanded by the Institute's examination."[106] One operator complained to the institute that "it is now generally recognised that the standard set by the Institute is too high." "Three times I have taken the examination," he recounted, "and three times received the same brief reply, stating that I have failed to satisfy the examiners . . . May I suggest that . . . the present method of entry be reorganised by division into 'grades,' each with its own standard of examination to be passed? . . . The manager and/or foreman of the smaller works (who, I venture to state, often carries more responsibility than the 'specialist' on a large works.) would have some qualification."[107] The desire for increased recognition and status on the part of operators had collided with the presumed qualifications for implementing scientific process control. Professionalization developed from the conflicting desires of sanitary officials for control and operator desires for independence and recognition. These tensions were ultimately resolved in favor of the public health agencies and control. Starting in the 1960s, mandatory licensing became increasingly common in the United States[108]

For operators to manage the increasingly complex plants and to pass the increasingly rigorous certification exams, sanitary engineers, operator associations, and state departments of health all thought that operators would need to be trained in the fundamentals of the bacteriological process, laboratory techniques, and the ideology of process control as a means of operating the plants. As it was in the organization of the professional associations, one of the overriding goals of operator training was the concerted establishment of "process control" as the dominant paradigm of sewage treatment.

Training began in an informal way at various conferences for sewage treatment plant workers that were established by many states in the United States. In these meetings, engineers from established plants or companies that produced equipment presented papers on theory and operation of sewage processes. In addition, operators had a chance to talk with each other and compare notes on operations and ways of overcoming problems. Although these conferences were valuable, many sanitary scientists saw the need for more formal instruction.

In the United States, state and regional organizations of sewage treatment plants organized formal courses—"short schools"—for sewage treatment plant operators. Texas established the first short school in 1926, piggybacking on an already established school for water treatment plant operators. In New Jersey, the first course was given the next year, in 1927. It consisted of laboratory work, lectures, and round-table discussions, and it covered practical issues such as the design and construction of treatment plants as well as principles of chemistry and chemical analysis.[109] Finding that engineers who were attending the course tended to dominate the discussions, the organizers split the course into two sections, one for operators and one for engineers.[110] By 1947, at least twelve states provided these programs, often in conjunction with state universities.[111] In England, the Association of Managers of Sewage Disposal Works debated how to set up formal courses in sewage treatment. It ultimately rejected correspondence courses or existing trade schools, and argued for setting up its own course of lectures.[112]

Teaching operators about process control was one of the main efforts of these various sewage schools and other educational efforts targeted to sewage treatment plant operators. In order to develop a labor force skilled in the laboratory control of sewage treatment processes, the short courses emphasized not only practical skills but also more theoretical scientific training (figure 3.4). Organizers included a good deal of basic science that might have been familiar in college-level

Figure 3.4
"Class Sessions at the February, 1965, Institute for Water and Sewage Works Operators."
Short schools taught basic chemistry and bacteriology as well as practical techniques for oper-
ating treatment plants. The schools also gave operators the opportunity to share experiences
and knowledge. Courtesy of North Dakota Water and Sewage Works Conference. *Source*:
Official Bulletin, North Dakota Water and Sewage Works Conference, 1965

chemistry and bacteriology classes. The short course offered by the University of
Illinois in 1941 was typical. Mornings consisted of lectures on general topics, such
as "General Duties of the Operator," practical sessions on "Operation of Activated Sludge Plants" and "Arithmetic of Sewage Treatment," and more basic scientific concepts such as "Theory of Oxidation." Afternoon sessions were devoted to
training in laboratory technique and procedures for keeping records (figure 3.5).[113]
In New Jersey, short course exams consisted of questions in basic chemistry, such
as "What is oxidation . . . what is reduction?" and "What is the relation between
ammonia, nitrates and nitrites?," as well as practical calculations needed for process control.[114] An Oklahoma short course had lectures on basic chemistry, including atomic and molecular weights, valence, reactions, and equations.[115] The 1950
course offered by New York University, although designed for operators with only a
high school education, covered a number of sophisticated concepts. It covered basic
microbiology, describing the morphology and life processes of bacteria, and their

Figure 3.5
"Laboratory Session—Water and Sewage Works Operators School, Fargo, N Dak.,1962." A central focus of the sewage treatment short schools and conferences was the training of operators in the philosophy of scientific process control and the laboratory techniques essential to its implementation. Courtesy of North Dakota Water and Sewage Works Conference. *Source*: *Official Bulletin*, North Dakota Water and Sewage Works Conference, 1962

biochemical composition as well as more practical work on specific sewage treatment practices.[116] While the goal of the courses was clearly to train labor in the particular techniques desired by the engineers—to control the labor process—they also helped develop new skills among the workers. In this sense, the emphasis on process control led not to a deskilling of the workers, but rather increased their skills.[117]

"If Only We Could Measure It"

To the mainstream engineers and scientists, "process control" meant "laboratory control." "The laboratory is one of the most essential units in an activated sludge plant. Proper operation is impossible without some laboratory control" was a typical exhortation.[118] Engineers emphasized over and over the importance of laboratory tests of each stage in the sewage treatment process and the careful recording of data. Textbooks and early operations manuals all stressed the importance of a laundry list of laboratory tests.[119] In 1931, the Federation of Sewage Works Associations tried to standardize laboratory tests and established a committee to recommend a minimum set of records to be kept by sewage treatment plants.[120] By the time the American Public Health Association drafted a review of operations practice for activated sludge plants ten years later, it recommended measurement of a standard list of parameters—biochemical oxygen demand, suspended solids, volatile suspended solids, organic nitrogen, nitrate and nitrite nitrogen, ammonia nitrogen, settleable solids, pH, and dissolved oxygen—all from several different points in the activated sludge plant.[121]

Despite, or perhaps because of, these copious laboratory testing and record-keeping regimes, engineers were almost universally disappointed by the failure of operators to apply laboratory tests to process control. "Too frequently operators and officials in charge of sewage works fail to recognize the value of full and accurate records of the performance of sewage treatment works." According to the *Sewage Works Journal*, operators felt that "if the effluent appears satisfactory . . . the daily log and weekly report sheets are simply time-consuming inventions of little use to anyone."[122]

Part of the problem, though, was that the scientists were requiring the wrong tests. The search for process control tests had quickly devolved into an effort to measure seemingly everything; the relationship between the measurement of conditions in a sewage treatment plant and appropriate operator actions was lost. "There now exist at many laboratories elaborate routines containing the more tedious determinations, which give, when completed, very little information or data of use in the actual operation of the plant" complained Wagenhals in his survey of treatment plant operation.[123] Effective process control needed tests that could provide information on the current state of the process. A biological oxygen demand test that typically required five days to complete, while offering information on the status of the sewage treatment plant, could not provide data useful for the hour-to-hour management of the process. Carl Nordell, arguing for his own patented control test, criticized the BOD test: "The values derived from this test afford little other than historical interest."[124] Speaking of biological oxygen demand and other tests, one researcher made the distinction between control tests and measures of overall efficiency of the process: "Valuable as these indices are in determining the end product of the process they are of very little help in intelligently controlling the process and in overcoming the troubles, when they occur."[125] Other tests simply didn't work. "The experience of all operators," noted an editorial in the *Sewage Works Journal*, "indicates that the results of volumetric tests of sludge concentration are too variable to serve as a basis of control."[126] Measurement was pursued for its own sake, a seeming requirement of the modern, scientific plant.

The goal of scientific process control and its failure to be achieved, at least by the 1950s, was clearly spelled out in an English review of activated sludge control tests:

Almost since the inception of the activated sludge process . . . plant operators have been on the look-out for methods of control which, ideally, and from analysis, would give accurate information relating to the performance and load-carrying ability of the activated sludge process and, at the same time, could be carried out expeditiously in the laboratory. Such a method would give practical and up-to-the-minute assistance in control of aeration processes, instead of just producing the purely "historical" statistics associated with the analytical analyses at present in use as a means of control . . . This desirable state of affairs has not been attained.[127]

A New Jersey sewage chemist, in similarly dismissing the utility for control of many of the tests, concluded that "undoubtedly the most important factor in the regulation of the activated sludge process is the condition of the sludge itself. Herein lies the seat of all activity." Delineating the principle of process control, he wrote, "Proper control depends on evaluating the condition of the . . . sludge rather than the effluent. If the mysteries of the sludge are solved, the effluent can take care of itself." He argued that the condition of the activated sludge process "must be reflected in some way in the condition of the sludge. . . . If we could only measure it," he lamented.[128]

The Visible and the Invisible

The failure to fully implement laboratory control, however, did not mean that operators neglected the regulation of the plant's treatment processes. Rather, from their experience running sewage treatment plants, operators had developed their *own* measures of the condition of the sludge. Instead of the chemical tests, laboratory analyses, and reporting forms developed by sewage scientists, many operators relied on more observational methods. Their methods were rooted in what can be called "craft knowledge," the knowledge developed by workers involved in the day-to-day operations of a production process. Operators, within the context of the process control strategy as outlined by sanitary scientists, had developed their own understanding of the processes and the tests appropriate for controlling them.

Early in the development of the activated sludge process, there was an emphasis on visible and olfactory properties of sewage and sludge, as researchers noted correlations between the color or odor of the sludge and its ability to treat sewage. An early study from England, for instance, noted the relation between sludge color and structure and settling qualities. "When the tank nearly went out of action," reported the manager of the Bradford, England, works, "the warning was given when a change in the colour of the sludge took place from brown to dirty grey and the size of the 'flocks' of sludge diminished noticeably in size."[129] Another British plant manager remarked that he could "tell instantly by the peculiar odour when the process is going off, the lustre in the bubbles on the surface of the liquid is indicative of the state of purification, the colour of the sludge is also a useful guide. Again the formation of various fungi on the edge of the walls will give indications as to progress." These observations, he acknowledged, might not be scientific, but they worked. "It doesn't matter whether you can spell or pronounce the names, the appearance is enough. This may not be soaring in the realms of science, but they are good practical guides for those who are not fortunate in having a chemist on the works."[130] Harrison Eddy, consulting engineer and founder of Metcalf & Eddy, one of major sanitary engineering firms in the United States, stated: "The character of the floc varies greatly under different conditions. That produced by thoroughly aerating sewage for such a period of time and under sufficiently favorable conditions as to produce a highly purified effluent will usually be of a golden-brown color, relatively compact and heavy, and will settle with surprising rapidity. On the other hand under-aeration will produce a light, voluminous floc which will not settle as rapidly and which will form a very voluminous sludge of high water content and correspondingly difficult to handle."[131] These kinds of observations, both scientists and managers learned, could serve as a basis for control and did not require extensive sampling and chemical analysis (figure 3.6). As operators worked in the plants, they too noted the relationship between visible and olfactory characteristics and process control. In a disposal works at Bolton, England, for instance, the plant manager noted that an "intelligent labourer" identified an operational problem in his plant by the septic character of the sludge and made appropriate adjustments.[132]

Because the new laboratory methods of control often failed in actual practice, operators fell back on observational methods of control they were developing as they worked in the plants. The U.S. Public Health Service concluded that "plant operators can best judge the condition of their sludge by its visible appearance and properties."[133] Other observers close to actual practice continued to note that visual

Figure 3.6
"Aeration Tanks and Operating Gallery," at the Milwaukee activated sludge plant, 1925. Foam is collecting on the right side of the tanks. By observing the color and kind of foam in the tanks, skilled operators could judge the status of the biological process. This photograph was taken soon after the Milwaukee plant began operation by an engineer from Grand Rapids, on an inspection tour of midwestern sewage treatment plants. Courtesy of Grand Rapids City Archives. *Source*: Environmental Protection Department, Wastewater Treatment Plant, Miscellaneous Historical Records, Grand Rapids City Archives

and olfactory clues were quite useful to control. Asked what tests were "used in keeping an activated sludge plant at its best efficiency?," Ralph J. Bushee, a sewage equipment manufacturer, "pointed out that appearance and odor of activated sludge should not be overlooked as important indices of its condition and that an experienced operator can detect very slight changes which furnish guidance in plant control."[134] Thus, at the same time that professional organizations were developing and encouraging scientific control, operators were learning to use their senses for controlling the process. As a result, qualitative process control techniques, based on experience, developed and persisted side by side with scientific laboratory control.

While early descriptions of actual day-to-day decision making are rare, a set of diary entries published in the *Sewage Works Journal* give some idea of how laboratory tests and more qualitative measures were used in actual practice of controlling the activated sludge process. W. W. Mathews, superintendent of the Gary, Indiana, sanitary district, documented the day to day details of operational decision making during the startup of that city's new activated sludge plant in 1946. By this time, the American Public Health Association had already published its review and recommendations for scientific process control, and laboratory control was institutionalized in sewage treatment practice. Yet, among descriptions of broken gas seals, pump stoppages, and construction garbage in pump lines, Matthews reported on how he used the smell and appearance of his tanks and effluent to manage his process. After starting the plant, he noted that "the return activated sludge possessed a somewhat stale and septic odor" and "concluded that the 25 per cent return rate was a bit low and decided to increase it." On adjusting the return sludge flow, he noted that "the final settling tank effluent now has a rather milky appearance." A few days later, "the effluent, turbid and cloudy heretofore, is beginning to clear up," and a week later he noted the overall success of the adjustments when "the plant effluent is clear and sparkling and low in B.O.D. but carrying a bit more suspended matter than is desirable."[135]

Despite his success in using sludge odor in control, Mathews showed an almost ideological commitment to scientific control. "A visual record of operating statistics has been of great assistance in noting trends in the treatment process." Plots of volume and pH of sewage, temperature, sludge index, and solids concentration "can be lined up for convenience in checking 'across the board.' Any trend or variation that may effect [sic] the treatment process is instantly apparent." Mathews concluded that "a daily inspection of the various tanks and samples along with a study of the graphs and control sheets has been of material assistance in controlling our plant operation." Judging from his reports, however, it appears that the laboratory measures recommended for control were not particularly useful and that the "inspection of the various tanks" was the measure actually used in control. For instance, three days after describing the importance of these daily records in controlling plant operation, Mathews reported a twofold spike in the sludge index from 155 to 315 over the course of one day. The effluent, however, remained in good condition, and he reported that "whatever has caused the high sludge index hasn't hurt the plant operation—yet!"[136] Over the next several months, he reported large variations in this index. The scientific literature on sewage treatment suggested that abnormally high values were usually associated with poor treatment. Yet the effluent at the plant remained satisfactory, and there is no evidence that the operators made any adjustments based on this control variable.[137]

The chemist for Chicago's Southwest activated sludge plant similarly reported on its initial months of operation. "The activated sludge effluent became dark and murky, and a film of grease appeared over the surface of the final settling tanks," reported Pointdexter.[138] Control tests did not indicate the failure of the process; rather, it was the visual features of the tanks. Laboratory tests were primarily used to analyze the event after the fact rather than give guidance to control.

Despite the success of these observational techniques, though, many scientists and engineers discouraged process control using sensory tests and insisted on what they considered to be more scientific approaches. This reflected an ideological commitment to notions of modernity and industrial organization as much as an accurate reflection of the relative success of scientific and craft approaches to control.[139] Sanitary scientists viewed the activated sludge process as superior to other forms of treatment because it was the direct outgrowth of scientific research rather than the result of "trial and error," by the "practical worker."[140] For these researchers committed to the application of science, the control of the process must therefore also be the outgrowth of scientific research as well. Researchers thus rejected observational techniques for controlling the activated sludge process and developed theories of scientific process control based on less tangible properties, such as oxygen demand, suspended solids concentration, or settling volume, that could only be discerned through laboratory analysis. "By chemical analysis we thus measure the invisible," wrote W. J. Dibdin at the turn of the century.[141]

At the same time, engineers discouraged the techniques relied upon by the operators that they considered to be outdated and unscientific. With each new advance in laboratory control, engineers tried to persuade operators to give up observational tests in favor of laboratory-based methods. "The routine observations normally performed to determine the relative characteristics of activated sludge, namely color, odor, settleability, and enumeration of the grosser microscopic life are of such a qualitative nature that little information of value can be expected," wrote one sanitary engineer.[142] Heukelekian criticized the use of color and odor as indices of sludge quality. Although noting that color could be useful as an indicator of conditions such as under-aeration or overloading, he quickly asserted that "it is obvious that color in itself cannot be a useful index of the condition of the sludge."[143] "Operating routine cannot be based upon mere visual inspection . . . but must be based upon the results of tests and chemical analyses," concluded a 1932 editorial in the *Sewage Works Journal*.[144]

These statements criticizing observational control were more assertions of ideology than statements of fact. Operating routine could be and indeed was based on "mere" visual inspection. Rather, sanitary engineers and scientists were committed to the ideology of scientific management and the quest for professional control. To assert control of the operators, to claim jurisdiction over the sewage treatment plant, engineers had to centralize decision making through the implementation of techniques that they could control. This insistence on laboratory analysis highlights the ideological nature of process control. Noble, in his study of automation of the machine tool industry, describes this ideological commitment to scientific control as an "obsessive drive for omnipotence" or a "queer quest for a perfectly ordered universe." In the machine tool industries, he showed, there was evidence that numerical control was more expensive and produced lower-quality products. Yet businesses continued to implement it, precisely for the control over workers that it provided.

In sewage treatment, laboratory control often failed where craft succeeded, yet scientists and engineers continued to insist that successful scientific control was just around the corner and that operators should abandon their craft techniques. This insistence was often placed in the context of modernity (figure 3.7). The laboratory of the Northside Treatment Works in Chicago, for instance, was highlighted as part of this "outstanding modern example of applied sanitary principles."[145] "Modernity," notes Peter Dickens, "is characterized by the attempt to rationalise all aspects of life, to organise production processes and forms of administration in such a way that their results are predictable."[146] Laboratory control was an essential element of this drive for modernity in the sewage treatment plant.

Complexity and Alienation

Despite the concerted efforts of the short schools, the *Sewage Works Journal*, and professional associations to implement laboratory-based scientific process control in sewage treatment plants, operators developed their own techniques based on sight and smell. And despite repeated exhortations to abandon these techniques, operators persisted in their use. The very complexity of the biological and physical processes defied the attempts of the engineers to rationalize operations. Shop floor control remained with operators. The very characteristics that resisted patenting and alienation of the bacterial processes also helped the workers resist the alienation of their work process.

In brewing, brewers had generations of experience in controlling the fermentation, even without knowledge of the biological conditions. Modern methods of

Figure 3.7
"The Electrical Brain" of the North Side plant, Chicago. "Operators in this main switch balcony control the entire mechanical equipment of the vast works." This view illustrates the modernist tendencies in the control of sewage treatment plant processes. *Source*: *Formal Opening, The North Side Sewage Treatment Project*, Sanitary District of Chicago, 1928

brewing that relied on more careful process measurement using thermometers and other instruments were imposed on these successful craft techniques. As a result, process control lagged. "The empiricist is still far too much in evidence," Chaston Chapman complained of the brewing industry.[147] In biological sewage treatment, however, scientific process control was not imposed on older, craft techniques. Rather, both developed simultaneously. Workers in sewage treatment plants began to understand the relation between success of the treatment and visible and olfactory characteristics of the process at the same time that scientists developed laboratory methods of control. The persistence of craft techniques in sewage can't be attributed to conservatism on the part of the workers and reluctance to adopt new methods. Rather, we need to look at what the advantages of craft techniques might have been.

The success of craft approaches to process control in sewage treatment can be explained in part by the biological complexity of the treatment processes themselves.

Activated sludge was a complex, incompletely understood physical, biological, and chemical process. Unlike a chemical industrial process, or even an industrial fermentation that, like sewage treatment, was based on biological activity, biological sewage treatment was highly variable. Sewage is both a biological and social product. Sewage, the raw ingredient of the process, differed from place to place, depending on the local industry and population. It changed over the course of hours, days and seasons, depending on social and economic activity of the city as well as local patterns of rainfall and temperature. "Sewage is very elusive," wrote the manager of the Stafford, England, works. "It alters from day to day, even hourly. . . . The production of a final effluent which will satisfy all interests is very difficult, sometimes impossible."[148] To the question "What is sewage?," William Dibdin quoted an "eminent authority" that it is "variable, very variable; complex, very complex."[149]

Further, the microbial community—derived from the sewage itself as well as from the local air, water, and soil—differed among different plants, composed of different species that were constantly changing. "Each individual sewage had its proper sludge," commented H. T. Calvert.[150] In a series of photomicrographs, J. A. Reddie documented the different communities of bacteria in sludges from different sewage treatment plants.[151] "Activated sludge, until recently, has been considered by many workers to be a definite entity," argued scientists with the Wisconsin Laboratory of Hygiene. In contrast, they reported, "The recent work of several investigators has indicated that unique bacterial sludges from separate sources might have widely different characteristics." They treated a uniform sewage with the activated sludges of four different New Jersey communities and showed that the sludges were distinct, with characteristic behaviors. But they could not determine the cause of the presumed biological differences.[152] Finally, because every plant was unique, with different-sized tanks, pumps, blowers, pipes, and other apparatus, no simple rules could be applied for process control. What was effective in one plant might not be in another, and no single strategy of process control was guaranteed to give good results. For effective process control, operators needed to understand how their particular plant responded to the constantly changing sewage and microbes.

Ardern, the codeveloper of the process, expressed the need for this understanding in decidedly unscientific terms. "Adequate control required intimate knowledge," he explained, "of the nature of the sewage treated."[153] This kind of knowledge could only be obtained by working at a plant and observing the changes in the sewage and process. The language of intimacy and intuition used here emphasizes the distinction between "scientific" control and the methods developed by workers in the plant.

Experience was the key element of process control as practiced by many operators. In fact, sewage plants organized their work structure to take advantage of this experience. Operators learned about craft techniques from other operators, and plant superintendents organized work so that new operators could learn from more experienced operators. At a plant in Mishiwaka, Indiana, new operators typically spent one day a week with experienced operators. The other days were spent on janitorial and mechanical tasks. A week later, the new employee would be back with the experienced operator. This allowed the new operator to "experience variable operating conditions due to weather, flow, equipment operation, sludge disposal" and other factors, allowing him to see the effects of his operating decisions on the plant process.[154] If he spent an entire week apprenticing with the experienced operator, he would not see the range of conditions experienced by the plant. Here, managers found they had to abandon scientific management's goal of centralizing knowledge. This kind of operating knowledge could not be written down. Rather, it had to be passed from operator to operator, if no longer from father to son.

While mainstream engineers recognized the importance of experience, they held out the hope that their research could make experience obsolete, replacing it with rational, definite scientific knowledge. Scientific process control, they felt confident, could substitute for experience. The superintendent of Chicago's North Side plant, for instance, recognized that it was "impossible to overlook the value to a plant operator of experience gained only by living with his plant and its own particular problems."[155] But once activated sludge control could become "definite and positive," he predicted, that experience would no longer be necessary. Other scientists defended the advantages of scientific training over experience. "When everything is going easily," wrote Heukelekian, the observational approach of the operator might be sufficient. But "how can you tell when your troubles are going to come; how are you going to detect them, overcome them; what are the things that cause these troubles? For that purpose we will have to have some understanding of the fundamental processes involved, and you will have to have chemical tests and bacteriological analyses."[156]

This contrast between science and experience was also reflected in arguments over the qualifications of operators to manage sewage treatment plants. Operators and officials of the New Jersey Public Health Department sparred over the credit given to experience in licensing sewage treatment plant operators. Licensing criteria emphasized written exams heavy on technical material. One operator argued that "some men have a good deal of trouble to write down answers to a lot of questions, but I don't think that should be a hindrance to their ability to operate a plant." Conversely, technically trained applicants "might very easily pass the examination to operate one of these large and complicated plants and still not have the necessary experience," another operator asserted. "I am very strong for Old Man Experience," he concluded.[157]

This craft knowledge, like scientific experimentation, was based on a constant iteration of observation, adjustment, and observation, and it allowed workers to develop a deep understanding of their local ecosystem. One operator, discussing problems at his plant, insisted on a very qualitative description of plant processes that could only come with the great familiarity he reached with his plant. He rejected the term "bulking" in favor of his characterization of the problem as "periods of unrest and uneasiness." For scientific control, a term like "uneasy" defied definition. How could an operator tell if a sewage plant was uneasy? According to prevailing ideas of process control, a value of the sludge index over 150 indicated a bulking sludge. This operator, however, developed a much more nuanced view of the sludge, based on his observation of its settling characteristics and familiarity with normal operating variation. "With all due respect to the persons who concocted this term [bulking] . . . the terms, as I have applied them, more adequately describe the situation that actually exists. During these semi-annual periods, the floc is finely divided, very light and fluffy, and almost slimy in appearance and in fact. The settling rate is slow and the general impression is one of uncertainty as to whether it will settle or stay in suspension."[158] For this operator, the feelings of "uncertainty," "unrest," or "uneasiness" were far more diagnostic than the term "bulking" and the sludge index.

In this context of a continually varying sewage flow and activated sludge process conditions, operators become investigators in their own right, correlating operating conditions with influent characteristics and effluent quality. Inspection of the tanks and other processes gave immediate feedback to the operators, allowing them to make process changes. "It is not necessary for him to make any of these intricate chemical tests," wrote Heukelekian, almost wistfully. "They don't even have to make a methylene blue test; they only have to go to the clarifier, take a look at the weir and see that everything is all right."[159] Observations based on sight and smell took no time to analyze, and if an operator could correlate these observations with effluent quality, over time he would be able to make informed process control decisions. The ability of operators to develop this craft knowledge, however, relied on their long-term experience in their plants. As one scientist noted, the long-time operator has "the opportunity to observe the effect, first, of the normal fluctuations in quantity and quality of sewage, and second, of experimental modifications which he is in position to make looking toward the betterment of his output."[160]

The qualitative indicators used by operators were often integrative, incorporating measures of both physical and biological condition. Examination of the physical characteristics of the activated sludge floc, or presence and type of foam in the treatment units, integrated all of the growth processes of the activated sludge organisms, influenced by biodiversity, dissolved oxygen, food availability, and population growth. Because treatment success is based to a great degree on the physical characteristics of the activated sludge flocs and their ability to settle, direct and instantaneous measures of the condition provided information that was often more applicable to process status and control than indirect laboratory-based measures such as dissolved oxygen, mixed liquor suspended solids, or mean cell residence time.[161]

While many scientists dismissed the craft techniques as outdated and insufficient, others struggled to understand their apparent success. Some recognized that this parallel system of craft knowledge represented a kind of technical knowledge but stopped short of acknowledging it as science. One engineer, in criticizing prerequisites of formal schooling for short school courses, argued that operators could master sewage treatment without formalized scientific education. He argued that sewage treatment involved "more practical mechanical genius than a highly technical training." Sewage treatment plant operators could produce a clean effluent without knowledge of the fundamental sciences. Instead, this engineer argued for the importance of technical expertise developed on the job. "For centuries mariners have successfully navigated ships on the seven seas," he argued by way of example. "Very few ship captains have ever seen a campus, let alone attended a university, and yet the art of navigation is based on such deep subjects as astronomy, spherical trigonometry and other technical and mathematical procedures." Nevertheless, he concluded, this kind of expertise was not, in itself, science. "The average navigator learns the mechanical function . . . without delving too deeply in to the basic principles of the sciences involved."[162]

Yet other scientists recognized that this knowledge gained from experience was, in fact, science and that operators (at least good ones) were indeed scientists. "Every plant operator has an opportunity which is the envy of those scientists who are advancing the cause of sewage treatment by laboratory and small-scale experimental studies," wrote Earle Phelps, the prominent sanitary engineer. "By cultivating habits of observation, of testing and of recording he may become the investigator and thereby in addition to doing well his set task, contribute scientific information of value in the development of the art."[163] For Paul Reed, sanitary engineer for the Indiana State Board of Health, this systematic observation was synonymous with being a good operator, and was not dependent on laboratory analyses. "A good operator will continuously study his plant in an attempt to increase the plant efficiency. Although a laboratory is desirable to provide information on efficiency of treatment, many times the matter is obvious to the naked eye. One does not need to have laboratory results to know that the efficiency . . . is greatly reduced when sludge is allowed to remain in that tank until is becomes septic and floats to the surface." "Be an operator," he exhorted, "not a valve turner."[164]

This observation echoed an earlier article by Abel Wolman, writing about water plant operators. "The function of science is no more than the function of every technical practical operator, namely, the observation and interpretation of facts. . . . If an operator sees facts and reasons as to their cause and effect, he becomes a scientific observer." The operator's duty, he continued, "has been enlarged from that of

Figure 3.8
Operators as valve turners. This device, at the Surbiton Sewage Purification Works, would automatically divert the sewage flow during storms to a greater number of percolating filters (seen in the background). Automation in sewage treatment plants was initially used primarily in controlling flow. Only recently has it been possible to automate control of the biological processes. *Source*: *Bacterial Sewage Treatment*, catalog from Adams Hydraulics, 1921, Milwaukee Metropolitan Sewerage District

a valve operator to an investigator."[165] Both Wolman and Reed contrasted the pejorative term "valve turner" or "valve operator" with that of the scientist. Valve turners were simply automatons, working at the direction of a plant manager, flipping switches or turning valves when indicated by instruments or laboratory measurement (figure 3.8). In fact, the most common use of the term "valve operator" was for a mechanical device, not a human operator. Yet a simple valve turner is just what managers in the chemical industry wanted from their labor force, the "simple machines" of Able's 1882 address to the Society of Chemical Industry. The advance of process control in industry was touted because it turned operators, who made independent decisions in regulating the chemical process, into simple valve turners, working under the direction of a plant manager who would look at the results of measuring instruments and laboratory tests and instruct operators when they should open or shut valves. It was a simple step from this conception to the replacement of the human valve operator with an automatic device. To elevate the sewage plant operator to the status of scientist and exhort him to be something more than a simple valve turner shows how some engineers, at least, understood the work of sewage plant operators to be intellectual as well as manual.

By establishing and maintaining their intellectual contribution to the control of sewage processes, operators were also able to claim jurisdiction over the plant. The goal of scientific management and process control, to "make it unneccesary for the workman to think," could not be accomplished by sanitary engineers in the sewage treatment plant.[166] Sewage plant operators had been at least partly successful in preventing the alienation of their labor in the sewage treatment plant. The biological treatment processes, and sewage itself, were far too complex to be successfully controlled by process control algorithms without the independent intervention of the experienced operator.

4

Profit vs. Purification: "Sewage Is Something to Be Got Rid Of"[1]

In 1927, the Milwaukee Sewerage Commission took to the airwaves to publicize an "epoch making achievement." "For the first time in the history of sanitation," the radio broadcast claimed, "the valuable plant food elements contained in sewage and trade wastes are being converted into a marketable fertilizer."[2] The Milwaukee fertilizer project was "being watched in all parts of the world," the Sewerage Commission declared. This was not hyperbole. Sanitary scientists internationally were paying rapt attention to Milwaukee's experiments in turning sludge into fertilizer, for they revived hopes held since the early days of sewage treatment that sewage could "be converted into gold."[3]

For almost sixty years, sanitarians had been pursuing the idea that the nutrients retained in domestic and industrial waste could make sewage treatment pay for itself by recycling sewage sludge into fertilizer. Towns could purify their sewage and make a profit too. Processes were patented, companies capitalized, franchises granted, all with the hopes of making money from sewage. But by 1916, when Milwaukee began experimenting with activated sludge, it had become clear to many, if not most, sanitarians that profit and purification were inherently contradictory goals. It might be possible to make money from sewage, they concluded, but only at the expense of the goal of sanitation. Conversely, sewage could be purified, but cities must expect to pay for it. It was not that sewage lacked valuable components. Rather, the expense of separating the valuable from the worthless was greater than the market value of the sewage. "The valuable constituents of sewage are like the gold in the sand of the Rhine; its aggregate value must be immense, but no company has yet succeeded in raising the treasure," stated German chemist A. W. Hoffman in 1857.[4] Yet when the originators of the activated sludge process measured the elevated nitrogen content of activated sludge, they revived the hope of profitable utilization of sewage. "We have the master key with which to unlock the treasure stored up in sewage sludge," according to an editorial in the *Engineering Record*.[5] Now, the Milwaukee Sewerage Commission had appeared to finally raise that treasure and was marketing it over the radio "at a very reasonable price put up in 100 lb. bags."

The "Search for 'El Dorado' in the Cities' Sewers"

Beginning in the 1840s, chemists and sanitarians began emphasizing the huge potential value of the fertilizing components of sewage. With the importing of guano from Peru, fertilizer had recently become an important component of farming in Britain. Agricultural chemists like J. von Liebig had developed the idea of limiting nutrients, and emphasized the possibility of artificial fertilizers to maintain soil fertility and increase crop yields, while British chemist John Bennet Lawes had invented and patented a process for making superphosphate fertilizer. These chemical theories, which had replaced earlier vitalistic concepts of soil fertility that emphasized humus, were applied to the analysis of sewage and suggested that the wastes of towns and cities might be worth fortunes.[6]

The high expectations, on the part of both cities and the firms they contracted with, relied on the chemical analysis of sewage. Applying the conceptual framework of the new agricultural chemistry pioneered by Liebig in Germany and by Lawes and Joseph Gilbert in England, chemists attempted to calculate the fertilizing value of sewage as they had analyzed the value of guano.[7] Chemists calculated the concentration of nutrients in feces and urine, measured the volume of excrementitious waste and "fluid voidings"[8] of a typical Londoner, and extrapolated huge quantities of nutrients in the sewage of cities. Using these techniques, analysts suggested various values of sewage, based on the equivalent fertilizing power of guano. Estimates ranged from less than 1d. to 2d. per ton of sewage or 6d. to 16d. per person per year. Lawes concluded that "the intrinsic value of the sewage of London, considered in this merely chemical point of view, is therefore enormous," equivalent to "nearly one-third of the wheat consumed by its population."[9] Other estimates of the monetary value of London sewage ranged from £1,385,540 as calculated by Corfield, £2,793,551 as calculated by Ellis,[10] or even £4,081,430 as estimated by Liebig.[11] There had already been culturally powerful arguments for recycling sewage on land. The process grew out peasant practices and had its basis in arguments of morality and natural philosophy. The new chemical studies added the argument of pecuniary gain.

Not all chemists were so taken by the value of sewage, however. "These theoretical calculations are . . . altogether fallacious," argued agricultural chemist J. A. Voelcker.[12] Lawes, even though he made these kinds of theoretical calculations himself, had similarly cautioned that the "merely chemical point of view" was simplistic. The problem was that sewage came not only with nitrogen, phosphorus, and potassium, but with abundant unwanted material as well, mostly water. Not only would all of that material need to be transported along with the sewage, but it would have to be applied to the soil as well. As one agriculturist calculated, sewage with the same volume of fertilizing components as about 100 pounds of guano would come with 160 tons of water.[13] Even if the water could flow downhill to the fields, reducing the cost of transportation, the soil would still have to be capable of absorbing every day twice the amount of rain that would typically fall in a year.[14] And the sewage would come every day of the year, when it rained and when it didn't, when crops were growing or the fields fallow. But if firms provided only the volume of sewage that a farmer actually needed, the sewage could not be "turned to any profitable account," criticized Lawes. A profit could not be made, he continued, "by sending a pipe into Oxfordshire, another into Bedfordshire, into Essex, and into Surrey, and selling it out by the gallon as you would ale or porter."[15] Rather than being of wide utility, these critics argued, sewage could only be effectively used under a narrow range of conditions on a limited variety of crops.

The profitable uses thus depended on the weight of the sewage, transportation costs, distance from a city, land values, and the kinds of crops grown and marketed. To surmount the difficulty and expense of transporting the liquid in sewage along with its fertilizing components, many firms looked to separate the liquid and solid components, concentrating the fertilizing materials in the solids. Such a precipitate, theoretically, would be economical to transport and could command a higher price. Despite the cautionary analyses of Lawes and Voelcker, and spurred by the optimistic determinations of agricultural chemists like Liebig and sanitarians like Chadwick, English cities were beguiled by the prospect of treating their sewage with little or no impact on ratepayers. Indeed, some cities hoped to profit beyond recovering the costs of sewage treatment setting off what a later commentator criticized as a "search for 'El Dorado' in the cities' sewers."[16]

Seeing the potential for large profits in these optimistic calculations of sewage value, dozens of companies organized to treat city sewage. A look at the balance sheets of proposed companies showed the promise of great profit. The Metropolitan Sewage Manure Company's 1846 plan to treat a portion of London's sewage by pumping it to outlying districts where it would be sold to farmers was to be capitalized at just over £120,000. The company predicted farmers would buy the sewage at 3d. a ton, and officials projected a profit of 11 percent per year on investment.[17] Over the following decades, companies of all sorts were formed to capitalize on the "mine of wealth" held in sewage; some for irrigation projects like the Sewage Manure Company and others for precipitation projects. Every one of the commercial efforts failed. For, as cities and firms both discovered, one could not both profit from sewage and purify it at the same time.

Crossness and Barking

The contradictions inherent in both profiting from sewage utilization and purifying the effluent can be seen in the parallel histories of two firms organized in the late

Figure 4.1
"Metropolitan Main Drainage, General View of the Southern Outfall Works at Crossness,"
1865. Sewage from the southern half of London was discharged to the Thames at this plant
at Crossness. The Native Guano Company erected its treatment works here in 1873. When
the works failed to demonstrate a profit, the Metropolitan Board of Works had it removed.
Source: *Illustrated London News*

1860s to capitalize on London's sewage; the Metropolis Sewage & Essex Reclamation Company and the Native Guano Company. These two companies exemplify the
effort to build commercially successful firms on the foundation of sewage recycling
using the leading technological alternatives of the period. These firms envisioned
profit through stock speculation, land speculation, and commercial sales of sewage. Their fate had enormous implications for how British cities approached sewage treatment, and through that experience, how American cities pursued it as well.

Built on opposite banks of the Thames, these schemes were parallel geographically as well. When engineer William Joseph Bazalgette and the Metropolitan Board
of Works built the system of sewers for draining the metropolis, there were two
main outlets across the river from each other eleven and thirteen miles downstream
from London Bridge: Barking on the north side of the Thames, and Crossness on
the south (figure 4.1). In 1869, 83 million gallons of sewage were discharged into
the Thames each day from these two outlets, increasing to 164 million gallons by
1881.[18] The Board of Works solicited plans to treat the sewage, with particular

interest in the profit that firms would guarantee the city. On the north side, at Barking, William Hope and William Napier were awarded the concession for London's sewage in 1865. Across the river, at Crossness, the Native Guano Company built a demonstration plant for its ABC process of sewage purification in 1871. Both companies promised huge returns on investment for stock holders, and each sought capitalization on the London Stock Exchange. As financial, as well as sanitary projects, both firms were failures, and their failure epitomized the difficulty in combining purification and profit in sewage treatment.

With the main drainage of London nearing completion, the Metropolitan Board of Works first advertised for proposals to treat the sewage at Barking on the north side of the Thames. In addition to details on the mode of treating the sewage, the board requested financial information on the "amount of capital to be invested, the number and value of the shares, duration of the contract" and particularly, the "royalty agreed to be paid to the Board."[19] Of the two tenders received, the Board of Works recommended the proposal of William Napier and William Hope.[20] Napier and Hope's proposal was both a fertilization plan as well as a land scheme. Sewage would be piped forty-four miles southeast from Barking to the coast. Farmers along the way would be able to buy the sewage for fertilizing their land, and the remaining material would be used to reclaim coastal flats near the mouth of the Thames. Profit would come both from fertilizer sales and the increased value of the newly made land. They estimated the sewage concession granted them by the Metropolitan Board of Works to be worth £1 million and annual profits to return 20 percent on investment.[21] To raise the estimated £2.4 million needed to capitalize the scheme, they sought the involvement of the International Financial Society, one of the leading finance companies of the City of London. International, with the help of Antony Gibbs & Sons, which until recently had held the British concession for marketing Peruvian guano, were to float the company in the capital markets. In the summer of 1865, they published the prospectus for the Metropolis Sewage & Essex Reclamation Company.

For a variety of reasons, however, the stock offering was a failure. None other than Liebig weighed in on the choice of concession, aligning himself with the competing proposal and severely criticizing Hope and Napier's plan in the press.[22] Perhaps because of this criticism, speculators in the scheme bet on a falling stock and shorted their shares. Two weeks after the prospectus was published, of the 21,000 shares offered, investors had applied for only a third. In order to prevent the collapse of the stock, International decided to actually issue only a fraction of even these shares, their resultant scarcity supporting the stock price. Its strategy helped defeat the "bears," but in doing so, International failed to raise the necessary capital to construct the project. With so few shares allotted, the London Stock Exchange refused to list the stock, leaving International and Gibbs to capitalize the project from their own funds. Major ownership fell to International and Gibbs, weakening Napier and particularly Hope's control of the company. With an eye toward minimizing costs, International greatly curtailed the scope of the project, first abandoning the land reclamation, and then even the sewage pipeline, leaving only a single sewage farm near the outlet at Barking. While the farm manager successfully conducted a number of experiments on the fertilizing power of sewage, the goal of utilizing the bulk of the sewage, not to mention making a profit, was never attained.

Hope charged the company with intentionally killing the scheme, and he tried to wrest back control from International, but without success. Whereas Hope appeared to have a true commitment to sewage utilization, it was clear that International viewed Metropolis Sewage as simply one among many possible avenues for investment. As argued by one of the stockholders in defending the company against Hope's charges, "They must not forget they were a financial company designed for making very large profits, and not for making scientific experiments which might fail to produce a return on the outlay."[23] International tried on several occasions to sell its interest in the scheme, but none of these plans were completed and they finally wrote off their investment in the early 1880s.

William Hope, bitter from his experience at Barking but undeterred, argued that "there is no doubt, for there can be none, as to the value, the bona fide intrinsic value, which exists in the sewage."[24] He bought land near Romford and contracted for the town's waste to conduct his own experiments on sewage irrigation. Among other sanitarians, G. V. Poore visited his farm and ate a dinner produced entirely from sewage irrigated food, declaring he had "never eaten better," with none of the "unpleasant results" that critics assured were inherent to sewage farms. (Ironically, Poore died in 1904 following a year-long decline in health "attributed to his having eaten sewage-fed oysters.")[25] Despite success in purifying the effluent, however, Hope suffered a "very heavy and ruinous loss in the conversion of the sewage."[26] Nevertheless, he continued to write missives to the press defending sewage irrigation through the early 1900s.[27]

While London's irrigation scheme was a failure, British cities continued to hold out hope for sewage irrigation. These other attempts to profit from irrigation fared little better. In 1879, the Royal Agricultural Society sponsored a competition for the best managed sewage farm. Farms were evaluated in terms of profit, productiveness of crops and stock, management of grassland, general upkeep and the mode of book-keeping. Tellingly, the success in purifying sewage was not an explicit consideration; there was no analysis of the effluents from the farms. Rather, the sanitary aspects of the farm were evaluated by noting if the death rate among residents was likely higher than other rural areas without sewage farms. One of the winners was Lt. Col. A. S. Jones, later champion of sewage irrigation and "natural" sewage treatment.[28] Sewage farms continued in England; in fact they were required by the

Local Government Board. But these farms were typically not intended as commercial profit-making ventures. Rather, they were built with the goal of purifying the sewage despite costs. If the products of the farm could offset some of these costs, all the better, but profit was no longer seen as the primary purpose of the farm.

As the Metropolis Sewage & Essex Reclamation Company was struggling in Barking, the Native Guano Company began work on the opposite bank of the river, at Crossness. One of at least thirty firms established from 1844 to the mid-1870s (fifteen in the early 1870s alone) to produce a manure from sewage, the Native Guano Company employed the ABC process of sewage treatment, patented in 1868 by William Sillar.[29] As Sillar described it:

> This process differs from the other so-called precipitation processes, in so far that it purifies the sewage before precipitating it. . . . The sewage as it comes down from the sewers to the works at Aylesbury, is met by a stream of, what we call, the A B C mixture. That is a mixture principally of charcoal, clay, and blood. The animal charcoal was the A, the blood was the B, and the clay was the C, and that gave it its name. When this joins the sewage, the sewage becomes turbid and black; a little further on, dissolved alum or sulphate of alumina is added to it. The moment the sulphate of alumina comes to it, it commences to precipitate; it is received into a tank, and in passing through the tank it precipitates as it flows, leaving the whole of the mud at the bottom. The water which flowed in dirty flows out at the other end clean; that is the beginning and the end of the process.[30]

Sillar relied on the fertilizing value of the precipitate, the "whole of the mud," to generate a profit.

The Native Guano Company was incorporated in 1869 to put Sillar's process to work. They issued a prospectus to raise £60,000, claiming "the extracted manure" needed "little more than drying to fit it for use." "The process is now in full and profitable operation," claimed the prospectus. The company proposed that towns should undertake the expense of building the works themselves, and give their sewage to the Native Guano Company without charge. The company would then split the profits with the towns. Aside from the anticipated benefits of "relieving towns of injurious refuse," and "restoring polluted streams to much of their primitive purity," the company promised "large and permanent profits."[31] It claimed to produce manure at a cost of 20s. 3d. and sell it for £3 10s., a markup of over 300 percent.[32] Given Great Britain's large annual imports of Peruvian guano, the company suggested that its works could spread throughout the country, replacing the imported fertilizer. A few months after issuing its prospectus, the company approached London's Metropolitan Board of Works with a proposal to treat the sewage on the south side of the Thames.[33]

Speculative stock gains were clearly one of the kinds of profit envisioned. Native Guano's shares, offered at £5, were initially traded privately, before listing on the London Stock Exchange. By September 1871, when the company was first officially listed on the stock exchange, the stock had increased in private trading to as much as £48 a share.[34] The speculation was no doubt driven by the potential agreement with the Metropolitan Board of Works. A concession for the London sewage would surpass by far the company's works that had already been erected or planned in Hastings, Leeds, Bolton, and Southhampton. It also seemed clear that unlike International Finance, which had resisted Hope's desire that they run up the shares prior to listing on the exchange, the directors of Native Guano had no such compunction. A later report by the London Stock Exchange Commission had characterized speculation in "Natives" as "of the most disreputable character."[35] The business section of the London *Times* questioned the "honesty of the management" of the company, and warned of the "danger of any speculative commitment." Despite the *Times'* cautionary statement, Native Guano's shares closed at £38 on the first day of public trading in 1871 (figure 4.2).

The speculation in Native Guano increased its visibility, and various official bodies investigated the process and its claims. Most, however, found it wanting, finding it neither profitable nor capable of purifying sewage. Frankland's Rivers Pollution Commission spent almost two years analyzing the results of the process in three cities, and concluded that while it could remove the suspended matters in sewage, it had no effect on the dissolved pollutants, and could not purify sewage enough to "render the effluent admissible into running water." Further, the commission concluded that the manure had only "a very low market value, and cannot repay the cost of manufacture."[36] Similarly, the Metropolitan Board of Works sent its chemist to Leamington to investigate the process. He found the plant producing so much nuisance, however, that he did not bother to investigate further. Given these poor scientific assessments of the process, the success of the stock—as opposed to the success of the process—baffled many of the leading irrigationists. J. Bailey Denton noted that Leamington, where the process had first been tried, had abandoned it and was now selling its sewage to the Earl of Warwick who was using it irrigate what became a model sewage farm.[37] Nevertheless, the company persisted in pursuing the London sewage concession. Defending itself against the poor results of official tests, the company complained of fundamental problems in the Leamington plant over which they had no control and claimed improvements to their process that made it more likely to generate a profit. Given the opportunity to prove itself in works built explicitly for the process, Native Guano assured London that it would be a success. To entice London to revisit the process, the company offered to build, at its own expense and with no future commitment from the Board of Works, a plant at the Crossness outfall to demonstrate the ability of the ABC process to treat London's sewage. The Board of Works accepted the proposal.[38]

William Hope, who still held the concession for the city's sewage north of the Thames, felt endangered by the floating of the Native Guano stock and the agreement at Barking. One of the major investors in Native Guano was none other than

Figure 4.2
Native Guano Company £5 stock certificate. Like many sewage treatment schemes, Native Guano was capitalized by listing on the London Stock Exchange. Speculation in "Natives" drove the share price as high at £48 prior to listing on the exchange. When the company failed to secure a contract to treat London's sewage, the stock crashed. With the rise of biological treatment, the company slid into oblivion and investors received only 2s. 6d. per share at liquidation. *Source*: Author's collection

the International Financial Society, with whom Hope was struggling for control of his Metropolis Sewage & Essex Reclamation Company.[39] Given the Metropolitan Board of Works' distress caused by the failure of Hope's scheme to "produce a pecuniary profit,"[40] Hope feared that Native Guano, with International's backing, would make a move for his concession on the north of the Thames. Hope wrote to the *Times* to remind both readers and the encroaching companies that the rights to London's "enormous wealth of manure" were under his control, and that neither the Native Guano Company, nor the International Financial Society could alter the plans for treatment without his or Parliament's assent, assuring the readers that "neither of these assents can they ever get." Hope wrote a scathing attack on the

ABC process, suggesting that the Native Guano Company was operating some kind of "juggle." He noted that the "company's works are always, everywhere, in a transitional state. The process never *is* but always 'to be' perfected. It was first 'to be' perfected by the introduction of centrifugal drying machines, then by new drying floors, and now apparently by larger drying floors. Again, it appears that they have improved it by the use of sulphate of alumina." Hope then mocked the company for justifying its ingredients by citing Moses's sanitary laws in the Old Testament. "I heard one of the promoters of the concern get up in the Institution of Civil Engineers, only a few months ago, and amid roars of laughter, declare that he had only adopted a receipt given by Moses in Leviticus." Hope added dryly, "I am assured by competent authorities that Moses does not even mention sulphate of alumina."[41]

As it turned out, Hope's fears of Native Guano's success were unfounded. Although Native Guano had received permission to run their experiments at Crossness, the company was long delayed in constructing the works. Combined with further skepticism from much of the scientific community and dissatisfaction from the other towns in which the company was operating, the failure to move forward in London caused the shares to plummet. Only a couple of weeks after its opening on the London Stock Exchange, shares in Native Guano began a precipitous slide, from £39 in October to £15 just nine months later.[42] At this time, Native Guano bought a patent from William Crookes, the eminent British chemist, who in return became consulting chemist and a director of the company. Crooks became the public face of the company, representing it in negotiations throughout Britain as well as Paris and Belgium. Crookes suggested changes in the process that would reduce expenses, and the ABC formula changed a number of times.[43] Despite these difficulties in getting the London experimental plant going, the chairman of the company assured the shareholders in early 1873 that they would be granted the sewage concession at Crossness and claimed that "they had succeeded in solving the great problem of the age." One shareholder, however, suggested caution. Before approving the optimistic report of the board, he argued, they should wait for the concluding report of the Metropolitan Board of Works, expected any day. After further pronouncements on the success of the process from company officers, this cautionary amendment was defeated, in favor of the wild optimism generated by the recently recovering stock price, licensing of the process in Belgium and Paris, and the expectations of the concession from London.[44]

The cautious shareholder proved to be the wiser. The Board of Works released its report on January 31, 1873. Even though it concluded that the effluent from the plant was "on the whole very good," and the entire process created no nuisance, it declined to grant a concession to Native Guano. For the board, the motive of profit took precedence over purification. Despite the evidently successful treatment of the sewage, the main conclusion of the report was that "it had not been proved . . . that the process could be adopted with any hope of profit to the ratepayers." The Board of Works considered there were two important aspects of their work: "sanitary and commercial." It declared "with the sanitary aspect they had nothing to do," for with the construction of the main drainage of the city, they had considered solved the sanitary question for the city's inhabitants, even if residents, fishermen, and boaters downstream suffered from the raw sewage. "So far as the main drainage scheme itself was concerned, it was perfect," it declared. The only consideration remaining was the commercial: "Could manure be manufactured at a paying price?" Bazalgette estimated that the cost of manufacture of the manure was over £6 per ton, yet "with the exception of a few shillings," no manure was apparently sold. Sillar disputed this conclusion, claiming that the whole of the thousand pounds of fertilizer produced was indeed sold, at the claimed price of £3 10s. However, because the firm to which it was sold was also owned by Sillar, lawyer Charles Norman Bazalgette did not consider this a sale on the open market, nor even a sale at all.[45] While "it was true the company had succeeded in producing effluent water," that was not the most important consideration, according to the board. Rather it was whether the company was able to make sewage profitable. Board member Newton, arguing that the potential for profit to the ratepayers was "a most important question," moved to cancel Native Guano's contract.[46] Given the choice of purifying sewage at a cost or doing nothing, the board chose nothing. Not only would Native Guano fail to get access to the northern outfall as Hope feared, but it wouldn't be allowed to keep its plant at Crossness either. The board passed the resolution and demanded that Native Guano remove the works from the Crossness site.[47] The Metropolitan Board of Works had demanded both purification and profit, but the company could deliver only purification. With this news, Native Guano's stock dropped below £5 for the first time since the formation of the company.

Native Guano survived the repudiation of the process at London, but for the next fifty years, its prospects continuously declined. Native Guano was unusual among sewage companies in lasting until 1926 but was otherwise typical of the sewage manure firms capitalized in the 1870s; they were all based on patented processes, they all promised large profits, and they all went bankrupt. In fact, the shares in sewage companies became notorious as speculative bubbles, and sewage company share offerings were widely seen as frauds and swindles.[48] The General Sewage and Manure Company, for instance, was capitalized in 1872 at £200,000, acquired the patents of Dr. Anderson for a process that added boiling sulfate of ammonia derived from coal shale treated with sulphuric acid. Tried in Coventry, the process was a financial failure losing £1,500 per year for every million gallons a day treated.[49] The company was liquidated in 1876 after only four years in operation.[50] The Sewage Disinfecting and Manure Company fared even more poorly. In early 1872, the company issued a prospectus to raise £120,000 in capital to purchase the patent for Hille's process, which used lime, tar, and calcined chloride of magnesium. The prospectus claimed the company had solved the competing commercial and sanitary demands of sewage treatment. "Unless manure obtained from any process . . . reaches a certain standard of chymical strength," it explained, "it will not pay commercially. Unless the effluent water attains the standard of purity fixed by the River Pollution Commissioners. . . no water will be allowed to flow into rivers." With Hille's process, it claimed, "both these ends have been successfully attained." The prospectus promised the company to be "most remunerative," but within the year the company had failed and petitioned for "winding up."[51]

In the 1880s, Native Guano attempted to revive its fortune, but it once again became caught in the dilemma between profit and purification. With the London concession gone, company director and chemist William Crookes declared that the company's life "depended on its process being adopted by towns and town councils." But "before such adoption," he continued, "they had to convince those with whom they were dealing that they were *bona fide* honest men, not speculators." The company needed to build a demonstration plant "where they could prove that the accusation of excessive cost or insufficient purification were false." The Native Guano Company managed to convince the council of Kingston-on-Thames to grant the company a thirty-year concession for treating its sewage with the ABC process. However, even before those works were completed, the company gave them up and hitched its future to works constructed in Aylesbury.[52] At the demonstration of the plant in 1880, representatives from Glasgow, Manchester, Southhampton, Dorchester, Luton, Margate, Croydon, and Bristol all attended, along with sanitary engineers and agriculturalists, including the prominent American sanitary engineer Rudolph Hering.[53] With the work at Aylesbury proceeding, the stock in the company climbed from its lowest point at about £3 to £7. The company's directors once again made wildly optimistic predictions. Congratulating the shareholders "on the present position of the company, as compared with a year ago," the chairman "believed that matters were now in such a position that they should be prepared for everything, and to undertake any work which might be placed in their charge."[54]

Despite this period of optimism in 1882 and 1883, the success of both the company and the process were illusory. The stock fell once again, and shareholders staged a revolt, charging the directors with "being asleep." The company was caught in the contradiction between profit and purification. To attract the interest of cities and towns, it needed to show that its process could successfully treat sewage. But to produce a high-quality effluent, they needed to use a ruinous supply of chemicals. "At Aylesbury, they were working at a great loss," the director reported, "as they wanted a place where they could show their process and deal with a large quantity of sewage." The problem with this strategy, he conceded, was that other towns would want terms similar to what Aylesbury was getting. Thus, while other towns approached Native Guano to treat their sewage, the company had to turn down these opportunities because it could not generate a profit operating as it did at its demonstration plant at Aylesbury.[55]

"A Mischievous Error"

Fifteen years after the completion of the London drainage scheme, and with the failure of both the irrigation and precipitation schemes at Barking and Crossness, the Metropolitan Board of Works was still dumping all of its sewage, untreated, into the Thames. In 1878, the steamer *Princess Alice* collided with another ship in the Thames, killing over six hundred people. In a series of sensationalist articles, newspapers described how passengers floundered and drowned in the raw sewage in the river. With the blame for many of the fatalities falling on the pollution of the river, a new Royal Commission was established in 1884 to investigate the causes and extent of pollution in the Thames. The commission determined that the effluent from the two London outfalls was causing horrendous pollution and convened a study to finally recommend a treatment scheme for the purification of the metropolitan effluent.

This study took place in an entirely different context from earlier commissions. By the 1880s, partly as a result of the failures of the two sewage schemes at the London outfalls, it seemed clear to most sanitary scientists and engineers that the twin goals of profit and purification could not be combined, and that they were, in fact, contradictory to each other. Backers of precipitation schemes claimed the incompatibility of purification and profit in sewage irrigation; irrigationists pointed out the impossibility of the two goals in precipitation. William Crookes, for example (and with the promotion of his precipitation schemes in mind), favorably quoted another sanitarian: "The difficulty of irrigation lies in the fact that the purification and the economical application of the sewage must go hand in hand. Purification *alone*, with a total disregard of expense, is quite practicable; economic utilisation alone is also attainable in favourable cases. But the irrigation farmer is compelled to take and to purify the sewage when he can make no use of it—in rainy weather and in winter."[56] Dr. Letheby put the contradiction more quantitatively: "One thing, however, is conclusive, and that is the diametrically opposite conditions of sanitary and agricultural success . . . the land must never receive more than from 1,000 to 1,500 tons of sewage . . . whereas for commercial profit it must never have less than 5,000 tons"[57] Achieving both profit and purification with precipitation schemes was similarly seen as impossible. Bazalgette, son of the engineer, described the impossibility of attaining profit and purification in the ABC process: "When, as at Crossness, the admixture of chemicals was ruinously extravagant," Bazalgette explained, "the quality of an effluent derived from dilute sewage was very good; but when, as at Leamington, the chemicals were sparingly apportioned, the quality of the effluent was notoriously bad." "Even assuming that the Company could produce a paying manure,"

Bazalgette wrote, "they could never do so concurrently with the production of a purified effluent. The two results of profit, or even solvency, and purification, are inconsistent and incompatible with each other."[58]

Sanitary scientists, finally calling the idea "nonsense"[59] that sewage could be profitable, now concluded that rather than a "mine of wealth," sewage was an evil that must be disposed of regardless of cost. In 1885, Robert Vawser, president of the Association of Municipal and Sanitary Engineers and Surveyors, declared the primary obstacle to sewage purification was "the idea that once so largely prevailed, that sewage purification would pay."[60] Chemist Tidy concluded in 1887 that "it is better to regard the sludge as a thing to be got rid of, and as a thing which, to be got rid of, must cost money and may not bring money"[61] Another chemist declared that "sewage is something to be got rid of, and that any question of pecuniary return must be absolutely subsidiary to this main point."[62] The experience of Dr. A. Dupre, London chemist and student of Liebig[63] in Germany encapsulated the shift in opinion among sanitarians toward the question of profit or purification. In 1886, he argued that "it was a mischievous error that had been propagated that sewage could be made to pay the community." Recalling that he had come "to England thirty years ago full of the theoretical notions inculcated by Liebig, who then compared England to a vampire sitting on the breast of Europe sucking out its life-blood, and discharging it through its sewers into the sea," Dupre concluded that "he had since become a sadder, and, he hoped a wiser man, and had come to the conclusion, which he thought most people had reached, that sewage was a nuisance to be got rid of, irrespective of expense or any idea of making a profit out of it."[64]

Reciting the long history of commercial failure in utilizing sewage, the 1884 Royal Commission argued that London must abandon profit as a goal for its sewage treatment. "In some very favourable cases . . . a profit may be made without purification, and very frequently the purification may be effected without profit, but the two cannot apparently be combined," it concluded.[65] In response, the Metropolitan Board of Works, having sought a profitable commercial scheme for decades, gave up the search and accepted its chemist Dibdin's proposal for precipitation. But Dibdin's plan differed from commercial precipitation schemes. Dibdin cast aside the goal of profit. London's sewage would be precipitated with the fewest and cheapest chemicals possible, and the resulting sludge, rather than being sold as fertilizer, would be shipped out to the Barrow Deep, ten miles offshore, and dumped.[66]

The Rise of Biological Treatment

Despite the recommendation of the commission in favor of Dibdin's scheme of precipitation with lime rather than its own ABC process, the Native Guano Company nevertheless considered the 1885 report as boding well for its fortunes, because it abandoned official support of irrigation and recommended chemical treatment for London's sewage. As the courts began increasingly demanding that cities treat their sewage, the company once again hoped to profit.[67]

By the 1890s, though, with the stock trading at a quarter of a pound, the shareholders were increasingly restive. The single town treated by the ABC process was now Kingston-on-Thames, from which the company claimed to make a profit of £740 in 1892.[68] One speaker at the annual shareholder meeting complained that "he had been waiting for about a dozen years for a dividend, and it seemed to him that they were as far off as ever from that result. . . . There must be something wrong somewhere if out of the 300 towns which had been communicated with . . . they could not succeed in getting even one or two of them to treat with the company." In response, the director of the company reported to the ever more restive shareholders "numbers of deputations came to Kingston-on-Thames, inspiring the company with the hope that they would adopt its process, but they generally went away and did not adopt any process at all." The director blamed weak river protection laws and lax enforcement, and continued in the company's tradition of groundless optimism: "There was, however, a vast field for the company's operations, and they felt that when the time came they would be in the best condition to deal with some of these towns."[69] The essential problem, however, was that once the promise of profit had been dismissed by municipalities, there were cheaper precipitation processes.

The Native Guano Company was finally doomed when the biological processes began to come to the fore. Although the 1898 Royal Commission on sewage disposal again evaluated the ABC process as undertaken at Kingston, it concluded that precipitation was only warranted when there was an unusual amount of manufacturing waste or the sewage was unusually strong. While it noted that the ABC processes could produce a good effluent, it was the most expensive process evaluated.[70] In addition, with the rise of biological treatment, the Native Guano Company found its fertilizer competing with sludge produced by these new processes. Manchester, UK, for instance, began producing a dried manure made from the sludge of its new bacterial filters. Requiring no chemicals for precipitation, this fertilizer was both equal in nutritive power and cheaper to produce. Lawyers for Native Guano, fearing comparison with these newer fertilizers, claimed copyright protection of their chemical analyses, and forced Manchester to destroy its advertising pamphlets that compared nutrient analyses between their new product and Native Guano.[71] Despite increased advertising for its manure,[72] and saddled with an expensive and technologically obsolete process, the Native Guano Company slid into oblivion. In 1908, Kingston-on-Thames, the only town still using the company's process, ended its relationship with the company, and in 1926 the company was liquidated, with shareholders receiving only 2s. 6d. per £5 share, one-fortieth of its original value (about one-eightieth counting inflation). As Bazalgette had already noted by

1876, "The A B C process has been more sensational than successful in its public career."[73]

Profit and Purification in America

American municipalities, engineers, and newspapers closely followed the arguments for sewage utilization in England, and frequently reported on English plans, reports of the various Royal Commissions, and statements of British engineers on the subject. But because sewage treatment in the United States lagged that in Great Britain, American engineers had mostly avoided the debate over profit or purification. Although many American cities had utilized night soil in agriculture, the first municipal sewage farms were not established until the 1880s, and they were never widespread. It was widely acknowledged that only in the arid West would sewage irrigation be profitable, and there primarily because of the value of the water content of sewage rather than its fertilizing capacity.[74]

By the time American municipalities were forced to treat their sewage, they regarded treatment as a necessary expense rather than a potential source of profit, and they sought the cheapest and most convenient means of disposal, without regard to utilization. In 1884, discussing sewage treatment plans for Boston, prominent physician George B. Shattuck argued that no system of sewage farming on a large scale had "proved actually profitable," in England or on the Continent, and argued that "where there was in the immediate neighborhood a large body of deep water" the solution was "simple, satisfactory, and inexpensive." Boston should dump its sewage into the harbor.[75] Similarly, in 1912, when Gilbert Fowler and John D. Watson traveled to New York City to consult on plans for sewage treatment there, their recommendations did not incorporate any means of utilizing the sludge to make a profit.[76] Engineers in the United States and Great Britain were in agreement that sewage was a nuisance that should be disposed of by any means, regardless of the potential to profit from its fertilizing power. But it was on this trip that Fowler observed the initial experiments that led to his development of the activated sludge process. Activated sludge promised to revolutionize not only the treatment of sewage but also the disposal of the leftover sludge.

The first chemical analyses of activated sludge revealed a high nitrogen content, almost twice that of the sludge from other biological processes. With these analyses, the hope of profitable utilization of sewage was revived. Manchester's Edward Ardern wrote that "one of the chief aims of the activated sludge process is to recover as much as possible of the valuable nitrogen." Because of this emphasis, F. R. O'Shaughnessy of Birmingham, UK, considered activated sludge "in fact a commercial proposition" in which the purification process was arbitrarily arrested to conserve nitrogen. The process could "be justified only on financial" rather than sanitary grounds, he concluded.[77] Fowler entertained such high expectations of the profitability of activated sludge that he declared "the fertiliser end of the process will come to be seen as the most important, sewage purification being, as it were, a by-product."[78] For American cities, which were just beginning to plan and construct large sewage treatment works, the promise of profitable disposal of the sludge was compelling.

Illinois chemist Edward Bartow conducted the first experiments on activated sludge as a fertilizer. Immediately on returning to the United States from his visit with Fowler in Manchester, he began his own investigations. After producing the first activated sludge in the United States, Bartow sent samples to various agricultural chemists to evaluate its potential as a fertilizer. Chemists replied that the sludge possessed a moderate to good value, comparable to other sources of organic nitrogen commonly used, particularly the waste products of the meatpacking industry such as dried blood and tankage. Bartow added the sludge to pots of wheat, lettuce, and radishes, and showed instead a remarkable increase in yield compared to dried blood.[79] His graduate student, William Hatfield, conducted a number of additional studies on activated sludge, analyzing its nutrient content, and most important the availability of the nutrients to plants. By comparing the fertilizing capacity of activated sludge to current nitrogen fertilizer available on the market, Hatfield concluded that activated sludge might easily be worth $20–30 a ton.

At this time, Fowler was trying to convince British authorities to take up the activated sludge process, and was particularly pleased with Bartow's pot cultures, because they emphasized the "economical reasons it is worth while to study the process carefully."[80] Encouraged by further experiments, Fowler concluded, "We have to do with a manurial agent of great possibilities."[81] Fowler hoped to profit from this manure himself by patenting a process for making fertilizer from activated sludge. Indeed, reflecting his prediction that activated sludge was more important for its nitrogen content than its ability to purify sewage, Fowler shifted the focus of his research to nitrogen chemistry and manures. Having moved to India, he spent the rest of his career working on the biochemistry of the nitrogen cycle.[82] Like Fowler, most sanitary engineers were excited at the possibility of finally combining sewage purification with profit.

Milwaukee and "Sewer Socialism"

Milwaukee, under the direction of T. Chalkley Hatton, became the first large city to initiate a sewage treatment plant based on the activated sludge process. From the beginning, the manufacture of a saleable fertilizer was an integral part of its sewage treatment plans. As with every city, Milwaukee's decision to build sewage treatment and the form that treatment would take was a complex combination of politics, available technology, economics, expediency, and the actual condition of the rivers, lake, and water supply.[83] Political pressures in the city for finally dealing with its sewage problem fortuitously coincided with the availability of the activated sludge process, making that technology a possible response. Further, because activated sludge required less land, the sewage plant could be built on a relatively small parcel near the harbor, a location that would not exacerbate political and ethnic tensions between the North and South Side wards of the city. Finally, the promise of revenues from fertilizer helped overcome objections to the relatively high initial capital costs. But the choice was not overdetermined. Milwaukee's socialist city government and the Milwaukee Sewerage Commission disagreed on the importance of sewage treatment versus water filtration, with city officials strongly favoring water treatment and cheaper sewage alternatives. The activated sludge process was by no means an established technology, and it took faith on the part of Hatton and the Sewerage Commission to pursue it. Because of Milwaukee's history of municipal utilities and the powerful political movement that put municipal trading at the center of its agenda, Milwaukee's decision to incorporate the sale of fertilizer as an integral part of its plans was a powerful argument in favor of sewage treatment over water filtration.

The notion that profit and purification were incompatible was also called into question by the movement for municipal trading that had begun in British cities in the mid-nineteenth century. The municipal trading movement sought to redefine the notion of profit to include nonpecuniary benefits to the broader public. In the rapidly expanding cities of Great Britain, competition between private firms had failed to regulate the provision of increasingly key services such as water, transportation, gas, and later electric lighting. Many municipalities began to provide those services directly in place of private enterprise. Birmingham, under Mayor Joseph Chamberlain, was one of the leaders of this movement, and he municipalized the city's water and gas works in the 1870s. Many British cities followed. As cities in the United States began to reform often corrupt municipal administrations, civic groups looked to municipally owned utilities and the experience in Great Britain. U.S. cities municipalized water, gas, and electric plants, but also pursued a wide variety of other commercial ventures—city-owned markets, slaughterhouses, or ice production. A city-owned fertilizer plant, as envisioned for Milwaukee, was solidly within this trend toward municipal enterprise.[84]

Private enterprises, like Native Guano or the Metropolis Sewage & Essex Reclamation Company, had sought traditional profit in the increased value of stock or surplus revenue earned from providing sewage treatment. Municipal sewage farms, although they may not earn a profit in this traditional sense, still provided benefits to the public. These benefits were characterized as a form of profit by Chamberlain. "The leading idea of the English system of municipal government is that of a joint stock or cooperative enterprise in which the dividends are received in the improved health and the increase in the comfort and happiness of the community." In this sense, there was no contradiction between purification and profit—they were, in principle, one and the same thing. Nevertheless, municipal enterprises often generated pecuniary income as well. Whether that profit should be a goal of municipalities, and where that profit should go, would be a continuing political problem.[85] The political calculus was not, in practice, a simple accounting of the public benefits of a municipal enterprise. Rather, pressures of the market came to dominate decisions that had important bearings on the environment, public health, and welfare.[86]

In the closing years of the nineteenth century there was a growing movement in Wisconsin, led in part by the Milwaukee Municipal League, to reform the system in which most city services were provided by private companies. Poor service, corrupt politics, and high rates all led to a sea change in the provision of electricity, gas, water, telephones, and transportation. Fed up with the interference of the corrupt Wisconsin Rendering Company in local Milwaukee politics, for instance, the city voters municipalized garbage disposal in 1898. At the same time, reformers were battling Milwaukee's private streetcar and electric power utility. The politics of reforming municipal utilities led voters to first oust the corrupt Republican administration and replace it with the ostensibly reform-minded Democrat David S. Rose in 1898 and, when his administration continued corrupt policies toward providing utilities, second to replace it with a Socialist administration that, with only a few exceptions, held the mayoral office into the 1960s.[87]

Municipal trading and public ownership of utilities comprised a central plank of the Socialist administrations in Milwaukee, particularly that of Daniel Hoan, who served as mayor from 1916 to 1940. The Milwaukee Socialist platform and its emphasis on city services was responsible for the later term "sewer socialism," but the relation between the Socialist administrations of Hoan and Milwaukee's sewage treatment plant was much more complicated. In fact, Hoan bitterly opposed both the building and later expansion of the activated sludge treatment plant.

Milwaukee had built a citywide water plant in 1875, replacing a haphazard collection of private providers. By 1893, the city plant had begun to generate a surplus of revenues over expenditures. In that year, the waterworks provided $15,000 to the city's general fund. By 1900, contributions to the city treasury had totaled $350,000. The city debated whether that surplus should go to reduce water rates, reduce property taxes, or be invested back into expansion of the city water system. While the director of public works maintained that "the department should never be operated as a money-making investment," successive administrations in fact did just that, using water works profit to fund not only general city expenditures but also projects in aldermen's wards designed to earn them votes. To raise money for these projects, Mayor Rose prioritized supplying water to communities outside the city limits, for the water plant could charge a 25 percent premium for this wholesale water while at the same time avoiding any costs for constructing a distribution system. Thus, at the same time that Polish wards on Milwaukee's South Side languished without city water service, the water works was building supply lines for suburban communities.[88]

The first reports on the sewage pollution of Milwaukee's rivers appeared in 1876 and again in 1889, describing the foul condition of the Milwaukee, Menomonee, and Kinnickinnic Rivers, which flowed through the central part of the city before emptying into the harbor and Lake Michigan, just a few miles from where the intake for Milwaukee's water plant was located. As a result of these engineering reports, the city built intercepting sewers and a pumping plant to flush Milwaukee's two main river channels of their polluted flow with water from Lake Michigan. While improving conditions in the rivers, this just displaced the pollution into the lake, near the water intake. In 1909, the city's health commissioner, Gerhard A. Bading, contracted three prominent American civil engineers, George C. Whipple, Harrison Eddy, and John Alvord, to investigate the problem. At the time, Socialists controlled the city council. They felt the charge to the committee was weak and advocated for a study that would identify means for treating the sewage. In response to the rampant corruption of the Democratic Rose administration, but partly on the promise of providing clean water, the Socialists won both the mayoral and council elections of 1910. The Rose administration had neglected basic city services like water supply to the immigrant Polish neighborhoods on the city's South Side. The report from Whipple, Eddy, and Alvord had also suggested a site on the south side for a newly constructed sewage treatment plant. Committed to a more democratic political climate, the new Socialist administration began funding expanded water supply to the South Side, and also expanded the scope of the sewage study to recommend alternate sites for the sewage treatment plant. The Socialist administration, however, was hesitant to build the recommended plant because bond debt would exceed its authorized limit, and prevent the construction of other priorities like a municipally owned gas and lighting plant.

With the corrupt Democratic administration gone, the Socialists were unable to maintain control in the next election. Bading, the former commissioner of health, was elected mayor in 1912. With no progress on a treatment plant, the State of Wisconsin stepped in and created the Milwaukee Sewerage Commission that would have authority to fund and build a plant. The commission named Hatton chief engineer in 1914, and he began a program to test various treatment methods on the local sewage before committing to the design of a plant. Late in 1914 Ardern and Lockett published their groundbreaking paper, and Hatton shifted the emphasis of the testing station to an evaluation of the new activated sludge process. In 1916, before a sewage plant was begun, there was an outbreak of typhoid in the city, which killed sixty people and sickened thousands, moving the public to approve a bond issue to build a sewage treatment plant. At this point, the sewage treatment plant became a major campaign issue in the battle between Socialists and "nonpartisans." Mayor Bading blamed the outbreak on Socialist failure to act on sewage, while the Socialists blamed the nonpartisans for failing to act on filtration.[89]

The conflict over the sewage plant reflected differing ideas on the best way to protect public health as well as intergovernmental rivalries and struggles for control. As Joel Tarr has shown, cities at the turn of the century saw water filtration and sewage treatment as accomplishing the same objective—clean drinking water.[90] Sewage pollution could be removed before it entered a river, or sewage-contaminated water could be filtered before being supplied to city residents. Cities of this period often preferred water filtration as the cheaper means to ensure a healthy water supply, and many public health officials concurred. In part, in a landscape where many cities on a single river or lake both withdrew drinking water and discharged sewage, the only way a single municipality could have control over its own public health was through water filtration or chlorination. Because it was difficult to control their neighbors' pollution, cities saw water filtration as more reliable. As Milwaukee's mayor, Hoan, later put it, even if Milwaukee's sewage treatment plant "operates 100% we will still be drinking Racine and Sheboygan sewage instead of our own."[91] Further, on rivers, the benefits of sewage treatment accrued to downstream water users rather than the city building the sewage treatment plant, making the economic calculus favor filtration as well. This choice was more complex in Milwaukee, however, where the city's sewage fouled its own water supply. Given the local conditions and knowledge in the field, there could be no engineering consensus on the best path for Milwaukee to take.

In Milwaukee, this debate between sewage or water treatment divided on the basis of politics. The Socialist mayor and aldermen supported water filtration while the nonpartisans supported sewage treatment. This conflict was, in part, related to an administrative divide. The City of Milwaukee controlled its water plant, while the sewage plant was built and controlled by the Milwaukee Sewerage Commission. Although the mayor could appoint some of the commissioners, the city did not control the commission, which was a separate and legally independent administrative unit. As the metropolitan area grew, conflicts between city and suburban interests on the commission intensified.

The Socialists regained control of the mayor's office in the 1916 elections. It was at this point that Mayor Hoan became a critic of Hatton and the activated sludge process. Sewage treatment versus water filtration became not only an engineering debate but a political one as well.[92] Hoan began a fifteen-year effort to block sewage treatment and build water filtration.[93] Hoan saw Hatton and the commission as political enemies, and he fought to have Hatton fired, the commission abolished as an agency independent of the Milwaukee city government, and ownership of the plant transferred to the city.[94] When equipment failure delayed the opening of the Jones Island plant, Hoan used the situation to force Hatton's resignation.[95] Hatton was an unlikely opponent, as he was by no means an opponent of municipal ownership or municipal trading. Although claiming to be no socialist, Hatton considered it "his best judgment that the public should absolutely own and control its own water supply and sewerage system."[96]

Milorganite®

It was in this political context that the plans for the activated sludge plant were developed. Hatton and the engineers at Milwaukee were the first to take up the technical challenge of producing a marketable fertilizer from activated sludge. If at first the goal was simply to offset some of the costs of sewage treatment through the sale of a valuable by-product, fertilizer production soon took on a life of its own. Experiments at Milwaukee's testing station showed that the sludge was up to 98 percent water, making it impossible to transport economically or apply efficiently to farm fields. Milwaukee's engineers used their new station to conduct experiments on dewatering the sludge, developing a three-stage process in which the sludge was first settled, then pressed in a filter press, and finally heat-dried to about 10 percent moisture, a level suitably dry that it could be marketed. Although heat drying the sludge cost more for energy, Milwaukee chemist William Copeland still considered it potentially profitable. Considering all of the expenses of production, including interest, depreciation, operation, labor, and fuel, Copeland estimated that fertilizer from activated sludge would cost $8–$12 to produce, but could be sold for $9–$15 dollars per ton. "The recovery of nitrogen in sewage sludge has at last been brought within the range of a commercially practicable problem," he concluded.[97]

The estimate of $9–$15 a ton, however, was entirely theoretical, because there was no existing market for sewage sludge. As Illinois's Hatfield had earlier commented, "It is useless to calculate the exact value of activated sludge since a market must first be established." Yet marketing was not one of the roles that sanitary engineers, or even municipal engineers, typically considered part of their purview, nor one in which they had much expertise. "It is not customary," Hatton remarked, "for an engineering organization engaged in constructing a plant to be interested in the work of marketing that which is to be produced."[98]

As Hatton recognized, Milwaukee would have to be actively engaged in marketing its sludge if it were to be successful. Despite their inexperience, engineers were initially sanguine. Hatfield, like proponents of sewage irrigation before him, considered truck farms the natural market for activated sludge; they were located near the urban centers where the sludge was produced, making transportation costs less of an issue, and they grew high-value produce that could potentially cover the cost of the fertilizer. In contrast, Milwaukee's engineers assumed that commercial fertilizer manufacturers were the most likely outlet. "In case the question arises as to the possibility of finding a market," reassured Copeland, fertilizer manufacturers would be happy to use it as a "base for making high-grade fertilizers."[99] Hatton canvassed fertilizer dealers and estimated that activated sludge could be successfully sold.[100]

As it turned out, however, marketing was not an easy task, and initial assumptions about the most lucrative markets for activated sludge were wrong. By building a sewage treatment plant that incorporated sludge drying, Milwaukee set itself on a path over the next eighty years that required it to continuously develop new markets for its output. With the exception of Milwaukee, cities' attempts to sell their activated sludge ended up failing. Milwaukee's success, on the other hand, resulted from its long and detailed effort toward innovations in building and sustaining a market. Some of these innovations were later adopted by large chemical companies and carried into the marketing of agricultural chemicals in general.

Before Milwaukee could "convince the manufacturer, dealer and consumer that we had something to sell which was of value to them,"[101] the city had to establish the value of the fertilizer for itself. The Sewerage Commission funded a fellowship in the School of Agriculture at the University of Wisconsin–Madison to study the fertilizing ability of Milwaukee's activated sludge. Professor of Agronomy Emil Truog selected O. J. Noer, a graduate student and former wagon salesman from nearby Stoughton, Wisconsin, to hold the fellowship. Unlike the mid-nineteenth-century agricultural chemists, the Wisconsin agronomists relied on more than chemical analysis to determine the fertilizer's value; rather, they conducted extensive field trials of sludge on Wisconsin crops (figure 4.3). Noer established his experiments under the assumption that the primary market would be fertilizer manufacturers who would use the sludge as a component of mixed fertilizers. Fertilizer makers took nitrogen, phosphorus, and potassium from various sources, both chemical and natural, and mixed them together in specific combinations for different crops and different regions of the country. Nitrogen sources ranged from inorganic sources like nitrate of soda or cyanamide from chemical companies to organic sources like tankage, the steamed offal that was a by-product of the meat-packing industry. Noer mixed sludge with inorganic and other organic fertilizers, and in conjunction with dozens of cooperating farmers he tested the various mixtures on cabbage, potatoes, sweet corn, barley, cucumbers, beans, onions, sugar beets, tomatoes, tobacco, pumpkins, peppers, eggplants, melons, squash, and raspberries as well as specialty crops like hothouse flowers and turf grass. These studies showed that activated sludge was effective as a component of mixed fertilizers. Because the nitrogen in activated sludge was released slowly, it was most effective on long-lived crops like potatoes or cabbage, rather than short-lived varieties like

Figure 4.3
"Potatoes on Plainfield Sand, 1926." O. J. Noer collaborated with farmers throughout Wisconsin to test the value of Milorganite on a wide variety of crops. Here, Milorganite ("sludge") was mixed with chemical fertilizers, because Milwaukee originally thought Milorganite's greatest market would be as an ingredient in mixed fertilizers. The plot treated with Milorganite produced the greatest harvest of potatoes, better than tankage, the waste from the meatpacking industry, which was thought to be Milorganite's biggest competitor. Courtesy Milorganite® Division, Milwaukee Metropolitan Sewerage District. *Source*: Turfgrass Information Center, Michigan State University

lettuce or beans. Armed with these results, the commission set out to market the sludge to fertilizer manufacturers.

For manufacturers to use sludge, however, they had to overcome a powerful prejudice against sewage materials. One large manufacturer, for instance, refused to prepare the mixtures for Noer's experimental plots after laborers in the plant refused to work with it "because the peculiar odor made them sick." Noer then arranged for the Armour Fertilizer Company, a subsidiary of the meatpacker, to make the fertilizer mixtures. Noer overcame any squeamishness on the part of the workers there by handling all of the material himself. The plant superintendent reported that the "mixing foreman was simply astounded to see the way you got in and handled the stuff." Partly as a result of these experiences, advertising for Milwaukee's activated sludge downplayed its origins in sewage and emphasized its "earthy" odor, while reiterating that it was a "sterile manufactured product." But until that marketing was successful, Noer warned that "the difference between activated sludge and raw sewage" would not be recognized by manufacturers.[102]

To help overcome this prejudice, one final detail needed to be completed before Milwaukee was ready to market its sludge. To attract the trade from fertilizer manufacturers, golf clubs, and flower growers, the markets identified as the most promising, "it appeared necessary to adopt a 'trade name,'" reported Hatton. The engineer felt that it was critical that the name should suggest not "sewage sludge," but rather "its ingredients appropriate for plant food." The sewerage commission sponsored a contest for naming its sludge and advertised in national fertilizer journals. The winning entry was submitted by a fertilizer dealer in South Carolina, who coined the name "Milorganite," for Milwaukee organic nitrogen.[103]

The Sewerage Commission was new to marketing, and its early efforts showed a poor grasp of how to market a product to a specific industry. The Sewerage Commission asked Alexander M. McIver & Son, the South Carolina fertilizer dealer that had named Milorganite, to provide a private critique of its first marketing materials targeted at the fertilizer manufacturers. In response, the firm cautioned Milwaukee to remember who its intended market was: the fertilizer manufacturer. The firm saw a booklet heavy on data and the kinds of evidence that had been appearing in the scientific literature, and lacking an appreciation for how fertilizer manufacturers would respond. McIver & Son urged particular caution with regard to Milwaukee's claim that, in contrast to Milorganite, high-analysis fertilizers might injure crops. As the fertilizer industry had "been doing everything possible to raise the percentages in their mixed fertilizers," McIver & Son thought a booklet making contrary claims "might not appeal to them, especially if they thought a copy of this booklet might accidentally get in the hands of the farmers." Further, the sales manager, V. H. Kadish, had included photos and data in the booklet showing that yields on certain crops were comparable between Milorganite alone and chemical ammonia. McIver pointed out that these data would appear to the manufacturers as critical of their products. Rather than encouraging Milorganite use, these results would alienate Milorganite's intended market. Instead of comparing Milorganite to chemical nitrogen, Milwaukee should stress the comparison between Milorganite and other organic nitrogen sources, particularly packinghouse tankage. Finally, they recommended that the signs in the photos that labeled which fertilizers were used should somehow be changed to improve advertising copy. Produced before the name Milorganite had been coined, they simply said "sludge."[104]

In the first years of production, shipments of Milorganite to fertilizer manufacturers dominated the sludge sales. But at the same time that Milwaukee marketed its sludge to fertilizer manufacturers, it began to develop a niche market for Milorganite in sales to golf courses (figure 4.4). It turned out that the chemical composition

Figure 4.4
Milorganite promotional brochure. The Milwaukee Sewerage Commission targeted its marketing and product development at the turf grass market. By 1932, Milwaukee claimed that Milorganite was the most widely used fertilizer on golf courses. Courtesy Milorganite® Division, Milwaukee Metropolitan Sewerage District. *Source*: Turfgrass Information Center, Michigan State University

of activated sludge was ideal for promoting turf grass growth. It possessed other advantages for turf as well. Because the sludge was heat dried, weed seeds were killed. In addition, many greenkeepers liked to avoid using potash in their fertilizers because potash promoted the growth of clover. Unlike manure, activated sludge contained negligible potash. What was a disadvantage in the mixed fertilizer market was an advantage in the specialty golf market. Finally, the potential profit was much higher for golf courses, because "activated sludge could be used straight instead of as just one ingredient" in a mixed fertilizer. Prices to fertilizer manufacturers were based on the concentration of nitrogen, and fluctuated with changing prices in the commodity market. In contrast, prices to the golf market could be set by the Sewerage Commission and Milwaukee was able to sell to golf courses at a substantial premium over fertilizer manufacturers.[105]

Milwaukee began to emphasize the use of Milorganite in golf courses and developed an infrastructure for marketing the fertilizer to this specialty market. When Noer had finished his studies at the University of Wisconsin, the commission hired him as a salesman for Milorganite, particularly to promote the fertilizer to golf courses. Noer had had mixed experience as a salesman in the past. Before returning to graduate school he had worked for Stoughton Wagon, a trailer manufacturer, with apparently little distinction. As it turned out, though, Noer was an inspired choice for the role of sludge salesman. With the Sewerage Commission, he not only managed to double his salary through commissions but was the driving force in the success of Milorganite. His position was later changed to agronomist for the Sewerage Commission, a post he held until his retirement in 1960. With his developing experience in the culture of grasses for golf courses, Noer became perhaps the leading expert in the United States on turf grass culture and was widely admired among greenkeepers.

To market to the golf course trade, the Milwaukee Sewerage Commission essentially established an agricultural extension service for golf courses. Named the Turf Service Bureau, it provided a number of services to golf courses, including soil testing. But most prominently, it offered the services of Noer. He and the sales manager Kadish traveled to golfing conventions, consulted on golf course turf problems, and answered queries on all nature of turf problems, including diseases and other issues not necessarily related to fertilizer. Noer traveled constantly from golf course to golf course. He prepared special fertilizer mixtures for particular golf courses. As a result of Noer's efforts, by 1930 Milwaukee was earning more money from sales to golf courses than to fertilizer manufacturers, even though it was selling a greater tonnage to manufacturers. By 1946, it stopped marketing to fertilizer manufacturers altogether and concentrated solely on the golf course and home garden market.[106]

Part of Milwaukee's success was related to its efforts to present a "scientific" and "objective" face to the golf course market, so that course managers could divorce the advice they received from a more crass attempt to market Milorganite. The original alliance between the commission and the Agricultural College at the University of Wisconsin was critical in this regard, as it gave the impartial imprimatur of the state's agricultural establishment to the Milorganite studies. Noer and others at the Sewerage Commission continued to cultivate an air of scientific impartiality to their promotional materials as well. Noer published a series on golf course grasses entitled "The ABC of Turf Culture" in *The National Greenkeeper*. These columns were later reprinted as one of the first books devoted solely to turf culture.[107] Covering topics such as photosynthesis, limiting nutrients, and soil structure, it was a scientific explanation of growing turf. In this book, Milorganite appeared only once, in passing, as one source among many of organic nitrogen. Tellingly, the one ABC column that appeared in the magazine but not the book was the one devoted to describing Milorganite experiments in detail. By distancing his agronomic advice from the interests of the Sewerage Commission, Noer's professionalism was enhanced. Later, the Sewerage Commission published a series of monthly advertisements in the magazines *Golfdom* and *The Greenkeepers Reporter* called "Timely Turf Tips." Many of the tips were designed to tout the advantages of Milorganite for specific turf problems, but others made no mention of Milorganite. This emphasis on impartiality contrasted strongly to the role of scientists in pushing sewage fertilizers in the previous century. Sir William Crookes, for instance, was one of the leading chemists of nineteenth-century England but also an officer of the Native Guano Company. Because of his conflicting roles, many questioned his claims of impartiality in his studies of Native Guano.[108]

All of these strategies were designed to help the Sewerage Commission create a trusted name in golf course management, allowing it to sell more sludge. By 1932, only six years after its introduction, Milorganite was heralded as the most widely used golf course fertilizer in the country. This success was notable, as Milorganite, despite great effort on the part of the Sewerage Commission, had not received the support of the Green Section of the U.S. Golf Association (USGA). Kadish had repeatedly tried to encourage the USGA to conduct trials with Milorganite, and when they finally did, they reported that were "unable to notice any results that would indicate the value of this product." The USGA, like many other potential users of sludge, was concerned over the sanitary character of sewage. It favored compost as an organic supplement and recommended that any sludge be composted for at least a year before being used on greens.[109]

Milwaukee ran into a number of problems in marketing Milorganite, however, and quickly found that maintaining fertilizer sales required an enormous amount of institutional effort, blurring the roles of the Sewerage Commission in purifying sewage and operating a fertilizer factory. Soon after Milorganite's introduction in 1926, the stock market crash and the Depression ate into the market for both the mixed fertilizer, and particularly the specialty sales to golf courses. Golf clubs significantly reduced their spending on upkeep as they lost members to the Depression. Milwaukee was competing against chemical fertilizer manufacturers as well as companies like Armour Fertilizer Company and Swift and Company that were marketing their meatpacking by-products as fertilizer. To counter these trends, Milwaukee continued to advertise in almost every issue of the magazines devoted to golf turf management. To compete against Swift's "Vigoro" which had a huge advertising budget compared to Milorganite, Milwaukee argued in advertisements and the turf press that Vigoro was unnecessarily expensive. Vigoro was marketed as a balanced fertilizer. Milwaukee countered this marketing with an appeal to "economical and rational" analysis and a critique of the word *balanced* as no more than an ad man's phrasing. Even though "the charm of the words 'complete' or 'balanced' carry an especial appeal," they argued, "except in rare instances there is very little evidence supporting the plea for so-called 'complete' or 'balanced' fertilizers on fairways." If the greenkeeper did not take into account the nutrients already supplied by the soil as well as the requirements of turf grasses, he was bound to waste money, if not provide for the growth of unwanted species like clover.[110] Milwaukee also emphasized the lower cost of Milorganite compared to inorganic fertilizers, since it required application only twice a year.

To counter the reduced spending on maintenance by golf courses, Milwaukee produced advertisements to convince golf course managers that the state of the turf could help attract and keep members. "Every Greenkeeper Now a Committee of One on Membership" proclaimed one ad. "It's largely your responsibility. . . . this holding of memberships. And *good fairways* is the answer," the ad continued. "Oh the Crime of it! letting him starve," another ad headlined, complete with pictures of "Mr. Starving Turf," that might have resonated with reader's images of bread lines. "Milorganite's Big Economy Plan," the ad declared, "*permits* even the most distressed clubs to continue a sound improvement program." In an economic environment where even the best clubs were under financial strain, Milwaukee tried to sell fertilizer by convincing smaller courses that attention to their greens could leapfrog them over their competitors. Milorganite "*creates* new Leaders out of Secondary Courses," ads proclaimed. "Milorganize your course," the commission enjoined.[111] By the end of 1932, it was apparent that the Sewerage Commission's marketing plan was successful when it declared that "More Golf Clubs use Milorganite than any other Fertilizer."[112]

Milorganite dominated, but it dominated a shrinking market. With the fertilizer market depressed, officials in the Sewerage Commission considered abandoning the production of fertilizer to incineration. The chief engineer was adamantly opposed, however, and produced an economic analysis that showed if Milorganite could be marketed at $8/ton, it would be favorable over incineration. Given that the average sale price of Milorganite in the previous three years was over $16, Townsend concluded that abandoning Milorganite production appeared "both illogical of thought and unwarranted." Yet this conclusion depended completely on Milwaukee's continued success in marketing the material.[113]

Rather than switching to incineration, Milwaukee initiated marketing plans that prefigured some of the vast changes about to take place in the agricultural chemical industry as a whole. Like others in the agricultural industry, Milwaukee sought to diversify its product line to increase sales. The granular nature and low water content of Milorganite made it an excellent carrier for other agricultural chemicals and Milwaukee agronomists had encouraged golf course managers to mix Milorganite with lead arsenate (an insecticide in common use) and other chemicals used in weed and disease control. The success of Milorganite as a carrier led the Sewerage Commission in 1940 to begin marketing its own specially formulated combination of fertilizer and herbicide. They combined Milorganite and sodium arsenite, a weed killer, premixed at the Milorganite production facility rather than by their customers. First described in advertisements simply as "Milorganite-Arsenite mixture," the Sewerage Commission quickly ramped up the marketing, coining a new name, "Milarsenite," and producing special bags, with Milorganite's trademark symbol adapted for the new product (figure 4.5). The sodium arsenite in the mixture killed the weeds while the fertilizer induced the "grass to spread and fill the voids left by the weeds." By combining the two products in one formulation, the Sewerage Commission pioneered the production of what later would become known as "weed and feed" applications. By the 1990s weed and feed products dominated the home fertilizer market, bought by more households than all other product types combined.[114] The development of fertilizers premixed with herbicides has been associated with Scotts' first marketing of "Weed and Feed" in 1947. But the development of Milarsenite preceded this by seven years.

In their tests, the Sewerage Commission found the weed control properties of Milarsenite to be "astonishing" and "almost perfect" with only faint or no discoloration of the grass.[115] Indeed, Noer treated his own front lawn with the mixture and photographed the results. Milarsenite was introduced in 1940 in the *Greenkeepers Reporter*, both in an advertisement and an announcement. The advertisement, under the slogan "kills weeds, saves grass" reported the results of Milarsenite's use for killing clover and other weeds and proclaimed "because results this fall were so promising, a limited amount of Milarsenite will be offered again in 1941. If you have a weed problem it will pay you to try Milarsenite next spring."[116] Milwaukee produced a leaflet instructing greenkeepers on the use of Milarsenite warning that it was poisonous, although "no worse than mercury." The commission recommended that workmen wear coveralls, waterproof gloves, and a mask, or at least a "pad of folded cheesecloth worn over the mouth and nose." While filling the spreader,

Figure 4.5
Milorganite and Milarsenite. Milarsenite ("kills weeds, saves grass"), first marketed in 1940, was the first commercial "weed and feed" product. The Milwaukee Sewerage Commission has had to continually develop new products and marketing to maintain its sales of Milorganite. *Source: The Greenkeepers' Reporter*

workmen should "stand on the windward side." Despite these warnings, photographs of workmen applying the product showed that they did not heed them.[117]

With the postwar explosion of the agricultural chemical industry and the apparent success of Milarsenite, the Sewerage Commission received a number of inquiries from agricultural chemical companies about combining their weed and disease treatments with Milorganite. In 1945, the commission began experimenting with the newly developed herbicide, 2,4-D, from Dow Chemical. Noer reported promising results to Dow of 2,4-D combined with both Milorganite and Milarsenite, but Milwaukee never developed a commercial Milorganite 2,4-D product. In ads beginning in 1945, it offered guarded criticism of the usefulness of 2,4-D in weed control without simultaneous fertilizer treatments. And the next year it criticized 2,4-D more strongly, recommending Milarsenite in its place. At the same time, however, the Department of Agriculture had begun tests with fertilizer, 2,4-D combinations, which eventually led to Scotts' Weed and Feed.[118]

In addition to 2,4-D, Milwaukee experimented with a number of other pesticide formulations. Also in 1945, B. W. Bellinger, of the U.S. Phosphoric Products company, had heard of Milarsenite from a fertilizer dealer and wrote to the Sewerage Commission about the possibility of incorporating sodium fluosilicate, an insecticide used to kill the mole cricket in the South. "Our interest, of course," he added, "is that we are the largest producer in the United States of Sodium Fluosilicate." Noer arranged for some tests through one of Milorganite's Florida dealers. The mixtures showed some promise, particularly because the physical nature of the Milorganite facilitated its formulation with the sodium fluosilicate, but there is no further record of developing this product.[119] Similarly, American Cyanimid was interested in combining potassium cyanate with Milorganite in 1951.[120] By this time, however, Milwaukee had ceased production of Milarsenite, likely because of the huge popularity of 2,4-D and its combination with fertilizer in Scotts' Weed and Feed.[121] Milwaukee wouldn't pursue formulations again until a marketing crisis in the 1980s.

"Plus Values"

In the meantime, Milwaukee continued to conduct research on other potential uses of Milorganite that could become the basis for marketing campaigns. In 1947, hops growers in the Pacific Northwest notified their Milorganite dealers that acres they had been treating with Milorganite survived an outbreak of blight that had affected neighboring fields. The dealer contacted the commission, wondering if there were some micronutrient in Milorganite that might give the plants immunity.[122] Reports such as these, and others from the Florida citrus industry, led Milwaukee to sponsor researchers at Rutgers University to investigate potential benefits of Milorganite in preventing disease.

This search for micronutrients and other potential benefits of Milorganite harkened back to a nineteenth-century debate over measuring the value of sewage. A lengthy and testy exchange between the Royal Commission on Metropolitan Sewage Discharge and Sillar, of the Native Guano Company, reveals the gulf between agricultural chemists who thought they could identify, with rational analyses, all of the valuable constituents of sewage, and those, like the Native Guano Company, who thought there was some holistic benefit to sewage, above and beyond its separate constituents, that was not measurable using current methods. As agricultural chemists identified limiting nutrients to crop growth—like nitrogen, phosphorus, and potatassium—they also established the monetary value of fertilizer as no more nor less than the separate value of these constituents. Thus, while Native Guano valued its fertilizer at £3 10s., agricultural chemists, based on the concentration of nitrogen and other chemicals, valued it far lower, perhaps one-sixth the amount.

The Native Guano Company insisted that the agricultural chemists were missing some constituent that promoted plant growth independent of the nitrogen or phosphorus, but they could not identify it. "Looking at the figure you name, there must be some other element of great value, besides the ammonia; can you tell us what it is?" asked the commission. Siller could only point to some near mystical property: "I attribute it to the fact that it is the natural substance which was intended to manure the earth, and that the earth has a natural liking for it independently of its actual ammonia or phosphate strength." The commission was having none of it. "We know the substances necessary for the growth of plants are ammonia, phosphoric acid, lime, magnesia, silica, and so on; without all those being present the plants would not grow. Now, which of those do you consider to be the efficient substance which gives this extra value to your product?" Sillar and the commission continued to argue, the commission insisting that rational analysis of the sewage could measure its value while Sillar insisted on some unnamed growth-promoting characteristic. When Sillar stated that most of the fertilizing power of sewage came from urine, the commission replied that it knew exactly what urine contained. "Something more than nitrogen and phosphates," Sillar replied. Exasperated, the commissioner responded that yes, it did contain more than nitrogen and phosphates. "It contains sulphates also, and chlorine." "But," he added, "we want to know to which of those ingredients belongs the peculiar fertilizing power."[123]

While Sillar's defense of native guano bordered on charlatanism, many sanitarians nevertheless did believe that sewage possessed some unknown "virtue." J. A Voelker (son of Augustus Voelker) noted that, although he didn't believe it, "it had long been an article of belief, even among scientific men, that there was some 'virtue' attributable to the use of materials of this nature, . . . which virtue was not to be represented merely by the contents of nitrogen, potash, phosphoric acid, etc., as expressed in analytical figures."[124] None other than Gilbert Fowler was one of these scientific men: "I am clearly of opinion that chemical analysis alone by no means defines the character of a manure."[125]

Sewage scientists continued to claim additional properties for sludge into the 1940s. "Users of sludge for agricultural or horticultural purposes have repeatedly called attention to the fact that the results obtained could not be entirely accounted for by the mineral or plant food constituents," wrote New Jersey's chief sewage scientist.[126] Agronomists in California termed these unmeasured values "plus fertilizing values." "It has long been recognized by those who have used . . . sewage sludge as a fertilizer that the value of the sludge in that capacity appears greater than the sum total of its fertilizing constituents."[127] When Milwaukee started looking for the plus fertilizing values, analytical science and the understanding of plant physiology had advanced to the point where scientists could begin to identify some of these growth promoting substances as vitamins, hormones, and other constituents. As a result, sewage scientists again argued that value should not be measured on the same basis as fertilizer. "It has long been our opinion that sludges should have applied to them a different yard-stick of value than the ordinary fertilizer analysis alone," argued the editor of *Water Works and Sewerage*. This yardstick would also allow treatment plants to increase the price of their sludge and tip the balance between profit and purification toward profit: "This matter of growth promoting substances in sludge has a definite sales appeal, which sewage works operators and managers can put to good advantage in this day of popularity of hormones and plant vitamins with florists, gardeners and horticulturists," continued the editorial.[128]

In addition to their search for "plus values" for their fertilizer, the Milwaukee Sewerage Commission began to search for industrial uses for Milorganite independent of its fertilizing power. Activated sludge, and thus its product Milorganite, was essentially composed of the cells of bacteria that grew on the sewage. As a bacterial biomass, it was composed of proteins, fats, amino acids, and other organic materials that might have a variety of industrial applications. Other waste products, like tankage from slaughterhouses, had proved to be good sources of adhesives, foaming agents, protein supplements, and the like. Milwaukee commissioned an industrial research company, the Miner Laboratories of Chicago, to explore the potential of Milorganite "as an industrial raw material."[129] The Miner Laboratories had made its reputation in developing a process for extracting furfural, a feed supplement, from waste oat hulls for the Quaker Oats company.[130] In its work for the Sewerage Commission, it was to conduct similar research on activated sludge. Miner used a variety of different solvents to extract various fractions from Milorganite and sent these fractions to industrial companies throughout the United States, like the U.S. Gypsum Company, International Minerals and Chemical Corporation, and the Commercial Solvents Corporation, for evaluation.

One of Miner's first discoveries was that various extracts of Milorganite could provide a previously unknown growth factor for yeast and bacteria that accelerated a variety of industrial fermentations, including ethanol, bread dough, lactic acid, and acetone-butyl alcohol.[131] The centrality of biological sewage treatment to the development of the industrial ecosystem was recapitulated when Miner sent samples of the extracts to the Commercial Solvents Corporation to test in its industrial operations. This company had licensed Weizmann's patent for the acetone-butyl alcohol fermentation. Activated sludge and acetone-butyl alcohol, processes that had both come out of Manchester University in the teens, were now reconnected. Also illustrating the shift in attitude toward patenting, Miner patented the fermentation acceleration and assigned it to the Sewerage Commission. Gone was the principled opposition to patenting natural processes.[132]

While activated sludge could activate commercial fermentations, its mode of action was unknown. Miner suspected that Milorganite provided a previously unknown B-complex vitamin. In searching for the specific activation factor, they discovered that activated sludge "was one of the richest sources yet discovered for Vitamin B-12,"[133] which had only recently been discovered and identified as a key animal nutrient in 1948. The presence of B-12 in activated sludge suggested additional uses for Milwaukee's sludge as an additive to animal feeds, and it became a focus of Miner's work.

Marketing products from sewage sludge continued to run into deep-seated antipathy toward using human waste in products for human consumption. The Milorganite yeast activator was used in beverage alcohol as well as industrial alcohol fermentations. On learning of this, the Food and Drug Administration issued a stop order, declaring "the idea of using such a substance in the production of alcohol for beverage purposes is so repugnant that the proposal should not be considered."[134] While untroubled by the use of Milorganite for industrial alcohol, the FDA threatened to take action against any use of the fermentation activator for beverages. The commission was thus surprised when the following year, the FDA indicated it might be willing to approve Milorganite as a feed supplement. Nevertheless, wary of the public perception of such a supplement, the commission sought the advice of the major feed producers in the United States on their willingness to consider Milorganite as a vitamin B-12 supplement in animal feed. Research directors for Quaker Oats, Allied Mills, and the Ralston-Purina Company all agreed that, regardless of the federal approval of Milorganite as a supplement, they would be unwilling to use it directly in their feeds. There had been a recent scare when bonemeal used in feeds had caused an outbreak of anthrax in the Midwest, and feed mixers were becoming increasingly cautious in using suspect ingredients. If they used Milorganite, they feared being "accused of using a sewage product." These fears applied only to using bulk Milorganite directly as a feed supplement. These advisors suggested that the industry would be quite willing to use an extract of Milorganite in which the B-12 had been concentrated, provided it had "a proper name" and the "source of the extract would not have to be given."[135]

Miner Laboratories began work on developing a commercially viable extraction process and presented plans for building an extraction plant at the Jones Island sewage treatment plant. As more manufacturers might enter the B-12 market, however, Milwaukee feared the effects of increasing competition on the potential price of the vitamin and would only consider building a plant if its costs could be amortized quickly, within three years. Commercial prospects changed, however, when Merck, the main producer of B-12 filed patents that covered a critical step in its production. The primary source of B-12 was from bacterial fermentations. Merck had patented a process for adding cobalt to the fermentation medium that substantially increased the yield. Merck apparently refused to license this process and thus looked to become the sole producer of vitamin B-12. Because Milwaukee's process would not infringe on Merck's patent, the Sewerage Commission anticipated being the only other source of the vitamin. "Merck & Company, using a fermentation process, and the Milwaukee Sewerage Commission, using the activated sludge process, ultimately may be the only important sources of Vitamin B_{12}," reported Miner. With the price thus prevented from falling due to increasing competition among producers, the potential profit for Milwaukee increased substantially.[136]

Producing vitamins from the sewage presented a different situation for the Sewerage Commission. "Our principal business is to purify the raw sewage that comes to the disposal plant," declared J. F. Friedrick, chairman of the commission. Producing Milorganite was a fundamental part of the disposal process, providing a way to dispose of the sewage. But commissioners did not see vitamin B-12 extraction as necessary to the treatment process. Rather, it represented "a new commercial undertaking" for which they felt it was inappropriate to expend public funds. For the commission, vitamin sales were not a legitimate form of municipal enterprise, because they were simply too far removed from essential municipal services. The commission thus sought a private company to build and own the plant, with Milwaukee providing the space. In return, Milwaukee would get 35 percent of the net profits.[137]

The Sewerage Commission contracted with the Verne E. Alden Co., a Chicago engineering firm, to build a vitamin B-12 extraction plant at Jones Island, selling the vitamin for use in poultry and animal feeds. The commission also hoped to produce B-12 for humans as well, making history as the first "by-product from sewage for human consumption."[138] After several delays, the Alden Co. finally built a pilot extraction plant in Milwaukee's fertilizer warehouse in 1956. When anticipated orders for the extract failed to materialize, however, the plant closed temporarily in 1958. Restarted in 1959, it closed permanently after only a few months in operation.[139] Despite the hopes that "Milorganite may well turn out to be another coal tar in the biochemical field,"[140] none of the industrial applications for Milorganite panned out.

"A Fancy Name . . . Helps a Lot"

When Milwaukee had begun Milorganite production, it had been the only sewage treatment plant producing commercial fertilizer. Seeing its success, however, several other cities entered the market. Pasadena started selling its sludge in 1928, Houston in 1932, and Chicago in 1939.[141] Other cities, however, did not take on the vast marketing program that Milwaukee did. Chicago, for instance, had no marketing staff. Rather, the city sold its entire production to one company that contracted for the sludge. In turn, the contractor assumed "all expense for selling, promotion and credit." Most of the sludge was resold to fertilizer manufacturers in the southeastern states for use on tobacco and cotton.[142] Receipts for Chicago's sludge grew throughout the 1940s, until 1948 when Chicago sold over $1 million worth of sludge.[143] McKee and Son, the fertilizer brokers who helped Milorganite get its start, had lost access to activated sludge materials when Milwaukee stopped selling to the bulk market. In 1948, it took over Chicago's activated sludge contract, marketing the product as variously "Vitorganic" and "Chi-Organic." Other sources referred to Chicago's activated sludge fertilizer as "Chicagrow" or "Nitroganic Tankage."[144]

Chicago's lack of a consistent name for its fertilizer suggests that marketing was much less of a priority than it had been at Milwaukee. Whereas Milorganite became a well-known brand, none of the other cities was able to develop a consistent marketing plan. Houston's market was restricted by the limited production of fertilizer material. The Houston treatment plant produced only about 2,000 tons per year of activated sludge, compared with 50,000 tons in Milwaukee. Similarly, Pasadena, despite the production of extensive marketing materials, did not stay in the market long, abandoning its fertilizer production in 1942.[145]

The success of cities in marketing their sludge depended on a number of factors. The most salable type of sludge was the heat-dried activated sludge produced by Milwaukee, Pasadena, Houston, and Chicago. This sludge had the highest concentrations of nitrogen and the lowest water content, making it economical to transport long distances. Of lower value was digested activated sludge, which had lost much of its nitrogen content through the anaerobic digestion process. Finally, of even lower value was the settled sludge from Imhoff tanks. The equipment required for heat drying was expensive, and thought to be economical only for the largest cities. Experts considered it impossible for small cities producing Imhoff sludge to sell their sludge as fertilizer. They could successfully dispose of their sludge to local agriculture, perhaps, but it would have to be given away and loaded for free. Marketing, however, was essential to even give away the sludge. P. N. Daniels, superintendent of the Trenton, New Jersey, plant, noted that of the three hundred plants in New Jersey, "the sludge cannot be given away, not to mention selling it." Plant managers needed to give greater attention to the condition of the sludge, he argued, making it more palatable for users to obtain and spread without nuisance. He suggested cheap machinery for grinding and preparing the sludge, as well as investment in fortifying the sludge with nitrogen and phosphorus to make it competitive with Vigoro and the other commercial fertilizers. Finally, he suggested, cities should give away samples to golf course superintendents and to other potential users and develop a network of sales agents.[146]

What most cities agreed on was the need for a catchy name for their product. "A fancy name such as 'Super-humus,' 'Grozall,' or 'Fortified humus,' helps a lot," suggested the superintendent of the Newark, New Jersey, sewage treatment plant.[147] Most names incorporated some version of the city name in conjunction with a phrase indicating growth or organic. Thus, Toledo produced "Tol-e-gro, " and Akron, "Akra-Soilite." Perhaps the most unfortunate of these names came from Clearwater, Florida, who called its product "Clear-o-sludge."[148] "Hou-Actinite," (presumably for Houston Activated Nitrogen) was a play on Milwaukee's trademarked sludge. "One of the tricks in advertising," according to Los Angeles's plant manager, involved writing the name of their product, "Nitrohumus," by sprinkling it on a potential customer's lawn. "In a week or ten days," he recounted, "the name of the sludge is written in bright green across the yard."[149]

To protect its own investment in developing the trademark Milorganite, Milwaukee tried to prevent other cities from coopting its marketing by developing similar names. It sued a number of cities for adding the suffix "-organite" to their product names. The Kelly Agricultural Products Company, located near Pittsburgh, had named a fertilizer derived from waste leather "Pittorganite." Milwaukee took the company to court, claiming a trademark on any name combining the phrase "-organite," and forced it to change the name (the company apparently chose "Groganic"). While the settlement netted the commission only $2,500, the lawyers pointed out that having its trademark validated in court was worth many times that amount.[150]

Despite the conclusion that heat drying would be economical for only large cities, several smaller cities tried to make a go of the fertilizer business. Grand Rapids, Michigan, started its sewage plant with high hopes of turning a profit. "Efficient utilization by the city of by-products produced at the new sewage disposal plant may make the enterprise an industry which is virtually self-sustaining," claimed the city's service director. But before it could turn a profit, the director admitted that he had to "seek a market." Grand Rapids marketed its sludge as "Rapidgro" and began its fertilizer production in 1931, in the midst of the Depression (figure 4.6). To get the project started, the superintendent James R. Rumsey used a variety of strategies for building the plant and marketing the fertilizer. He used scrip labor to build the fertilizer plant. Workmen "from the city's unemployed were paid in scrip which they exchanged at the city store for food or clothing." With no market initially, Rumsey turned the city into its own customer. All of the plant's sludge was used by various city departments. The sewage plant used the fertilizer and scrip labor to beautify its grounds. It also used the fertilizer to build up "waste land" and turned it into a city garden that provided vegetables to the city's welfare department, providing food for the indigent. It also enlisted county farm agents to advise local farmers on the use of Rapidgro, and it distributed samples to local garden clubs and fruit growers. Finally, the sewage plant, following the example of Pasadena, enlisted unemployed men to act as commissioned salesmen for Rapidgro.[151]

Selling mostly to a limited local market, Grand Rapids nevertheless managed to sell a thousand tons of sludge per year, earning a net profit on fertilizer sales of between $1,000 and $10,000 per year. They were embarrassed, however, when it

Figure 4.6
Sludge-drying beds at the Grand Rapids sewage treatment plant. From 1931 to 1965, Grand Rapids produced its own brand of activated sludge fertilizer, "Rapidgro." Unlike Milwaukee's product, which was heat-dried, Rapidgro was made from air-dried sludge. Although a much cheaper process, Rapidgro did not have the high nitrogen concentration of Milorganite. Courtesy of Grand Rapids City Archives. *Source*: Environmental Protection Department, Wastewater Treatment Plant, Miscellaneous Historical Records, Grand Rapids City Archives

became known that the Parks Department, rather than using its own Rapidgro, had ordered eight tons of Milorganite for use on the city's parks and cemeteries. When asked to explain his order for Milorganite, the parks superintendent explained that "the Milwaukee product has a faster action than the fertilizer we make at the sewage disposal plant." "To save face," the city announced that it would enrich Rapidgro with extra nitrogen for use by the Parks Department. "It will cost an extra $10 a ton to bring the local product up to the potency of some of the nationally-distributed brands," stated the *Grand Rapids Herald*.

A 1936 report of the American Public Health Association surveyed sewage treatment plants to determine the use of sewage for fertilizer. Of the 452 plants surveyed, most (62 percent) disposed of at least some of their sludge as fertilizer, ranging from 43 percent of the plants in the Rocky Mountain states to 75 percent in the East. However, only 63 plants reported that they were able to sell their sludge, 19 of which were activated sludge plants; the remainder gave their sludge away.[152] With the exception of Milorganite, most of the plants disposed of their sludge locally, with no national market. But by the 1950s and into the 1960s, most producers of fertilizer material from sludge ceased operations. Finding markets drying up and production more trouble than it was worth, city after city abandoned their sludge recycling programs and opted for either incineration or landfill for their sludge.[153] Despite the continued, albeit modest, success of Rapidgro, for instance, Grand Rapids abandoned the manufacture of fertilizer in 1966 and began taking all of its sludge to the town dump. Chicago was finding that the companies that had contracted for its sludge were forced by market conditions to withdraw from bidding. With no outlet for its sludge, Chicago briefly looked into packaging its fertilizer and selling it directly, as Milwaukee did, but these plans never materialized.[154] Chicago then began dumping its fertilizer on the banks of the Sanitary and Ship Canal and encouraged the public to remove it free of charge.[155] In 1965, Chicago stopped making fertilizer and planned to demolish the plant in 1969.[156] By 1960, when the Water Pollution Control Federation updated its 1946 *Manual of Practice* on the utilization of sewage sludge, it concluded that "the sale of sludge can never be expected to provide a net return over processing costs."[157] Selling sludge might be less costly than other disposal methods, but the hopes of "unlock[ing] the treasure stored up in sewage sludge" were not realized.

"We Are a Sanitary Facility"

As the Native Guano Company had discovered in the previous century, profit and purification remained as contradictory as ever. If Milwaukee operated its plant to minimize pollution, its profit was reduced. If it operated to maximize its profit, purification suffered. This inherent conflict between profit and purification initiated a long-running controversy between Milwaukee's sewage plant manager and chief engineer. James Brower, plant superintendent from the 1930s to the 1950s, recounted how he believed the aims of the sewage treatment plant were being subverted by the need to market fertilizer. "I have always believed and still do, that the plant was designed and built to treat sewage, that the main purpose was to keep the rivers and lake as free from sewage polution [sic] as was humanly possible, regardless of what happened to the fertilizer." Nevertheless, concerns over marketing caused the commission to release untreated sewage into Lake Michigan during storms even though the plant had the capacity to treat it. But, because the capacity of the fertilizer plant was lower than the capacity of the activated sludge tanks, the commission sacrificed treatment to limitations of fertilizer production. "This was mainly done," Brower charged, "to maintain a high percentage of nitrogen so that the sales department could meet the guarantees as printed on the bags." Dilution of the sewage by large rainstorms would decrease the percentage of nitrogen in the fertilizer. Even though the health department and water purification plant complained to the Sewerage Commission, James Ferebee, the chief engineer at the time, refused to alter the practice. Brower, concerned over the pollution, demanded a written order from Ferebee before he would bypass the plant. In 1941, Ferebee issued that order and declared that "until further notice, the quantity of sewage taken in by the purification plant shall be reduced to the lowest possible operating quantity during times of quick, intense rain storms. The purpose of this order is to decrease the dilution of organic material." After continued complaint, though, Brower was able to convince Ferebee to take in the purification plant's capacity during storms. That decision was quickly reversed, however. Brower claimed that "someone got to him, stating that after a storm it took a considerable amount of time to bring the fertilizer back to the standard guarantee."[158]

As Gail Bradford notes about the history of public authorities like the Sewerage Commission, entering the market often meant structural pressures to emphasize revenue at the expense of environmental concerns. "Agencies . . . are impelled by their very nature to behave much like private businesses. They have to pay bondholders, staff, and vendors from revenues gained mostly by charging for the goods and services they provide. Thus, they must make revenue growth and cost reduction the central criteria for deciding what to produce and how to do so. Only with difficulty can they give goals such as environmental protection . . . much weight."[159]

Milwaukee's history of Milorganite production showed that sewage treatment plants could not be run at a profit. Even in its best years, when World War II had elevated the price of nitrogen by diverting chemical nitrogen into the production of explosives, sales of Milorganite barely covered the cost of its own production, not to mention the cost of other parts of the plant.[160] Yet repeated economic analyses showed that once Milwaukee had sunk its initial capital investment in its fertilizer plant, continuing production of Milorganite was a cheaper means of disposing sludge than either landfill or incineration. But even this depended on continuous innovation in product development and marketing. Milwaukee could not simply assume a market and sell to it. By pursuing an outlet for its sludge through the marketplace, Milwaukee was caught in not only the contradiction between profit and purification but also the contradictions of capital as well. The pressures under capitalism for increasing competition, falling profits, and reduced market share demanded continuous attention to the business aspects of sludge disposal. Only relentless innovation in product development and marketing allowed Milwaukee to dispose of its sludge. Over time, however, the Sewerage Commission thought of itself less and less as a producer of fertilizer. "We are not in business as a manufacturer," stated Raymond Leary, chief engineer in 1960. "We are a sanitary facility to collect and treat sewage."

That a sewage commissioner needed to reiterate this points to the distortions in public policy that were created when Hatton and the commission made sewage treatment subject to the relentless demands of the market. Half a century earlier, the potential to profit from its sewage had lured the Milwaukee Sewerage Commission and its engineer Hatton to construct a fertilizer plant as an integral part of its sewage treatment facility. Now, in a newspaper article headlined "Milorganite Moves from Profit to Loss," the commission regretted its decision: "If not saddled by the 10 million dollars tied up in the plant necessary to produce Milorganite, it would be better for the district to burn the waste."[161]

"Would plant officials recommend another fertilizer operation if Milwaukee had to start all over?" asked Kadish, Milorganite sales manager in 1951. "Probably not," he answered. It was still economical to manufacture Milorganite at the Jones Island plant, "with most of the facilities built years ago."[162] But when Milwaukee decided it needed a new plant to meet sewage demands on the south side of the city, there was no preexisting capital investment. When the commission built the new South Side treatment plant in 1968, it answered Kadish's hypothetical question. Despite Milorganite's widely known name and its well-developed marketing network of agronomists, salesmen, and distributers, the Sewerage Commission decided it would be cheaper to digest its activated sludge and incinerate it than to produce Milorganite at its newly constructed plant.[163]

5

The Contradictions Continue: Sewage Treatment since the Clean Water Act

By the late twentieth century, the biological sewage treatment processes that had been developed a century before had become widespread and had become the international convention for dealing with sewage.[1] Activated sludge was practiced with very little change from the basic process as it was laid out in 1914. A trickling filter built in 1990 would have been entirely recognizable to sanitary scientists of the 1890s. Certainly there were changes and refinements of the technology. Scientists had developed a much better understanding of the biological basis for sewage treatment, and, recently, modern genomics techniques have allowed a more thorough characterization of the bacterial communities than ever before. Yet the same contradictions and dilemmas that faced the sanitary scientists during the previous century persisted. Engineers, scientists, and operators continued struggling to resolve these contradictions. At the same time, as society began relying more and more on the biological sewage treatment plant to solve increasing pollution problems, the contradictions threatened to collapse the ecosystem.

During the 1950s and 1960s, sewage treatment in England and the United States faced a deepening crisis.[2] The population was becoming more urbanized, and the pace of sewage treatment plant construction severely lagged behind the required capacity. Further, plants already operating were failing more frequently from increased loading, detergents, and industrial waste. The federal government in the United States began a program to finance the construction of new sewage treatment plants that culminated in the passage of 1972 Clean Water Act and its subsequent amendments in 1977 and 1987, while Great Britain completely restructured its water and sewage services in 1974, nationalizing the hundreds of local authorities, consolidating and reorganizing them along watershed boundaries. However, the potential responses to the sewage crisis were conditioned by the fundamental contradictions of the sewage treatment plant ecosystem. In turn, the sewage crisis reworked those contradictions into new forms.

Not only was it the technology that was rooted in the early nineteenth century. The discourses surrounding sewage treatment revolved around identical issues as well: whether sewage treatment processes were artificial or natural, the role of science and craft in process control, whether sewage treatment should (or even could) be profitable, and whether it should be run for private gain or public good (figure 5.1). As scientists, engineers, public officials, businessmen, and sewage treatment plant workers addressed each of these contradictions, they became inextricably intertwined. Whether a process could be privatized depended on whether it was natural or artificial. Whether treatment plants could turn a profit became connected to whether they were public or private. Because privatization centered on sewage treatment plant operation as more amenable to profit, the control of the labor process inherent in process control became more important, and the demands for profit led to the increased automation of sewage treatment plants. Yet, as automation failed to deliver pure effluents, engineers found they needed to incorporate the craft expertise of operators into automated process control technologies. That this same set of contradictions persisted for over 150 years of progress and change in sewage treatment technologies testifies to the strength and irresolvability of these contradictions They are intrinsic to the industrial ecosystem.

Profit vs. Purification

During a 1972 meeting with some of Milorganite's largest customers, sales manager Charlie Wilson painted a rosy picture for Milorganite. Two years after the first Earth Day and barely a month after passage of the Clean Water Act, the Milwaukee Metropolitan Sewerage District was selling every ton of Milorganite it produced, and Wilson thought if the current "ecology kick" continued it could probably sell twice that. The chief concerns were expanding production and maintaining the ability to keep profit margins "undoubtedly the highest in the industry."[3] There were some clouds on the horizon, however, ironically from the very same "ecology kick" that Wilson hoped would spur sales. By the middle of the decade, Milwaukee faced a deepening crisis in disposing of its activated sludge.

Having held a virtual monopoly in marketing sludge for over forty years, Milwaukee was suddenly faced with increasing competition from other cities that, because of the provisions of the 1972 Clean Water Act and other legislation, began marketing their sludge as well. A new batch of sludge products like "Metrogro," "Biogro," and "Gardenlife" appeared.[4] In 1988, the federal government instituted a ban on the dumping of sludge in the ocean. Major East Coast cities like Boston and New York that had been ocean-dumping their sludge were forced to find other outlets. Rather than dumping it at sea, these cities began dumping it on the market,[5] eating into Milwaukee's market share and profit.

In addition, there was increasing concern over the safety of sewage sludge as a fertilizer, reducing the ability of any city to sell its sludge. In the 1970s and 1980s, there were a number of scares involving the safety of Milorganite, resulting from a

Figure 5.1
Using Nature's Way to purify sewage still holds important rhetorical importance into the twenty-first century. Waste from portable toilets, however, is usually pumped to municipal sewage plants, and treated with whichever process is used there. *Source*: Photo by author

fundamental contradiction of sewage treatment. Sewage contained both industrial and municipal waste, including heavy metals and other contaminants. The more material removed by the sewage treatment process, the cleaner the effluent from the plant, and the lower the pollution of rivers, lakes, and ocean. But it also meant a greater concentration of these pollutants in the resulting sludge. As long as nutrients were the main component of sewage, this led to a more marketable sludge. For instance, when the widespread use of phosphates in detergent led to an increase of phosphorus in the sludge, Milwaukee trumpeted the higher concentrations as an improvement in the fertilizer. But when sewage contained toxic waste, either from industrial discharges or household waste like solvents, paints, or other material washed down the sewers, the cleaner the effluent, the more toxic the sludge.[6]

In 1974, recognizing the potential for sludge to contain high concentrations of pollutants, the EPA began to develop regulations limiting heavy metals in sludge, much to the consternation of many municipal sanitary districts. The EPA's proposed regulations would have excluded from use over 90 percent of the municipal sludges from Wisconsin and Indiana and 80 percent from Illinois. Individual states also began to develop regulations. Following the EPA's example, Wisconsin published guidance on toxics in sludge in 1975.[7] The Sewerage District, fearing the effects on Milorganite, complained to the state about rigid limitations on heavy metals.[8] At the same time it approached members of its "Milorganite Advisory Council"—fertilizer distributors and other large users of Milorganite—to contact their congressmen to help prevent the USDA and EPA from "endangering" Milorganite with demands for health warnings.[9]

With municipalities in Wisconsin reluctant to address the issue of heavy metals, a Milwaukee environmental organization, Citizens for a Better Environment, began a campaign against the use of sewage sludge contaminated by heavy metals. In 1978, it issued a well-publicized report on the high cadmium levels in Milorganite.[10] Milwaukee, to its credit, responded to the publicity by placing a prominent label on its Milorganite bags: "WARNING: DO NOT USE ON VEGETABLE GARDENS, OTHER EDIBLE CROPS OR FRUIT TREES. Eating food grown on soil containing Milorganite may cause damage to health." While the warning was limited to Milorganite's use on edible crops, consumers understandably began to shy away from it for all uses. "Even though the notice pertains to home grown vegetable production, many homeowners do not want their children or even family pets walking on lawns that receive Milorganite 'heavy-metal cadmium' applications," stated the marketing staff.[11] The warning label caused an almost immediate 30 percent drop in bag sales.[12]

In 1987, the health effects of Milorganite were further questioned when a former quarterback for the San Francisco 49ers was diagnosed with ALS, or Lou Gehrig's disease. At the same time it was revealed that he was only the most recent member of his 1964 team to contract the disease; two others had already died. Public health officials speculated that this unusual cluster might be due to the heavy metals in Milorganite that had been reportedly used on the 49ers' practice field. A Milwaukee physician made further news by claiming that an unusual number of his ALS patients had used Milorganite as well. Researchers investigating the link between Milorganite and ALS found no evidence that the fertilizer was responsible for the disease. Yet, as the Milorganite marketing staff recognized, the cadmium warning label raised the question among users "'If Milorganite does not cause ALS, why do you have the warning on the bag?'"[13]

To reduce the cadmium content in Milorganite, Milwaukee had passed an ordinance in 1980 requiring industries to treat their waste before sending it on to the municipal plant. As a result, the cadmium concentrations in Milorganite were more than halved within a couple years. As the Sewerage District reduced the concentration of cadmium and other metals in Milorganite, it received enormous pressure from the Milorganite marketing staff to remove the warning label. The label had never been required by either the Wisconsin or Federal government. Rather it had been placed voluntarily, albeit under pressure from environmental organizations. Because Milwaukee was competing against other fertilizers that had no label, even though they might have had equivalent cadmium concentrations, the marketing staff, heeding the advice of its large customers, repeatedly sought to downplay the warnings. In 1982, with cadmium concentrations in Milorganite further reduced from around 120 to 45 parts per million, the staff convinced the commission to downgrade the "warning" to a "notice" and move it from the front of the bag to the back. The Sewerage District also sought the approval of Citizens for a Better Environment for the label change. With the organization's approval, the label was changed to "NOTICE: Milorganite should not be used on vegetables due to possible uptake of the heavy metal cadmium. When home grown vegetables make up a large portion of the annual diet over a period of many years, cadmium may accumulate in the kidneys, resulting in possible dysfunction." "Safer Milorganite to be sold in snazzier bags," crowed the *Milwaukee Journal*.[14] Nevertheless, the cadmium content remained a problem; it was too high for Milorganite to be exported to Canada, and other states sought to ban its distribution within their borders.[15] Cadmium concentration continued to decline, though, as Milwaukee enforced its pretreatment regulations. By 1990, concentrations of cadmium were around 5 parts per million, and marketing staff successfully lobbied to get the label changed once again: "While Milorganite meets all federal and state regulations for fertilizer products, it should not be used on food crops. Use only in accordance with label directions on nonedible plants and grasses. Most natural organic and chemical fertilizers contain trace amounts of heavy metals. Between application, store in a dry area out of the reach of children and away from pets. Wash hands after use. Do not eat."[16]

With increasing concern over the safety of sewage sludge, the Environmental Protection Agency finally released its long-delayed regulations governing the application and disposal of sludge. Mandated by the 1987 amendments to the Clean Water Act, the regulations were finally promulgated in 1993. The regulations divided sludge into several categories based on the concentration of pathogens and ten heavy metals, including cadmium. If concentrations were below a critical value, the sludge (now termed "biosolids") could be considered of "exceptional quality" and distributed essentially without further regulation. These regulations provided incentives for sewerage districts that marketed their sludge, like Milwaukee, to ensure that their product met these criteria. It also helped ensure a market for the sludge by providing reassurance that sludge was, indeed, safe.[17]

Because of Milwaukee's pretreatment rules, the concentration of cadmium in Milorganite was below the EPA limit of 39 parts per million. Milorganite was classified as an exceptional quality sludge, and the Sewerage District sought to remove the warning label altogether. With the encouragement of the EPA, Milwaukee removed any reference to potential health hazards, stating simply that Milorganite meets "'exceptional quality' guidelines as defined by the Federal Environmental Protection Agency."[18] The EPA regulations had performed their function, as defined by the EPA—encourage the disposal of sewage sludge on land as well as the reduction in toxics concentration. Whether the regulations were sufficiently stringent to protect public and ecological health is still a source of controversy.

The market for Milorganite, however, remained difficult. Even with the changes in the label, dealers reported that homeowners were "still concerned about the metals content" of Milorganite, causing the loss of 60 percent of one dealer's market. Other dealers reported they had stopped carrying Milorganite after the ALS scare and had not carried it since.[19] With both increasing competition and the warning label depressing sales, the Milorganite division of the Sewerage District saw a critical need to improve the marketing of Milorganite, both by creating new products and by finding new markets. Yet many in marketing felt that the commission had ignored the importance of generating revenue through Milorganite sales. Rather, they charged, the commission was focusing solely on the quality of the sewage treatment plant effluent. Once again, sewage treatment was caught in the contradiction between profit and purification. Charles Wilson, O. J. Noer's successor, charged the commission with ignoring Milorganite and its marketing. "You have said that we are not in the fertilizer business," he wrote to William Katz, a director at the commission. But, he continued, "70,000 plus tons of yearly sales and a 4 million plus income could be considered good business by most folks."[20]

In response to the difficulty of selling Milorganite, the Sewerage District contracted for a number of marketing studies, in 1982 and 1985, in 1986, and again in 1994. Each study reached similar conclusions: there was a conflict between purifying sewage and seeking a profit from the plant's operation. The 1985 study argued that the Sewerage District would "have on-line an eighty million dollar fertilizer factory—a heavy capital investment in its fertilizer products." Yet, despite this investment, the commission had neglected basic business planning. The district, the report charged, had become "technology" rather than "marketing" oriented. The Sewerage District's "products are the responsibility of the operating, technical side of the organization with no product planning influence from the marketing staff," the report concluded.[21] Technology, here, meant sewage treatment. The Sewerage District was focusing too much on treating the sewage, and not enough on selling it.

The 1986 study renamed the contradiction between profit and purification one of customer vs. operations orientation. In the early years of the activated sludge process, the report argued, the district operated like a "fertilizer factory" focusing on the needs of the fertilizer customer. Recently, however, it had shifted its approach to one of "operations," or purifying the effluent. The report urged the Sewerage District to move back toward a marketing orientation.[22] In 1991, Milorganite staff proposed a reorganization that would strengthen the marketing division and allow it more flexibility to create new products. In response, district officials again raised the contradiction between profit and purification as a fundamental problem in their operation. "Isn't there the potential for conflict between one side of the MMSD which advocates reducing toxics at the source and the other side of the organization which hopes to maximize revenue?"[23] Some staff even wondered if "the MMSD should be engaged in 'business' at all."[24]

Perhaps, thought the Sewerage District, a resolution to this contradiction could be found in spinning off that side of its operations "which hopes to maximize revenue." When the Sewerage District began Vitamin B-12 production, for instance, these kinds of considerations over the proper role of the district in business ventures led it to structure the vitamin production plant as a private operation under contract to the district. By the early 1990s, officials began to investigate the privatization of Milorganite marketing. Houston, after decades of ineffective marketing of its sludge, had begun privatizing its sludge operation in 1990. Tampa, New York, Boston, and other cities were about to follow. Houston had contracted with a relatively new private firm, Enviro-gro, to take over its sludge marketing. With EPA regulations imminent, cities were desperate to find ways to dispose of their sludge. Enviro-gro sought to take advantage of this new municipal market and soon had contracts with Houston; it was also in negotiations with New York and Boston to both build and operate sludge drying and fertilizer production, as well as to market the resulting sludge. Recalling the discussion of sewage from nineteenth-century England, the environmental industry press asked, "Can cities turn profits from sludge?"[25] In seeming answer, the *New York Times* evoked echoes of the nineteenth century as well: New York and Enviro-gro were turning "sludge to gold."²⁶ With Enviro-gro's growth, Milwaukee began to investigate privatization of its own sludge operations.

The marketing staff opposed the privatization of its functions. What Milorganite needed was not a private operator and marketer, but rather the "blessing" of the Sewerage District to act like a private business in terms of responding to the market. The risks of actual privatization were too great. Staff predicted that a private operator would take the Milorganite name and apply it to sludges that were cheaper to produce. These other products would be "up-valued" while the Milorganite name would be eroded. When the contract terminated, "the District [would be] left with a bastardized product, no staff, and the continuing problem of how to dispose of the stuff."²⁷ The marketing staff convinced the Sewerage District to keep the marketing operations public, and its predictions about private sludge companies proved prescient. Many of the private sludge marketers generated terrible publicity and ran into enormous opposition in disposing of their sludge. A number of firms were tied to organized crime and others were prosecuted for bribing public officials, while their sludge was found to contain high concentrations of heavy metals and other pollutants.²⁸

Milwaukee's market at this time was roughly evenly divided among retail, institutional, and bulk markets. Milwaukee was selling two-fifths of its production to the Southeast, primarily Florida. The Milwaukee environs, the rest of the Midwest, and the mid-Atlantic received about a tenth of the Milorganite production each, while Texas and California regions accounted for about 5 percent each.²⁹ As Milwaukee worked to continue product innovation and compete in this national, and even international, market, the privatization of sludge processing and marketing dramatically increased. Enviro-gro and Bio Gro were both bought by the environmental services company Wheelabrator in 1993, itself owned by the waste processing giant, Waste Management. New York City paid Wheelabrator over four times per ton what it cost Milwaukee to produce Milorganite, so that Wheelabrator could sell the product much more cheaply, about one-third of the price of fertilizer shipped to Florida, Milwaukee's largest single market.³⁰ Milwaukee charged that Wheelabrator was dumping sludge on the market below cost.³¹ Pressures on Milwaukee have continued to increase as the private sludge processing industry has consolidated. Waste Management, owner of Wheelabrator, decided to leave the water treatment field and sold its sludge fertilizer subsidiaries to another firm, Synagro. Synagro is now the largest sludge management firm in the country, serving over six hundred communities and industries.³²

The alternative to privatization was an increased commitment to developing and marketing new products. The foremost priority was the development of new products, primarily through licensing of the Milorganite name to blends created by other fertilizer and chemical dealers. "So-Green," "Sta-Green," "Sunniland," and

"Mil-chem," were all Milorganite blends produced by regional distributers for use in their local areas.[33] One such proposed blend was a weed and feed product. Milwaukee proposed providing Milorganite to Parker Fertilizer who would blend it with an herbicide and market it as "Milorganite Weed and Feed." Milwaukee had actually blended 2,4-D and MCPP with Milorganite at the request of the Milwaukee Parks Department in 1971, but it was never marketed to the public. A number of commissioners were concerned, however, that Milorganite's reputation as an "organic" or "natural" product would be damaged by "associating the Milorganite name . . . with '-cides' that may be environmentally unfriendly." If Milorganite were to take advantage of the "ecology kick," it would have to avoid association with pesticides. The marketing staff was thus directed to pursue blending Milorganite only with other nutrients until the commission could investigate the "implications of Milorganite/'-cide' blends."[34] The Sewerage District had seemingly forgotten its own history in inventing this product in the first place. But marketing Milorganite as a natural product came under attack nonetheless in 1992 when the Federal Trade Commission (FTC) began an investigation into Milorganite marketing. The FTC believed that Milorganite advertisements were deceptive by portraying Milorganite as "natural, organic, and/or safe." By 2000, the Department of Agriculture explicitly banned all sewage sludge from use on crops labeled organic.[35]

Another arena for marketing a new product was suggested by customers who noticed that Milorganite apparently kept deer from browsing in their gardens. An increasingly damaging garden pest in suburban ecosystems, deer were repelled by the odor of Milorganite. EPA regulations on pesticides prevented the Sewerage District from marketing Milorganite for this purpose without first registering the product under federal pesticide rules. Milwaukee commissioned studies of the effectiveness of Milorganite and in 2006 applied to the EPA to register Milorganite as a deer repellant. The EPA denied the application in 2009 for lacking studies on potential health and environmental risks.[36]

In addition to the problem of new product development, changes in Milwaukee's economy were causing a steep decline in Milorganite production. Milwaukee was caught between two opposing forces. It needed to innovate new products to replace the revenue lost to other cities from its traditional golf course market, but at the same time it needed to prevent steep declines in production to provide material to the markets it was busy creating. Sewage treatment was like any other manufacturing process, relying on inputs to the plant to produce a product. These inputs—the sewage itself—depended on developments in technology, economic conditions, and social movements, and in turn placed constraints on the ability of Milwaukee to market its product. One example of the changing nature of these influences is in the role of phosphorus. With the replacement of laundry soap with detergents containing phosphates,[37] the concentration of phosphates in the waste stream increased substantially, elevating the phosphorus concentration in Milwaukee's fertilizer. In the 1950s, the commission advertised this increase to prospective buyers, claiming that Milorganite could now guarantee 4 percent phosphorus. However, as the role of phosphorus in fertilizing lake and river ecosystems and causing blooms of unsightly and toxic algae became clear, phosphorus went from being a benefit to fertilizer manufacturers to a pollutant that needed to be controlled. Wisconsin instituted strict controls on phosphorus in detergents, and the phosphorus content of Milorganite again decreased.[38] Milorganite's fortune might be changing again. Madison and other Wisconsin cities have enacted bans on phosphate fertilizer use to protect lakes. Municipal sludge products, like Milorganite, were exempted. With customers unable to use commercial fertilizers, they may turn increasingly to Milorganite.[39]

Industrial waste played a contradictory role in Milorganite production as well. The Sewerage District sought to prevent waste from the Milwaukee's many machining and metal factories from entering the sewers and contaminating Milorganite with heavy metals. Yet other Milwaukee industries were vital to the production of Milorganite. Milwaukee's investment in its Milorganite production capacity required a steady stream of sewage to produce the fertilizer. As Milwaukee industry began a long economic decline in the 1970s, the closure of polluting factories in the metal industry helped lead to a cleaner sludge, with lower concentrations of heavy metals. But as the giant breweries, Schlitz and then Pabst, closed, followed by malting and yeast production plants, the sewage treatment plant was robbed of much of the raw material for its product.[40] Commenters had often thought that Milwaukee's ability to market Milorganite was partly due to a fortuitous set of circumstances, including the particular mix of industries in the city. The city, famous for its beer, produced a great deal of nutritious waste that helped feed the microorganisms of the activated sludge process. When the breweries and yeast factories closed, all of that material was lost to the plant, and Milorganite production declined. Milwaukee tried to make up for it, in part, by pumping digested sludge from its South Side plant to Jones Island, to be made into Milorganite, as well as contracting with neighboring communities to process their sludge.[41] The latest solution to the problem of decreasing supply involves trading on Milorganite's name and reputation. For over seventy years, Milwaukee's marketing staff has built up and protected the name and brand of Milorganite and may find that the name is more valuable than its own sludge. Milwaukee is seeking to license its name to other cities' sludge production.[42]

Craft vs. Science

From 1957 to 1968, the Federal Water Pollution Control Administration invested $1 billion for more than eight thousand sewage treatment plants across the United States. While the federal government helped pay the capital costs of these plants, local governments remained responsible for their operation. With sewage treatment plants across the United States failing, the U.S. General Accounting Office (GAO) investigated the reasons for poor purification. Investigators mostly blamed sewage treatment plant operators; operators lacked appropriate training, they failed to perform laboratory analyses—in short, they were unwilling and unable to conduct appropriate process control.[43] The report outlined the same set of requirements for sewage treatment plant operators as in the 1910s and 1920s: "Operators should have a basic knowledge in the fields of chemistry, biology, microbiology, bacteriology, physics, and mathematics in order to accurately perform laboratory tests and interpret the results." Given the low salaries and lack of "social status" of operators, the report was not surprised that sewage plants lacked "competent individuals."[44]

For many, the problem was that operators were still using sight and smell in process control rather than laboratory tests. "The activated sludge system requires food, oxygen and living organisms in a delicately controlled environmental balance. The stereotype system design, operated by visual inspection and assisted by infrequent analysis, will not meet the needs of today's and tomorrow's quality control standards,"[45] wrote the assistant plant superintendent of Decatur, Illinois, in 1970. A renewed emphasis on operator training and scientific process control became the solution to the operations crisis.

As a number of investigators had determined in the 1950s, part of the problem was a lack of adequate process models and timely laboratory tests to run the plant. Developing new process control strategies was one of the priorities for the Federal Water Pollution Control Administration (soon to become part of the Environmental Protection Agency). The administration instituted a renewed program for addressing the process control issue. In 1969, it established a "task force" on pollution abatement. This task force was a group of ten engineers, biologists, statisticians, and economists that would be deployed "instantly" to sewage plants that were not performing well.[46] The team traveled to Denver, St. Louis, Chicago, Washington, and other cities, where its members would observe operation and make recommendations for adjusting the process, improving operation, and training operators. The task force thus tacitly acknowledged the importance of local "intimate" knowledge of sewage treatment that Ardern had intimated and that operators possessed. The engineers could not hope to understand the operation of a plant without spending a substantial amount of time "to discover capabilities and idiosyncrasies of the specific plant." The team spent up to four months at a plant to improve its operation. The structure of the program also recognized the local variability inherent in sewage. No city was like another, no sewage plant like another. Sewage varied over space and time.[47]

Under the guidance of sanitary engineer Alfred West, this task force helped increase appreciation for the operator's craft. West's appreciation for craft originated in the tension between the search for universally applicable control strategies and the recognition that variability and local specificity would make such a search vain. West had trained under E. B. Mallory at the Lancaster Ironworks research laboratory where he was exposed to Mallory's patented process control techniques. Mallory had tried to patent a universal procedure for operating activated sludge plants, and he continued to refine his techniques. But Mallory, working in industry, and concerned with patents and secrecy, had apparently never revealed his proprietary techniques (to West or anyone else) for interpreting his control tests to provide specific guidance for managing the sewage treatment process under varying conditions.

As part of the task force, West sought to recover Mallory's ideas on process control and build a robust and universal scheme for managing sewage treatment plants. For West, the universal nature of the control algorithms was central, and he delayed publication of his research for years because he recognized he was not yet able to demonstrate control-response relations that applied in all instances. Both the local specificity of individual sewage plants and the complex biology of sewage treatment defeated this goal.

Instead, West developed a stronger appreciation for the observational tests that operators could make to control their plants.[48] Perhaps because he recognized the importance of the kind of knowledge that could only be gained by spending time at a plant and experiencing its "idiosyncrasies," a central aspect of his approach to scientific control was a renewed respect for the operator's craft. "Much can be learned," wrote West, "from simple but perceptive sensory observation of process features." The type and color of foam, its extent on the surface of the tanks, the presence of scum or floc particles in the clarifiers could all indicate the condition of the treatment process, argued West. "From such observations, a skilled operator usually can determine the basic phase his process is moving towards or is locked into." In his treatise on activated sludge operational control, West presented a key to different process variables that could be elucidated through direct observation, systematizing the evocative terms developed by operators over the years—"straggler floc," "pin floc," "clumping," or "ashing"—and differentiating between a "fresh crisp white foam," a "billowing white foam," and a "thick, scummy, dark tan foam."[49] West acknowledged that "skilled experienced operators" recognized the relation between the final effluent quality and sludge quality. But he was writing to help teach the less experienced operator how to recognize the process conditions at his plant.[50]

West's treatise addressed one of the key problems with relying on experience and operator's craft. There needed to be a way to pass on this experience in a context of the rapid expansion of sewage treatment. In the nineteenth-century industrial context, parents would pass this knowledge on to their children who worked in the same craft. With the rapid expansion of plant construction, there was now a shortage of skilled operators, and the problem became one of how to teach inexperienced operators the skills that only came with experience. One of the concerns for scientific management was collecting this kind of knowledge so it would be in the hands of management. In contrast, West tried to systematize the knowledge so it would be available to new operators, not just managers.

West also emphasized the craft nature of the laboratory analyses as well. Even the "scientific" tests of the engineers required a craft developed from experience to conduct them well. The "settleometer" test (in which the ability of a sludge to settle out in a quiescent cylinder was measured) had been a mainstay of scientific process control for decades. But this test, West argued, required more than simply adding a sample taken from the clarifiers and reading a value thirty minutes later. "Operators who set up a test cylinder, walk away, and then return for a single observation after 30 minutes, miss most of the information that actually defines sludge quality," he reproached. West instructed that the operators should carefully observe the behavior of the sludge, especially at the beginning of the test. "During this first five minutes, the conscientious operator will critically observe how the sludge particles agglomerate while forming the blanket. He will see whether the sludge compacts slowly and uniformly while squeezing clear liquid from the sludge mass or whether tightly knotted sludge particles are simply falling down through a turbid effluent," West explained. "The importance of conscientious, critical, perceptive observation during the first five minutes cannot be over emphasized," he concluded.[51] West, like Abel Wolman and others before him, had a respect for the skills of the operator and blurred the marked distinctions between operator craft and scientific analysis that were common in the engineering profession.

Despite the efforts to improve the training of sewage plant operators, a 1977 GAO report once again laid the blame for failing sewage treatment plants on inadequate operation. Sanitary engineers once again focused on establishing scientific process control as the solution. When, in 1984, the federation instituted a new journal exclusively for operators, titled *Operations Forum*, improved process control was listed as the highest priority. The journal published numerous articles for training operators in process control, encouraging them to adopt scientific strategies for controlling the activated sludge process. Once again, despite the work of West, the sanitary science profession tried to discourage operators from using their craft to control the processes. An editorial noted that "some of you have acquired the ability to determine what is going on by seeing or smelling and know that a white to grey thin foam can mean that there has been insufficient time to degrade common surfactants, and a thick, pasty or slimy foam could be nutrient deficient." Nevertheless, despite this acknowledgment of the skill of the operator, the editor urged operators to "use more than your senses for process control."[52] This exhortation appeared in the federation's research journal as well. "Laboratories are no longer a mere appendage to a treatment plant," declared Karen Carter, the journal's associate editor, "but are integral to successful operation. . . .The days when the adequacy of treatment could be assessed by the color and smell of the water in the aeration tanks are long past."[53]

At the same time, the conflict between scientific and craft approaches to operation manifested itself in an escalating controversy between engineers and operators in the primary professional organization for people working in sewage treatment. The struggle for jurisdiction over the sewage treatment plant that had begun in the early twentieth century continued. Although operations personnel had originally been involved in establishing the Water Pollution Control Federation (called the Federation of Sewage Works Associations at the time), they found the organization, now dominated by engineers, scientists, and manufacturers, to be increasingly indifferent, or even hostile to their interests. In England too the sewage treatment organizations became more and more dominated by consulting engineers, academic researchers, and officials in governmental public health agencies, and the profession there became divided between supervisory personnel and operators.[54]

Plant operators found themselves in a situation common to many technical workers. In many industries, the knowledge systems of technicians are subsumed in the governing science or engineering fields, placing them in a subordinate position in professional societies. Scholars of professions and technical work have thus labeled these technicians as "uniquely passive" in the construction of their professions.[55] Yet this oversimplifies the situation in sewage treatment. Because of the large numbers of operators relative to more highly educated and credentialed members of the profession, technicians also possessed greater power in professional societies than has been previously thought. In sewage treatment, operators demanded and received changes in the federation.

Starting in the 1950s, operators in many states began organizing "operator associations" to fill the void left by the changing priorities and makeup of the federation. Operators began to develop their own organizations, focusing on training, exchange of operations experiences, and fellowship. Over the succeeding decades, this trend accelerated, and by the 1980s it threatened the dominance of the federation in the pollution control field. Indiana's experience was typical. Operators became dissatisfied with the Indiana Water Pollution Control Association (IWPCA), the local affiliate of the federation. Operators complained that meetings of the IWPCA "which started out as simple primers on plant operation gradually branched out into areas of more scientific and technical concern, leaving behind the operators' basic needs." Publications, like Indiana's *Sewage Gas*, that focused on "simple, instructive, unsophisticated" advice for plant operations had morphed into research journals for academics. The state organization, these operators concluded, "was primarily for engineers." In 1969, operators in northern Indiana formed the first regional splinter group, with groups representing the state's other regions proliferating over the next decade. In reaction, the IWPCA formed an ad hoc Operators Unification Committee to respond to the threat from these independent operators associations. But by the early 1980s, the IWPCA found its membership declining.[56] Experiences like Indiana's, repeated across the country, threatened the position of the federation. Officials of the national federation began to warn that the over 100,000 operators in the country would form their own national organization, threatening the federation's finances as well as its prestige and political clout.[57] It might also threaten the sanitary engineering profession's control over the sewage treatment plant and its operation, which had been contested since the beginning of the century.

Of the 30,000 members of the federation in the early 1980s, only 6,700 were operations personnel. But even these were mostly managerial or supervisory staff, the federation president warned, not the "on-line, valve-turning, laboratory, collection systems, maintenance personnel" that represented the vast majority of workers in the sewage treatment field.[58] These operators, the federation found, believed that "the Federation is not for operators. The Journal was not for operators . . . The Member Associations are not for operators. There is no operator recognition." The federation concluded that in the face of national organizing by operators, there was not enough time to change operator perceptions; they would have to provide the services the operators demanded.[59]

To get the operators back in the fold, the federation recognized the importance of professional identity and established a Professional Wastewater Operations Division (PWOD) in 1984. It got off to a rocky start. The federation was, from its origin, a federation or association of state or regional organizations; members joined a state organization first, which provided access to the national federation. These "member associations," dominated by engineers and managers, worried that the PWOD would weaken their influence in the national organization, and many refused to allow operators to bypass their local association to join the national organization through the unaffiliated local operators' associations that had arisen to represent operators' interests. Rather, they would have to join the already-affiliated member associations that they controlled. Operators in many regions were angered, protesting the exclusion of their operators' organizations, which were in many cases better organized than the member associations they were forced to join.[60] With membership of the new PWOD much lower than expected, the federation was forced to allow operators to join the federation directly, without going through the local associations.[61]

Operators also complained that the main journal of the federation was irrelevant to their needs. The journal, which had previously included a great deal of material for the operator, including operations reports, "tips and tricks" for operating a plant, and a question and answer section, was now dominated by research that was perceived to ignore "the practical every day real world problems" of the operator.

The federation was publishing a popular newsletter for the operator, called *Deeds and Data*, that included much of the information on plant operations that had previously appeared in the journal. To convince operators of the value of federation membership, it expanded *Deeds and Data* to a glossy magazine expressly for the operator called *Operations Forum*.

Another popular innovation was the "Operators' Challenge," or "sewage Olympics," as it was widely known. The federation established this competition at its national meetings. Teams of operators from various cities would compete in a "pentathalon" of events that highlighted the various skills needed by operators to successfully manage their plants. In the first competition, operators competed in laboratory techniques like calibrating a pH meter or identifying "activated sludge microorganisms using a microscope." They also had to demonstrate mechanical skills by repairing a pump and patching a chlorine leak. Finally, they had to demonstrate their ability to perform process control. They were required to "troubleshoot an operation problem in an activated sludge wastewater treatment plant given appropriate available information."[62] Designed to "increase visibility of and appreciation for the individuals and the expertise involved in wastewater treatment plant operations," the sewage olympics were extremely popular among operators.[63] But while the competition offered respect to the mental and manual labor of operators, it also simultaneously transformed their work into spectacle, and it reveals the complex position of the operator in the professional society.

The expanded efforts of the federation to stem the loss of operators was successful. However, in recent years, with the advancement of scientific process control, automation, the application of computers to process control, and the outsourcing of operations, operators are finding themselves in a weaker position that may also be reflected in a retrenchment in the respect garnered by their craft approach to process control. The most recent *Manual of Practice* (the comprehensive and authoritative publications of the federation) on the operation of activated sludge plants, for instance, has deleted its recognition of the value of observational methods of process control. The 1996 edition of the federation's *Manual of Practice* declared, in discussing color, odor, foam, and other visible qualities of the sludge, that "qualitative measurements of such characteristics may often better reveal process performance (and be more timely) than some quantitative measurements." Even this carefully qualified statement acknowledging the value of the operators' craft was removed from the most recent 2008 version.[64]

In addition, even the widely popular *Operations Forum* was discontinued. There was widespread disappointment when the federation, now the Water Environment Federation (WEF) canceled its publication, and the long history of suspicion between engineers and operators resurfaced. An open discussion on the federation's web forum on operation and maintenance raged for almost three months as operators responded to the loss of *Operations Forum*. One operator contrasted "those of us who actually operate the plants" with the managers and supervisors "that some operators think of as largely parasitic." Another charged that "for years, engineers, salespeople, consultants and managers have stood on the backs of operators." Another remarked that "WEF was geared to consulting engineers first, equipment manufacturers second, and operators a distant third . . . With the exception of Operations Forum and this web site the WEF has done little for operators. Even the operators' challenge at WEF exists, in my opinion, because of corporate sponsorship." The rifts and resentments between the operators and other elements of the federation had quickly resurfaced, and the professional identity of operators was central. "Since these utilities, almost without exception, are run by engineers," wrote another comment, "it's not surprising that the prevailing attitude continues to be disdainful of the operator's profession. This is despite the fact that many operators scattered throughout the country DO have 4-year college degrees, some with advanced degrees (they're just not 'engineers'); hence, operators are not 'professionals'(?)"[65]

Even though operators had made strides within the professional organization, it was clear that they did not feel completely empowered within it or respected by it. For this operator, and many like him, the professional identity of the operators was key, and the key aspects of professionalism were respect and autonomy. Here, autonomy meant autonomy of the professional organization, but it might well have meant autonomy of the operator himself. Central to their conception of professionalism was the ability to make independent decisions, precisely the ability that was being increasingly eroded.

Since the 1980s, as the design of the activated sludge process became more complicated with the addition of nutrient removal systems, as engineers and sanitary scientists developed a greater understanding of the biology of the sewage treatment processes, and as limited municipal budgets and privatization of sewage treatment plants pressured for a reduction in operating budgets, research increasingly focused on the development of automated approaches to process control.[66] As the sewage treatment establishment was removing its recognition of observational methods of process control, it was simultaneously appropriating these same methods to improve its ability to automate process control.

Automatic control relies on instrumentation for measuring process conditions and an automatic controller for adjusting sludge and air pumps to keep process variables within predetermined values. Automation had been tried much earlier in a very limited capacity in sewage treatment plants, mostly to adjust flow. In the 1930s, engineers at Chicago's North Side plant had installed an electric eye in the clarifiers to detect the height of the sludge blanket (the depth to which the flocculent sludge settled). Wasting rates were controlled by this electric eye to keep the blanket at a fixed height. In general, however, process control was based on laboratory or visual assessment of conditions and manual adjustment of air supply or flow, and automatic control lagged behind many other process industries like petroleum refining.[67] The chemical industry had adopted automatic control of their processes long before sewage treatment, despite the similarities of process between the bulk processing industry and wastewater.

Even as engineers began developing automated approaches, there were a number of factors inherent in the treatment process itself that limited the penetration of automation into the sewage plant.[68] Like most ecosystems, sewage treatment plants are not at equilibrium. There is constant variation in key state variables, so that bacterial populations and activity are constantly adjusting to new conditions. Second, the sewage treatment process is complex and poorly understood from a theoretical perspective. Even theoretically sophisticated models are necessarily approximations, based on incompletely understood bacterial kinetics. Finally, the objectives of process control are context-dependent. For instance, the objective during normal plant operation may be the production of the highest-quality effluent, but during large storms or other events operators may change the objectives temporarily to deal with the transient conditions. These difficulties applied to complex industrial ecosystems in general. The fermentation industry likewise encountered difficulties in automation. "Why has the fermentation process been so difficult to operate?" asked several process control engineers in 1983. "The answer is that the living organisms and their metabolic processes add another level of complexity."[69]

Most attempts at automation in sewage treatment focused on the simpler hydraulic or chemical processes within the plant, such as adding chlorine or adjusting flow rates, but little in the way of control of the more complex biological processes. Early researchers thought that automatic control was hampered primarily by technology—development of reliable automatic sensors for pH, dissolved oxygen, or nitrate, for instance—and "suitable engineering understanding" of processes. Once the sensor technology was available, they thought, it would be fairly straightforward to implement automation. But even with improved technology, automation failed to take hold. By 1995, reported one leader in automation, the "use of quality parameters for the automatic control of sewage works is still not widespread."[70]

Labor issues were one reason. Automation had "not been fully accepted by the user community," wrote one commentator, over concern over job loss and the lack of skilled technicians required to maintain process equipment.[71] Others blamed the lack of automation on the public nature of sewage treatment works, immune from the competition of the private sector.[72] But perhaps the most important reason was that automation had an extremely poor track record; it simply didn't work very well. A number of early systems failed spectacularly. Los Angeles spent over $11 million for control systems specially designed and built for four of their wastewater treatment plants. One system failed after only three years of operation because parts were no longer made for the computers. Only by scavenging working parts from the failed system could a second system stay in operation. Yet a third system never worked at all. Operators had never accepted the systems because they quickly learned that they could not trust the data produced by the equipment.[73]

There was also a growing recognition among researchers into automation that their automatic systems were failing not because of technical shortcomings of equipment but rather fundamental shortcomings of their whole approach. Automation failed because it did not take into account the craft knowledge of the operator. "Any operator is aware of the dynamic character of a wastewater treatment plant simply by observing time-varying flow rates, concentrations and compositions," remarked a leading researcher in automation. "A person's ability to look, smell, feel and hear a lot of phenomena in a biological process is superior to any sensor," he continued.[74] Developers began to look to the behavior of human controllers—the operators—and sought to incorporate the qualitative nature of their decision-making processes in the structure of their automatic control models. Engineers designed systems to explicitly take operator expertise into account. One group of engineers developed numerical control programs that would mimic the ability of human operators to learn from past experience and "re-use" knowledge.[75] Another group recognized that even though human operators' "know-how is essentially qualitative, empirical and incomplete," operators were still more effective than automatic controllers, because conventional control approaches required accurate models of the complex systems, a characteristic they considered "uncommon in real processes."[76] Other engineers developed what were termed "fuzzy logic" models to incorporate the experienced operator's mode of analysis.[77] "Fuzzy logic" is a term that exemplifies the engineer's complex relationship to the operator's craft. It is a term at once both perjorative (as in "fuzzy thinking") and complimentary, a recognition of the more complex capacity of human reasoning compared to a computer's on-or-off logic. Yet these researchers recognized that just such fuzzy reasoning was required to control complex systems.

Not just the operators' reasoning processes but also their observational skills were recognized by yet another automatic control system. This system incorporated "visual information describing the state of the plant such as water color, foaming, and odors." Operators were to input data on these "critical parameters" into the computer database that the control system would use.[78] Nevertheless, even these newer approaches to automatic control have not been universally successful. By one 2001 estimate, fully half of all automatic control loops in U.S. plants that year were set to run in manual mode.[79]

As engineers have developed a better understanding of the microbial processes, sensors have become more reliable, and computing power has become essentially free, automation has been making rapid inroads in wastewater treatment plants. Nevertheless, because of the wide variation in the quality of the influent to a wastewater treatment plant, there are commonly disturbances that exceed the ability of the automatic systems to control. In these cases, "other kinds of control actions are mostly needed, where the operator makes manual adjustments of the process."[80]

Process engineers continue to search for computer tools that can replace the skills of the plant operator. Their recognition of the role of the operator is clear. "In reality," one group of researchers wrote, "human operators observe the values of several variables obtained from a WWTP, and then perform integration and analysis . . . via heuristic reasoning based on their own every day operational experiences." The very subjectivity of the decisions, how they are based on skill and past experience, is both their strength as well as, according to these engineers, their weakness. The role of human experience is something to be "overcome." Yet, as they acknowledge, "the technology is not available for the automation of inference procedures based on empirical knowledge of the human operator." Their inability to do this has become the "bottleneck" in implementing automatic control.[81] The flexible, context-specific decision making that is a hallmark of operators' craft processes remains a necessary part of sewage treatment plant operation.

Most recently, WEF's conferences and the pages of WEF's journal and web site have been filled with the problem of "succession planning" or "knowledge management." As baby boomers are approaching retirement, sewage treatment plants are facing the loss of a generation of knowledge about how the plants work. Many recognize that computers and information management systems cannot replace this skill and knowledge, and encourage plant management to implement mentoring programs and on-the-job training. Faced with the loss of decades of experience, plant managers recognized the importance of craft knowledge. "A lot of the work that goes on is based on a little tweak here and a little tweak there to get the plant to operate at its absolute optimum. . . If you're not sure what those tweaks are—especially seasonally or when upsets [occur]—when a new person comes in, they've got to learn that all over again, because a lot of it is not well documented," explained Steven Agor, president of the California Water Environment Association.[82] Yet, despite acknowledgment of operators and their craft skills, the ideology of process control and rationalization of the plant persists.

Public vs. Private

By the 1970s, with the promise of huge spending by the federal government for pollution control, a number of firms went into the market for supplying equipment. One such firm was Envirotech. Envirotech was formed specifically to take advantage of the increased interest in and spending on pollution control that was evident in the 1960s and early 1970s. Formed in 1969 "from the corporate flotsam and jetsam" of conglomerates seeking to divest themselves of marginal divisions, it quickly became one of the largest wastewater equipment manufacturers in the world, and analysts singled it out as one of the most promising investments in the pollution control field.[83]

There were a number of obstacles however, to the long-term growth of pollution control firms. Most municipal wastewater plants were bid on a fixed-cost basis. As the lead time for constructing plants increased, profits were eroded by inflation. Further, while enormous sums were promised by the federal government for construction, a far smaller amount was actually committed. Finally, the long-term growth of pollution control firms was limited. Once the huge capital expenditures under the Clean Water Act had been made, there was a declining rate of construction of new wastewater treatment plants.[84] As a result, manufacturers and industry analysts alike began to view the wastewater equipment field as a difficult area for investment, despite the huge sums of money being spent by the federal government for treatment plant construction. "Cleaning up is a grubby business with low profit margins, low return on equity and high demands on capital," commented one analyst.[85]

One final difficulty derived from the struggles over patents and the public interest in the early part of the century. While the patentability of sewage treatment processes had long been settled, there remained lingering antipathy toward patents in sewage treatment. The Clean Water Act of 1972, for instance, specified that no bids for sewage plants built with federal funds could require "proprietary" processes or equipment, except under extraordinary circumstances. Even then, the bid had to specify at least two trade names and the phrase "or equal." These regulations restricted the ability of patent holders to profit from their patents since they couldn't be uniquely specified in contracts. Financial analysts cited these provisions in cautioning investors about the limited profit potential of the environmental control industry. Testifying before Congress in 1979, an Envirotech official blamed these procurement regulations for the company's own withdrawal from the sludge incineration market, which had accounted for 10 percent of its revenue.[86] Indeed, soon after his testimony, Envirotech began divesting itself of its pollution control equipment businesses.

To bypass both the structural problems of the pollution control industry as well as the restrictions on proprietary processes and equipment, Envirotech emphasized a new arena for privatization and profit: the operation rather than the construction of sewage treatment plants. Riding the fallout of the GAO reports on the poor operation and maintenance of wastewater treatment plants, Envirotech started a division to provide operation, maintenance, and training to municipal sewage districts. Envirotech Operating Services (EOS), as it was called, argued that it could operate plants more efficiently than public agencies, train operators more effectively, and save municipalities money as well. EOS signed the first private operations contract in the United States in 1972, to operate the Burlingame, California, sewage treatment plant. This was the first instance of the privatization of sewage plant operations. Over the next eight years, EOS continued to grow, and by 1980, it was operating over eighty plants across the United States.

Operating contracts had a number of advantages from a structural perspective. Even as the construction of wastewater treatment plants slowed in the 1980s, there continued to be a large and growing market in operating already-constructed plants. Further, operations contracts could conceivably continue indefinitely. In 1987, public expenditures on treatment plant operation and maintenance first exceeded capital expenditures in the United States; by 1995, they were almost double.[87] Financial analysts emphasized this innovation in marketing operations services rather than construction as one reason to invest in what they considered to be an otherwise lackluster industry. Other equipment firms also entered the services market including Wheelabrator, CH2M-Hill, OMI, and others. By 1997, 736 wastewater plants in the United States were privately operated.

Entrepreneurs in the private operations industry linked the contradictions between craft and science and public and private. "Plants are in the hands of highly skilled, private-industry professionals," they boasted.[88] This phrase sought to contrast the operator under privatization with the historical image of the sewage plant worker as a former gas station operator[89] or "high school dropout." In fact, because of both the power of municipal worker unions, and the continued need for the locally specific expertise developed by the operators, the "private-industry professionals" touted by the private operations firms were the exact same operators as before, rehired by the new company. Only there were fewer of them.

Other companies that sought to enter the private operations market were stymied by the same kinds of conflicts between operators and engineers, craft and science, that had arisen in the Water Pollution Control Federation. Engineers at CH2M Hill, a prominent consulting engineering and construction firm, were trying to enter the private operations field, but the firm initially resisted the idea because of the potential for "internal conflict between 'professional engineering' and 'blue-collar' businesses." Part of this concern was the fear that if it had to hire operators as part of the firm's services, organized labor might gain a "foothold" within the company.[90] Engineers have often held a tenuous position in the workplace. Most engineers are employees of large organizations, with many of the same interests as other employees.[91] CH2M Hill feared that "blue-collar" workers might infect these engineers with working-class sentiments.

Privatization of sewage services had been a common approach to sewage treatment in the 1850s, when cities, anxious not to increase taxes, turned their sewage problems over to firms like Native Guano and others. But the private sewage companies of the nineteenth century had floundered on the dilemma of profit or purification, and ever since, municipal sewage treatment had been considered a fundamental responsibility of local governments. In all the discussion and conflict over municipal trading, there was little controversy over municipal sewage plants. It was considered a given that municipalities would construct and operate their own sewerage and treatment operations. As late as the 1950s, sewage treatment experts considered it impossible that municipalities might privatize their sewage treatment responsibility, because they saw it as a fundamental responsibility of the public sector. The editors of *Wastes Engineering* were "practically unanimous" in the "disapproval of the idea that private investment in sewage works projects might be . . . feasible and workable." They "assailed" the privatization of sewage treatment as "a means of dodging such social obligations by government."[92] However, as cities faced increasing budget constraints, and the ideology of the free market began to expand, the opposition to privatization began to fade.

Despite the increasing privatization of operations through firms like EOS, municipalities in the United States continued to own their sewage treatment works. In England, however, the plants themselves were privatized in a mass liquidation of publicly owned water and sewage treatment works under the Thatcher government. In the early 1970s, all of the separate, locally run sewage and water services in England had been nationalized and geographically reorganized according to large watersheds. Following nationalization, though, the British government had severely limited investment in infrastructure. By the late 1980s, there was over £24 billion in delayed investment needed. The Thatcher government, ideologically committed to neoliberal policies, saw privatization as a solution to the capital needs of the industry. By selling the assets of the water and sewer service authorities, they could transfer the investment needs to the private sector. In 1989, the public water and sewage authorities were floated on the stock market, where they sold for £3.6 billion.[93]

Within a decade of privatization, however, the British water service companies faced a period of restructuring, with several companies seeking to abandon private ownership of water and sewage plants. In her analysis of the political economy of the water industry privatization, Karen Bakker points to the multiple contradictions created by the inability to completely commodify water and sewer services. She attributes the pressure for restructuring on the British government's review of price caps for the industry in 1999, after which they forced the water and sewerage companies to reduce their prices. Up to this point, the industry had been able to achieve profits far higher than anticipated and greater than similar industries in other parts of the world. Because of public outcry over the high prices and high profits, the British regulators were forced to cut prices charged by the private companies. But these price cuts struck at their bottom line, pushing the industry to find other sources of profit. "The political unacceptability of higher profits and commercial unacceptability of high losses introduces a contradiction," Bakker argues, that threatened the structure of the privatized industry.[94]

The contradictions of capital inherent in the privatization of sewage treatment plants propelled the implementation of automatic control. Pressures on profits forced the British companies to lay off experienced operators and rely more on automation as a cost-reducing strategy. In the ten years following privatization, the water and sewerage sectors of the companies reduced their workforce by over 20 percent.[95] Yet automatic process control was not yet very successful; control still required the years or even decades of experience possessed by just the operators being let go. The water service companies attempted to resolve this contradiction by appropriating operator experience, finally implementing in sewage treatment Taylor's first principle of scientific management—the "deliberate gathering in on the part of those on the management's side of all of the great mass of traditional knowledge."[96] At two of the water service companies, Thames Water and North West Water, researchers interviewed experienced sewage treatment plant operators to incorporate their site-specific knowledge and experience into expert systems for process control.[97]

The British water companies had limited opportunities for growth, with nearly every household in the country connected to centralized water and sewer services. Further, each company's licensed boundary for operation was defined by watershed boundaries and could not be expanded. Because natural boundaries placed limits on profitability, increased profits had to come from increasing rates rather than increasing customers.[98] Yet with the new Labor government reducing the rate of return, the water and sewer companies were squeezed. One response to this contradiction was for the water companies to diversify their holdings to remove themselves from the conflicts between price and efficiency incentives. Their options were either to internationalize to countries lacking price controls or diversify into activities that were not regulated by the central government. Some water and sewer companies diversified into other unregulated utilities, like electricity and gas, others into solid waste. Still others, recognizing the structural advantages of contracting operations, entered the private operations market. Severn Trent entered the U.S. market in 1990, when it established a subsidiary to operate U.S. sewage treatment plants. Other British firms quickly followed, and by 1997, the private British water and sewer service companies operated over 150 sewage plants in the United States.[99]

The British firms were following the same line of reasoning as the American equipment manufacturers and engineering firms. By the early 2000s, this logic also pushed several of the British firms away from owning their sewage works. These companies had found their financial situations untenable and sought to return their regulated water and sewage treatment functions back to public control. In these proposals, the water industries proposed to "hive off" the unprofitable infrastructure assets and sell them back to the public. They would maintain, however, the profitable contracts to operate the water and sewage treatment facilities. Although profit margins would be lower, the capital requirements would be substantially reduced because they would have few assets.[100]

In the United States, President George H. W. Bush sought to emulate the UK privatization program. However, the situation in the United States was different, posing a number of obstacles. Prior to privatization, the British had already consolidated and nationalized their treatment plants run by some 1,300 local councils into ten regional water authorities in the 1970s, making the transfer to private hands much easier.[101] In the United States, there were over 15,000 publicly owned treatment plants in 1992, each operated by local governments.[102] Further, financial regulations that required repayment of federal funds used to construct the plant made them poor targets for acquisition. In response, in 1992 President Bush issued Executive Order 12803 that relaxed the repayment requirements, making it easier for municipalities to sell public assets like sewage treatment plants. Subsequently, a number of U.S. cities began investigating the sale of their treatment plants. In 1995, Franklin, a small Ohio city, took advantage of the executive order, and sold its sewage treatment plant to its long-term private operator, Envirotech, by this time a subsidiary of Wheelabrator. For a variety of reasons, however, Franklin has been the only city in the United States to take advantage of the incentives for full privatization. The city of Wilmington, Delaware, investigated selling its treatment plant, but ran into obstacles because of the complex nature of the local jurisdictions served by the plant. Although the city owned the plant and would profit from its sale, most of the sewage came from the surrounding county, which would receive no benefit from the sale and objected.[103] Great Britain, which had regionalized its sewage authority thus ran into fewer of these kinds of problems.

Another disincentive for privatization was a provision of the Clean Water Act termed the "domestic sewage exclusion." Under this provision, industrial waste that mixes with domestic sewage on its way to a publicly owned treatment works (POTW) is exempted from hazardous waste regulations. The exclusion does not apply, however, if the treatment plant is privately owned. Private treatment works are subject to the far more rigorous hazardous waste treatment standards under the Resource Conservation and Recovery Act. To bypass this problem, Franklin actually maintained ownership of just enough of its plant to keep its designation as a POTW. The plant was thus a chimera of public and private ownership; some of the pipes were owned by the city, others by Wheelabrator.[104] But overall, losing preferred status as a POTW prevented most cities from selling their sewage treatment works assets.

In 1995, Milwaukee began investigating the possibility of either selling its sewage treatment plants or contracting out their operation. Milwaukee had just completed a $2.1 billion construction project, and district officials thought the time was right to sell the assets: "It's at its maximum cash value," said one official.[105] The Sewerage District decided, however, that rather than sell its assets, it would contract out the operation of its treatment works. Milwaukee chose United Water Services, a subsidiary of Suez Lyonnaiase des Eaux, a French multinational that had previously bought one of the English water and sewage systems. United Water claimed its contract would save the district over $13 million per year in operating costs by reducing the number of workers, automating the plants, and entering into long-term energy contracts. Concerned over layoffs, workers at the district opposed privatization until United Water guaranteed that there would be no layoffs, wages and benefits would not be lowered, and workers could keep their city pensions. When Milwaukee signed the privatization contract, the district became the country's largest privately run sewage treatment works.[106]

While privatization of operation did save the Sewerage District money, United Water itself was apparently losing money on the contract because of rising energy costs. With its profits being squeezed, there were concerns that the contractor was cutting corners on maintenance and other responsibilities.[107] Operations problems plagued the sewage treatment works. The district blamed United Water, and, after only four years, called for a review of its ten-year contract.[108] When the next year, hundreds of "spent" condoms were found floating in Lake Michigan, pressure on United Water to improve its operations increased.[109] The performance review concluded that, while United Water was, in general, performing well, there were not enough incentives for United Water to spend sufficient funds on maintenance of the system (which was still owned by the Sewerage District).[110]

When it came time to renew United Water's contract, the district rejected the company's bid. The remaining finalists were Veolia, another French multinational company, and the district's operators themselves in a bid to return operation to the public. In a controversial decision, the district selected Veolia. The decision to contract with Veolia was controversial in Milwaukee. Over a thousand residents petitioned the district to return operation to the public. But even though the unions' proposal was lower than the rejected United Water's bid, it could not come in below Veolia's.[111] Union officials charged that Veolia purposefully undercut the union by not including any profit for the company in their bid and questioned whether Veolia would be able to live up to the terms of its contract.[112] Milwaukee's operating contract was considered a plum, allowing bragging rights as the largest sewage treatment plant under private operation.[113] Just four days before Veolia took over operation of the district's plants, vandals apparently sabotaged the Milorganite production plant. Police had not determined whether it was related to the change in contractors.[114]

The contrasting fate of the Milorganite marketing division with plant operations reveals an irony between public and private operations. Milwaukee had rejected privatization of sludge marketing. It never hired a firm like Synagro to handle its sludge, and in the privatization agreements with both United Water and Veolia the Milorganite marketing staff remained public employees of the Sewerage District. In a curious contradiction, the treatment functions, most critical to the public interest and health, were privatized, while the business functions that most resembled private industry remained public.

While Milwaukee did not return its operation to the public, a number of sewage treatment plants have. In 1997, there were at least thirteen companies operating 736 wastewater treatment plants in the United States. By 2007, after a series of industry consolidations, there were only seven firms, operating 1,052 plants.[115] City managers, like those of a century before, were looking at the prospect of monopoly in urban technology. With the rapid consolidation of the services industry and the resulting lack of competition, they feared that costs would increase substantially. In 2008, both the Fairfield-Suisun and Petaluma treatment plants in California reverted to city operation. Yet the logic of privatization has permeated even public operation. When Charlotte, North Carolina, considered privatization of operation, employees submitted a successful bid themselves, but only "by taking the approach that they should think and bid as a private entity."[116]

Natural vs. Artificial

In the 1970s, with the ecology movement's developing critique of energy-intensive industries and an increasing interest in "appropriate," small-scale technology, sewage engineers revived the distinction between natural and artificial sewage treatment, so characteristic of the development of biological treatment in the late nineteenth century. The biological sewage treatment processes developed from the 1890s to the 1910s had successfully taken over the mantle of "natural" from the older land treatment processes. But by the 1970s, the industrialization of sewage treatment had obscured many of the natural characteristics of the biological processes.[117]

In their criticism of the industrial nature of the modern sewage treatment plant, late twentieth-century authors mimicked Poore's critique of the nineteenth- century sanitary apparatus. One author contrasted the "energy- and chemical-intensive processes" of conventional (i.e. biological) sewage treatment to natural systems. Conventional sewage treatment processes, she wrote, "need energy to power massive aerators, to run mechanical mixing and scraping devices, and to fire sludge incinerators. They need chlorine to disinfect the waste-water, or electricity to generate ozone or ultraviolet radiation. Finally, they need large volumes of concrete and steel" and occur at an "artificially accelerated pace."[118] "All waste management processes depend on natural responses," wrote other researchers. But in conventional systems, "these natural components are supported by an often complex array of energy-intensive mechanical equipment."[119] Proponents of land treatment in the late twentieth century recycled precisely the same arguments as irrigationists at the turn of the nineteenth century. Because they recognized, like earlier advocates of bacterial treatment, that activated sludge was based on natural processes, proponents of natural methods had to redefine what was natural and what artificial. Sherwood Reed, a leading exponent for the revival of land treatment, somewhat vaguely defined natural treatment systems as those processes "that depend primarily on their natural components to achieve the intended purpose."[120]

In phrases that might have also come directly from the nineteenth century, the administrator of the U.S. EPA reasoned from natural theology to support land treatment. "Wastewater management policies," he declared, should be "consistent with the fundamental ecological principle that all materials should be returned to the cycles from which they were generated."[121] As in the nineteenth century, "nature" was mobilized to support particular technologies. Scientists, engineers, and activists who were informed by the ecology movement began to increasingly see these biological methods as *un*natural and tried to steer sewage treatment toward processes they considered more ecological (and thus natural).

In addition to reviving older "natural" methods like land treatment, engineers influenced by the ecology movement began to develop a variety of systems for treating sewage that were explicitly ecological and "natural" in their motivation and design.[122] Using aquatic plants, artificial or natural wetlands, or various types of lagoons, these systems all tried to create complex ecosystems for treating sewage. The development and reception of one particular treatment process, dubbed "the Living Machine," reveals the complicated mobilization of the term "natural." Described in its current advertising as "treating wastewater nature's way,"[123] the Living Machine was originally developed by ecologist John Todd at the New Alchemy Institute in Cape Cod, Massachusetts. Coming directly out of the emerging ecology movement of the time, the New Alchemy Institute was founded in 1969 by John Todd, Nancy Todd, and Bill McLarney with the explicit goal of "looking to the natural world for the clues to develop a science and supporting technologies in its image, mimicking its materials, processes, and dynamics."[124] John Todd was trained as a behavioral ecologist but became disenchanted with what he considered to be conventional "doomwatch" ecological research. With his cofounders, he sought a way to apply his ecological training to help sustain human society rather than simply document its collapse. Informed by the back-to-the-land movement of the late 1960s, they founded a community in Cape Cod, where they worked on the development of aquaculture, organic farming, and solar energy designs.[125]

Their revival of "natural" waste treatment processes began with a critique of conventional biological sewage treatment as an essentially artificial process. Todd characterized the "conventional waste treatment industry" as "one of the major environmental destroyers," addressing one problem but in the process creating a multitude of others. Pollutants that the public sought to regulate, Todd explained, "were being treated with unregulated compounds that were far more dangerous. For example, when the citizens demanded that phosphorus be removed from waste water, industry poured in aluminum salts. That got rid of the phosphorus, but now we have everything from weakened forests to Alzheimer's because of this huge infusion of toxic aluminum in the environment." Adding chlorine, he continued, produced chloramines, "making every sewage plant, in effect, a carcinogen factory."[126]

Todd based his initial designs for waste treatment on the fish aquaculture systems the group had developed to produce food. After observing the ability of the aquaculture tanks to process waste, he began to develop sewage treatment systems that used a series of tanks filled with bacteria, algae, plants, and fish. Todd seeded his tanks with "hundreds of species—ranging from small trees to fish to microorganisms from the anaerobic world." He thought of the process as one of "assembly": creating a new ecosystem. He called his technology a "solar aquatic waste treatment plant." "Nature," said Todd, had an "extraordinary, dynamic purifying power . . . if the right organisms can be found to work in the right kind of concert together."[127]

Despite his grounding in the back-to-the-land movement, an important consideration for Todd was the commercialization of his technology. Todd wanted solar aquatics to be widely adopted, and recognized that it would not only have to be economically viable but also commercially feasible. "First, I was involved in establishing a business to take these ideas out widely. The old idea of just letting this information be assimilated slowly by the next generation of students, from whom it would spread into the academic world, engineering firms, and eventually into society, just wasn't fast enough—that takes at least twenty years. The only way I could think of compressing the process was through the corporate arena." Todd patented his design as the "solar aquatic apparatus for treating waste." He established a company, Ecological Engineering Associates (EEA) to commercialize his technology, to whom he assigned the patent.[128] Although not a partner in the firm himself, he had great ambitions for it. As described by Todd, EEA "has created what it calls a 'clean water service'—they build and operate a facility that turns waste into clean water, which it puts back into the town's ground waters. That's a whole new concept, and it's my hope that it will spread across the country very quickly—like McDonald's franchises."[129]

Solar aquatics and EEA ran into a number of problems, however, problems that centered on the continued tension between the natural and artificial elements in sewage treatment technology. The first municipality to install a solar aquatic plant was the small town of Ashfield, Massachusetts.[130] The ensuing controversy over the plant's operation and cost split the few hundred residents of the town over their vision for the sewage treatment plant and the central contradiction of the industrial ecosystem: was it, or should it be, natural or artificial? One of the primary proponents of the plant was town resident and landscape architect Harry Dodson. Dodson, a member of the town's Board of Health that recommended the sewage plant design, criticized conventional sewage treatment plants as part of the "sewer industrial complex" (figure 5.2).[131] Visiting one of EEA's pilot installations, Dodson was beguiled by using "the processes of nature to purify water."[132] Another proponent, writing in the town newspaper, laid out the advantages of a solar aquatic plant in terms that also incorporated a critique of the artificial nature of conventional waste water plants. "It does not rely on expensive, energy consuming machinery to process waste. It discharges clean water and does not dump pollutants into the soil. It's not ugly and doesn't smell or make noise."[133]

The Board of Health recommended that EEA be hired to design a solar aquatics plant, and the town voted overwhelmingly to build the plant. The town hired EEA to design the solar aquatics portion of the facility, but in a complex arrangement, it contracted with a conventional sewage engineering firm to design and build the overall project. Tensions between the ecological and conventional aspects of the design would create a number of future problems. With financial support from the Massachusetts Department of Environmental Protection and the U.S. Environmental Protection Agency for innovative sewage treatment technology, the solar aquatics plant was completed in 1996. It consisted of two greenhouses enclosing sixty-four solar tanks and four marshes, supporting a diverse community of water hyacinth, elephant ear plants, fish, freshwater clams, and bacteria and other microorganisms. The operator of the plant stocked the greenhouses with other animals and plants, fertilized by the wastewater, to give it a tropical feel.

From the beginning, however, the plant was plagued with construction cost overruns and operational difficulties. With only 160 households to share the cost of the plant, sewer charges quickly soared to become the highest in the state of Massachusetts, averaging $575 per year in 1998. With costs so high, over 40 percent of the town's total sewer bill went unpaid, and the town was forced to file property liens on households who were unable to pay their bills. Townspeople divided over the sewage controversy, with one group of citizens blaming the costs on the ecological nature of the plant and the other on the "sewer industrial complex"—namely, the conventional engineers who built it and the state regulatory agencies who imposed ruinous and unnecessary design constraints. In 1999, townspeople sought relief in the form of a grant from the state and federal governments for "corrective action." One of the conditions for the grant, however, was a reassessment of the plant's design and operation. As part of this "corrective action phase," the conventional engineers responsible for the facility identified the vegetation itself as the culprit for cost overruns and operational difficulties. "Higher than expected labor

Figure 5.2
"Filter cake cylinders, looking southwest, Jones Island Wastewater Treatment Plant." The waste treatment plant as the "sewer industrial complex." This machinery was used to dry the activated sludge as part of the Milorganite production process. The ecology movement developed a critique of the activated sludge process as too energy intensive, requiring power to compress air for the aeration tanks, to pump the sewage through the plant, and to dewater and heat the sludge, leading to a revival of "natural" processes for sewage treatment. This photo is part of the National Park Service's Historic American Engineering Record that documents significant engineering structures and techniques in the United States. *Source*: U.S. Library of Congress, HAER WIS,40-MILWA,37-29

and operation and maintenance . . . costs were associated with managing, harvesting, and disposing of the solar tank plant and sludge material [which accumulates on the surface of the water in the tanks]," they reported. In addition, there were a number of other ecological problems. Insects infested the sludge mats, plant material clogged pipes, plant roots reduced aeration, and plants obscured the tanks, making observational control difficult.[134] In 1999, in an effort to correct these problems, the engineers removed all of the vegetation from the treatment tanks.

Removing the vegetation, critics charged, changed the very nature of the plant. "Having funded that green technology," complained Dodson, "they're turning around and taking the green out of it."[135] Another supporter argued that the vegetation was "the heart of the technology," and charged that critics of the design were motivated by a "distrust of natural systems" and had organized "a misguided campaign against the use of plant-material."[136] As Dodson and others claimed, the town had turned an innovative plant into a conventional one. "Removal of the plants changes the overall system from solar-aquatic to 'activated sludge,'" stated supporters of the solar aquatics design. Plant proponents organized the Ashfield Solar-Aquatic Interest Group and flooded town meetings to defend the idea of a natural sewage treatment plant.

Engineers who studied the effect of plant removal concluded that, to the contrary, removing the vegetation had had no effect on the quality of the effluent. It did change the ability of the treatment plant to receive funding, however. Because the plant was funded through a program to support innovative technologies, town officials feared that the EPA would consider the plant, with the vegetation removed, to be a conventional sewage treatment system, as Dodson charged. In order to protect their funding, six months after spending $100,000 to take the plants out, officials replaced the vegetation.[137] "It will be a green machine again," headlined a local newspaper. The changes made to the plant, however, failed to improve the main concern, which was the high cost. Even after reducing the size of the plant, removing vegetation, and streamlining operations, sewer charges remained the highest in the state, increasing to $1,350 per year by 2003. The state and federal EPA commissioned a second study of the plant, by the New England Interstate Water Pollution Control Commission. The commission concluded that, contrary to the claims for the plant's natural operation, conventional plants "costs less to operate, have a much smaller footprint, use less energy, less (no) chemicals, use much less fuel, request less operation attention and are much easier and reliable to operate" than the Ashfield solar aquatic plant. "Other than the fact that the Solar Aquatics System can be aesthetically pleasing," the report concluded, "none of the other claims for the Ashfield Solar Aquatic System seem to hold up." As of 2009, Ashfield had abandoned its commitment to natural treatment but, without a ready source of funding, has not been able to replace the Solar Aquatic Plant.[138]

Figure 5.3
View of the "Living Machine" water treatment facility at the Findhorn ecological community in Scotland. This plant was built in 1995 to treat the waste of the 350 residents of the village. John Todd built living machines for a variety of clients, including food processing plants, visitors' centers, universities, and small communities. Waste is pumped through a succession of tanks that support different ecological communities of bacteria, algae, plants, and other organisms. *Source*: Photo by L. Schnadt, used under a Creative Commons Attribution-Share Alike 3.0 Unported license, terms at <http://creativecommons.org/licenses/by/3.0/legalcode>

As Ecological Engineering Associates was installing the solar aquatics technology at Ashfield, Todd continued to adapt his designs for larger waste treatment systems. He started another firm, Living Technologies Inc., filed for new patents, and was now calling his designs "Living Machines," a term he trademarked in 1991. Todd's new firm began building Living Machines to treat the waste from food processing industries, nature centers, and ecologically based communities, and in 1995 it constructed its largest system to date, to treat municipal waste in South Burlington, Vermont (figure 5.3).[139]

As revealed in the controversy in Ashfield, the conventional waste treatment profession had a complicated relationship with Todd's ideas. Martin Melosi draws the distinction between environmental engineering and the emerging discipline of "ecological" engineering. Environmental engineers had come out of sanitary engineering with a conventional approach to waste treatment that built on the century of experience with trickling filters and activated sludge. In contrast, "ecological" engineers were more likely to come out of the life sciences, particularly ecology, with roots, like Todd's, in a more well-defined critique of modern science, technology, and society.[140] These two trends in the profession collided when the EPA, which had itself funded Todd's demonstration project in South Burlington, commissioned an independent evaluation of its success. This evaluation revealed a divide among the proponents of the various "natural" wastewater systems. The authors of the EPA report were from the more mainstream environmental engineering profession. The lead author was Sherwood Reed, an engineer with the U.S. Army Corps of Engineers who was the author of textbooks and the WEF manual on "natural systems" for waste treatment.[141]

The EPA's evaluation of the Living Machine concluded that it was capable of producing an excellent effluent at costs comparable to those of conventional technologies. But it could do this, they concluded, not because it was especially ecological but rather because it utilized precisely the same processes responsible for treating waste as a conventional plant.[142] "The mechanisms responsible for the treatment are biological," the report concluded, but rather than being unique to the Living Machine, they were "common to the majority of 'mechanical' wastewater treatment processes (i.e. activated sludge)."[143] Rather than being a uniquely effective application of ecological principles to waste treatment, all of the processes functioning in the Living Machine could be mapped onto the familiar unit processes of the conventional environmental engineers. "The system," the report argued, "can be characterized as an anoxic reactor followed by extended aeration activated sludge, followed by clarification and filters for pathogen reduction and final polishing." In a cutting critique, the report concluded that "in effect, the system is a combination of conventional processes." The only difference was that in the Living Machine there were "plants floating on the water surface." But these plants, they concluded, were merely decorative, irrelevant to the Living Machine's function.[144]

The failure of the Living Machine to successfully harness solar energy and ecological principles, the report argued, was because the "ecological" design principles themselves had created their own set of contradictions for the Living Machine. "Having committed to plants and solar energy, the design logic then requires the use of a greenhouse when colder seasonal climates prevail—for protection and continued year-round growth of the plants." The requirement of a greenhouse in turn

causes a design dilemma since the high cost of the space enclosed by a greenhouse then requires deep high-rate treatment units for cost effective use of that space. Such high-rate units consequently minimize the surface areas available for utilization of plants, so the role of the plants is diminished along with the original and highly desirable intent to utilize plants and solar energy as major components in the system. A treatment system based on the use of plants and solar energy as major components must provide sufficient surface area so that the plants are in fact a major physical presence in the system. However, this does not appear to be the case.[145]

With no biological function, the report argued, the role of the vegetation was reduced to one of aesthetics, and the treatment plant was essentially conventional: "Instead of solar energy the system depends on the same energy sources commonly used in conventional wastewater treatment (i.e., electrically powered pumps and aeration compressors.)"[146]

According to the conventional waste treatment engineers, not only was the Living Machine not "ecological" but, according to Reed's definition of natural systems, it wasn't even natural. While he acknowledged that natural systems "might typically include pumps and piping for waste conveyance," the primary treatment processes would not rely on "external energy sources."[147] Thus, for Reed, land treatment, wetlands, and sewage ponds were "natural," but not activated sludge nor the Living Machine.

The EPA report recommended no further government investment in Living Machine technology. "Based on evaluations to date, it can be concluded that the AEES 'Living Machine' has not yet demonstrated reliable attainment of all of its process goals. . . .The floating macrophytes appear to contribute more to the aesthetics of the system than to the treatment performance. Also, the 'Living Machine' concept does not appear to offer any economic advantages over conventional technologies. In view of these conclusions, the continuation of Federal funding support for these demonstration projects is not warranted."[148] With the U.S. EPA demonstration completed and funding cut off, Todd began to look to the facility itself to generate funds. Students and scientists undertook a study of how brewery wastes could be managed to provide nutrients that "they could use to develop new economic products with commercial value. Our intent was to prove that ecological waste treatment could become an economic engine for a community." Like the nineteenth-century treatment plants, they looked to the nutrients in sewage to provide profit. "Nutrient farming has a rich future," they optimistically concluded. For several years the South Burlington plant "was a cornucopia of activity," as scientists sought to once again raise the treasure lying in a city's sewers. These treasures might be plants used as ornament, in perfume, or fish used as bait. Like the Miner Labs' search for "plus values" in Milwaukee's sludge, Todd and his coworkers looked for "new economic byproducts" from sewage.[149] By 2004, however, with resources at Todd's nonprofit foundation declining, he could no longer afford to run the plant in the winter, and it was mothballed.

In the meantime, an investor and business partner, Tom Worrell, had bought Todd's company, the rights to the name "Living Machine," and all of the company's intellectual property. Worrell Water Technologies now builds "Living Machines" based on Todd's ideas and several new patents for "tidal wetlands." Todd continues to design treatment systems through his environmental consulting company Todd Ecological Design, but he now uses the term "living machines" generically and calls his installations "Eco Machines."[150]

The solar aquatics system and the Living Machine both foundered on the skepticism that "natural" designs were substantially different from conventional biological wastewater treatment. The vegetation growing in the treatment tanks was the most obvious difference between the living machine technology and conventional processes, but the environmental engineering community concluded that the vegetation played at most a minor role in purification. By relying on bacteria, the Living Machine was as natural as conventional waste treatment; by requiring electricity for its pumps and compressors, it was just as artificial.

Biological Sewage Treatment in the Twenty-first Century

The "Living Machine" is just the latest manifestation of the threads that have been woven through biological sewage treatment since its invention in the late nineteenth century. After one hundred years of biological sewage treatment, citizens and engineers are still arguing about whether the technology is natural or artificial. The value of long-term skilled employees is being weighed against computerized control. In a technology crucial to the public health, corporations, inventors and municipal engineers are still fighting over the public interest. The persistence in sewage treatment of the key contradictions of the industrial ecosystem show just how sticky these contradictions are. Through it all, however, the basic importance of the living organism in treating sewage has stayed constant. The diversity and adaptability of bacteria and the flexibility of microbial communities have proved critical as sewage treatment plants have faced the task of removing more and more pollutants. As new challenges face sewage treatment plants, the living organism remains an irreplaceable resource. However, the ecosystems of the sewage treatment plant are threatening to collapse under the weight of the new demands and the contradictions of the industrial ecosystem.

Biologists and sanitary scientists have discovered that bacterial communities possess the capacity to deal with a huge variety of pollutants. By manipulating the environment of the microorganisms, researchers have been able to favor different microbial communities within the basic framework of the activated sludge process, allowing sewage treatment plants to adapt to meet new challenges. This has been most clear in the development of what is called biological nutrient removal. As phosphorus and nitrogen pollution led to overproduction of algae in lakes, rivers, estuaries, and oceans, sewage plants have been required to remove more of these nutrients from their effluent. Sanitary engineers first tried chemical methods for removing the nutrients, but these were expensive and made the precipitate unusable for fertilizer or land application. James Barnard, a South African sanitary engineer discovered that by manipulating the environmental conditions of the activated sludge plant, he could encourage certain bacteria to flourish that stored a superabundance of phosphorus in their cells. As these bacteria settled out with the sludge, they removed the nutrients from the effluent.[151] However, the bacteria that accomplished this so-called enhanced biological phosphorus removal (EBPR) required specific environmental conditions during different stages of growth, necessitating much more complex strategies for process control. Removing nitrogen could be similarly accomplished by manipulating conditions in parts of the plant, but the bacteria that performed denitrification required an even different set of environmental conditions from the bacteria that removed phosphorus or organic matter. The ecology of the wastewater treatment plant needed to be more finely controlled than ever, with rates of the various processes carefully synchronized. As the demands placed on the industrial ecosystem have increased, the stability of the system has declined; the EBPR process is subject to frequent upsets of the biological community and failure of the process.[152] Further, the plant is required to deal with newer and more varied sources of pollution, constraining the ability of the operators to adjust the process to meet what are becoming more and more conflicting demands.

The most recent challenge to the biological sewage plant is the problem of waste pharmaceuticals and other "micropollutants." These products enter sewage after being metabolized by the body and excreted in urine, or dumped down the sink. Antidepressants, steroids, ibuprofen, all can pass through a biological sewage treatment plant. Sunscreen, fire retardants, or the synthetic musk fragrances in shampoo and detergent are increasingly being found in natural waters and their sources traced to municipal sewage.[153] As *E. coli* was used as an indicator of sewage pollution for much of the twentieth century, caffeine is being studied as an indicator today.[154] Many of these compounds are biologically active, move through the food web, and are implicated in biochemical pathways for disease.[155] Triclosan, for instance, a bactericide added to bicycle shorts, socks, cutting boards, and kitchen sponges, has been found in fish downstream of sewage treatment plants, in earthworms in fields that received sewage sludge, and in human breast milk as well. Galaxolide and Tonalide, synthetic musks, are almost ubiquitous in sewage effluent and sludge, and have been shown to be endocrine system disruptors.[156] One proposed solution has been to introduce into sewage treatment plants the technology used for purifying drinking water. Ozonation, activated charcoal, and reverse osmosis are some of the chemical and physical processes that can remove micropollutants from plant effluents. These technologies, however, come with a large cost, both financial and in terms of energy use. Further, the degradation products of some of these technologies are themselves toxic.[157] Instead, argue some researchers, the ability of the activated sludge organisms to degrade these pollutants should be optimized.[158] Activated sludge bacteria can remove many of these compounds, but the conditions under which they might be most effective are not well understood. For instance, by increasing the volume of activated sludge returned to the aeration tanks, operators can increase the average amount of time that bacteria remain in the tanks. This allows slow-growing bacteria to survive and exposes pollutants to a greater diversity of bacteria and diversity of biochemical pathways that can degrade the pollutants. However, this can also favor bacteria that can cause foaming and other problems in the plant. Nor can activated sludge bacteria remove all the pollutants. Some chemicals are highly refractory and even for chemicals that are degraded, efficiencies vary widely.[159]

"The core of the sewage problem was really what was to be done with the solids," Gilbert Fowler had stated in 1908.[160] The problem of what to do with the solids remains. The standards used to exempt sludges of "exceptional quality" from further regulation have come under constant pressure. In a 2002 report, the National Research Council reviewed EPA's risk assessments of the land disposal of sewage sludge. They concluded that the EPA's efforts were inadequate and recommended a reevaluation of risk, the regulation of more contaminants, and studies of the health effects of long-term and multiple exposures. As a result, sewage plants will be under increasing pressure to reduce the volume of biosolids produced.[161] Balancing the needs for greater nutrient removal, degradation of micropollutants and the production of less sludge may be impossible.

The biological sewage treatment plant seems to be teetering as more and more potentially conflicting demands are placed on it. When the technology for biological sewage treatment was first developed, the key objective was to remove pathogens and as much organic matter as possible to keep the effluent from putrefying. Once the sewage plants were in place, they became the solution to other pollution problems as well, so that now they are asked to remove phosphorus, nitrogen, and other pollutants. At the same time, many of these pollutants become a source of other environmental problems, in the disposal of contaminated sludge. The remarkable thing about biological sewage treatment has been its flexibility and adaptability to meet these new demands. But its ability is not infinite, and society needs to look upstream of the sewage treatment plant at the sources of the various pollutants.[162]

From Sewage to Biotech: "What We Have before Us Is an Industrial Product"[1]

In 1978, Genentech announced the first instance of a human drug created using the new methods of recombinant DNA. Scientists had succeeded in splicing the gene for human insulin into the bacteria *E. coli*, turning the bacteria into "microscopic 'factories'"[2] for the production of medicine. This announcement was heralded in the press as the first major accomplishment of the new biotechnology industry.[3] But as historians like Robert Bud and Angela Creager have noted, the biotechnology industry had much deeper roots than simply the manipulation of DNA.[4] The use of *E. coli* as a model organism, the intellectual property environment allowing Genentech to profit from its new microorganism, the very conception of microorganisms as little "factories," are all connected to the role of sewage treatment in the creation of the industrial ecosystem in the late 1800s.

Despite claims that the techniques of genetic manipulation represent a revolutionary change in biotechnology, fundamentally altering the logic of production and "transforming the whole nature" of the industry,[5] the contradictions of the industrial ecosystem, dating to its creation in the late 1800s, have persisted in and continue to characterize genetic engineering. The manipulation of DNA did not alter the fundamental problem of using living organisms to produce industrial products. In turn, however, the new biotechnology industry has expanded the scope of the industrial ecosystem, extending it to encompass the natural systems that harbor the raw materials for genetic manipulation, the natural biodiversity of the earth's ecosystems. As technology like Genentech's hybridized human and bacterial genomes in *E. coli*, so it has hybridized industrial and wild ecosystems, helping to transform all of the earth's habitats into industrial ecosystems.

The scientific understanding of *E. coli*, the organism Genentech had modified for the production of insulin, has its roots in sewage and the sewage treatment plant. *Escherichia coli*—originally known as *Bacillus coli communis*, or the common colon bacillus—was first identified in infants' feces in 1885 by Theodor Escherich. Scientists studying the bacteria of sewage quickly isolated *B. coli* from both sewage and sewage-polluted rivers as well as from sewage treatment plants.[6] Because it was so commonly found in sewage, *B. coli* was used as an indicator of sewage pollution, and its removal during the sewage treatment process was taken as a measure of the success of purification techniques.[7] Interest in its role in sewage-polluted waters led to new culturing techniques that allowed it to be easily grown in the laboratory. Edwin Oakes Jordan, whose work was so important to the Lawrence, Massachusetts, studies of sewage treatment, began to focus on *B. coli* and its use as an indicator of fecal pollution.[8] Its presence was often ambiguous, however, and scientists began investigating bacterial variation and mutation in *B. coli*.[9] These early sanitary studies of *B. coli* started to reinforce each other. As scientists accumulated a greater understanding of the species, they continued to use it for new studies, and it became a model organism for genetic research, culminating in the 1940s in the work of Joshua Lederberg and Edward Tatum on bacterial conjugation.[10] In the 1960s, following on these studies, Herbert Boyer, the cofounder of Genentech, had identified restriction enzymes in *E. coli* that could cleave DNA, and with Stanley Cohen, used them to insert novel genes into the bacterial genome.[11] With the techniques for manipulating the *E. coli* genome established, this microbe became the natural choice for Boyer and Genentech's development of insulin-producing cells.[12]

As biotechnology companies like Genentech developed novel organisms, they extended the reach of the patent system to include the recombinant organisms themselves. In the landmark case *Diamond v. Chakrabarty*, the U.S. Supreme Court ruled for the first time that "a live, human-made micro-organism is patentable."[13] *Diamond v. Chakrabarty* enabled the growth of the biotechnology industry and established a new arena of privatization. Ananada Chakrabarty, working for General Electric, had created a strain of bacteria that could use crude oil as a growth substrate and could thus help clean up oil spills. He sought to patent not only the process of making his organisms but the organism itself.[14] The patent examiner, repeating the argument first articulated by Vernon Richard in the 1930s in the *Prescott* case, denied the patent. Chakrabarty appealed. The Patent Office Review Board affirmed the examiner's decision, but it was reversed by the Court of Customs and Patent Appeals, a reversal later affirmed by the Supreme Court. Sewage treatment and the sewage patent cases were instrumental in this decision as the sewage cases provided important legal precedent for this landmark case. But the cultural work of biological sewage treatment in denaturing natural processes was critical as well.

Daniel Kevles looks to the history of plant patents as the precursors to *Chakrabarty*, with the first patent protection for living organisms granted in the 1930 Plant Patent Act. The direct ancestors of the *Chakrabarty* patent, however, were not plant patents but rather the microbial process patents granted in *Cameron* and *Activated Sludge*. In a line of decisions over the twentieth century, precedents initially established in the sewage patent cases were applied to the patenting of industrial bacterial processes, like the production of butyl alcohol and acetone by *Clostridium*, patented by Weizmann, and by *Bacillus technicus*, patented by Prescott and Morikawa. All of these cases, resting ultimately on *Cameron*, were then mobilized to support the ruling in *Diamond v. Chakrabarty* that not only the microbial processes but the bacteria themselves were subject to patenting and privatization.

Nor was the nature of DNA critical in this finding. Kevles argues that the Supreme Court justices came to see bacteria as simply DNA: "What enabled the patenting of Chakrabarty's bacterium was the finding of molecular biology that genes are DNA."[15] Rather, the justices came to see bacteria not as DNA but rather as industrial products. Viewing bacteria as an industrial product rather than a living organism was critical. It was the denaturing accomplished by industrial microbiology that allowed the courts to uphold the septic tank and activated sludge patents, and it was this same denaturing that allowed the courts in *Chakrabarty* to see bacterial activity as a chemical rather than a vital process. When, at the beginning of the nineteenth century, the engineering opposition failed to convince the courts that the bacterial processes in sewage treatment were natural and thus public, the stage was set for finding bacteria themselves to be the result of artifice.

Although they were critical to the reasoning in the case, *Cameron* and *Activated Sludge* do not appear in the Supreme Court's opinion in *Chakrabarty*. Rather, their influence is clear in the lower court decisions that were affirmed by *Chakrabarty*. Chakrabarty's lawyers cited *Cameron* and *Activated Sludge* as well as *Guaranty Trust*, but the reasoning goes deeper.[16] The Court's arguments in the Chakrabarty case were first developed in reference to another patent application, that of Malcolm Bergy and colleagues.[17] Bergy had isolated a pure culture of a fungus that produced the antibiotic lincomycin. They sought patents not only on the process of producing the antibiotic, but, like Prescott's patent application, on the microbial culture itself. This was the precise claim that the Patent Office had successfully denied in the *Prescott* case, and it denied it for Bergy as well. Following the rulings in the early sewage cases, the Patent Office granted Bergy's process patents but denied the patent on the culture itself, reasoning that living organisms could not be patented.[18] Bergy appealed to the Court of Customs and Patent Appeals, which overturned the decision of the Patent Office and allowed the claim for the pure culture. When *Chakrabarty* later came before the same court, the court simply applied *Bergy* as ruling precedent without any additional analysis.[19] The commissioner of patents appealed both rulings. As *Bergy* and *Chakrabarty* moved up to the Supreme Court, the two cases were formally linked. Fearing that his part of the case was weaker and hoping for a successful test case of the Patent Office's procedures, Bergy withdrew, leaving only *Chakrabarty* to be decided. But the wording of the decision had been almost completely framed for *Bergy*.[20] To understand the Court's reasoning in *Chakrabarty*, then, we have to examine *Bergy*.

Relying on the decisions in *Cameron*, *Activated Sludge*, *Guaranty Trust*, and *Prescott* that found bacterial processes to be patentable, the Court of Customs and Patent Appeals ruled in *Bergy* that in addition to processes, bacterial cultures were also patentable "notwithstanding the employment therein of living organisms and their life processes." The court found it "illogical" to deny patent protection to a bacterial culture yet allow protection for the "functioning" of that culture, and it extended the patentability of a process to the organism itself.[21] In *Bergy*, the court equated the production of lincomycin using *Streptomyces vellosus* to any commonly understood industrial manufacturing process and declared that "what we have before us is an industrial product used in an industrial process." The industrial product the judge referred to was not lincomycin, the product of the microbial fermentation, but rather the organism *Streptomyces* itself. How had a natural, living organism become an industrial product? The court's justification was based on an understanding of bacterial processes as chemical processes, "the essential similarity of what we normally think of as 'chemical reactions' and the complex procedures wrought by the life processes of microorganisms."[22] This was the argument made in *Guaranty Trust* and *Prescott*. In *Guaranty Trust*, the court ruled that the bacterial process relied on an "agency of nature, the use of which for practical purposes can be patented," in the same way that "chemicals, minerals, or anything else" could be.[23] In *Prescott*, the court further equated the life process of bacteria with a "chemical process," calling the discovery of the bacteria, *Bacillus technicus*, simply "a new reagent."[24]

The industrialization and denaturing of bacterial sewage treatment had helped courts at the beginning of the century overcome the opposition to the patenting of bacterial processes. It was the denaturing inherent in the use of bacteria for sewage treatment that allowed the courts to uphold the septic tank and activated sludge patents, and it was this same denaturing that allowed the courts in *Chakrabarty* to see bacterial activity as a chemical rather than a vital process. This, in turn, initiated the transformation of patent law that allowed the patenting of living organisms themselves.

Sanitary engineers themselves were complicit in this industrialization of the bacterial processes. As we saw in chapter 1, engineers and scientists had thoroughly industrialized the biological processes of purification, making bacterial treatment appear more and more like a chemical manufacturing process and less like a biological one. The continuous nature of the process, the application of process control techniques, the biological simplification, all denatured the fundamental living character of the system. Thus it should have come as no surprise that the courts began to treat the bacterial processes more like chemical manufacturing processes than "nature's methods."

Emphasizing the view of bacterial processes as simply another example of an industrial, chemical process, the judge in the *Bergy* case stated, "The nature and commercial uses of biologically pure cultures of microorganisms . . . are much more akin to inanimate chemical compositions such as reactants, reagents, and catalysts than they are to horses and honeybees or raspberries and roses."[25] Because the patent appeals court reasoning relied heavily on the industrial nature of bacterial processes and the professed similarity between bacterial and chemical processes, the judges in the Bergy case viewed their decision as limited to bacteria and other lower forms of life. They dismissed warnings that their decision could lead to the patenting of animals as "far-fetched."[26]

The Supreme Court, in a narrow 5–4 decision, affirmed the reasoning in *Bergy* (but now applied only to *Chakrabarty*) and allowed the first patent on a living organism itself—the bacteria created by Chakrabarty. The decision has had far-reaching implications. Although the lower court decision was crafted for bacteria, as particularly chemical-like, the Supreme Court opinion applied to any life-form modified or created by people, "anything under the sun that is made by man." Once one form of life was patented, there was no logical obstacle to patenting others. In 1988, the cancer-susceptible mouse, "onco-mouse," was successfully patented by Harvard University.[27] Since that first animal patent, there have been well over a thousand patents granted in the United States on animals themselves, including fish, pigs, birds, sheep, and cows.[28]

For Genentech, *Chakrabarty* came too late to patent its recombinant *E. coli* that produced insulin, but it was not too late to take advantage of the new intellectual property environment that would result. Based on the success of its recombinant *E. coli*, Genentech was about to take the company public, and a ruling from the Supreme Court in favor of Chakrabarty would assure investors that it could protect its intellectual property, increasing the value of the company enormously.[29] Genentech submitted an *Amicus Curiae* brief to the Supreme Court in the *Chakrabarty* case, encouraging the Court to validate Chakrabarty's patent. In the brief, the company explicitly linked the industrialization of the living organism, the combining of the natural and the artificial, to genetic engineering. Genentech emphasized the status of microorganisms as industrial products: "For industrial purposes, bacteria that produce human insulin can be regarded as . . . machinery." But conceiving of bacteria as little factories had its roots in the denaturing accomplished in the sewage treatment plant, and it was a critical element in the creation of the industrial ecosystem.

Biotechnology and the Contradictions of the Industrial Ecosystem

This view of the living organism as an industrial product is central to the biotechnology industry. This is made explicit in the title of a new biological journal, *Microbial Cell Factories*, established in 2002. The journal promised results from both engineered and "natural . . . microbial factories," with apparently no awareness of the internal contradictions of the phrase.[30] What had made the microbe into a factory and, at the same time, made a factory natural? This characterization of living organisms as biological machines or "little factories" was tied to the discursive moves that had reconfigured natural processes like anaerobic digestion or aerobic oxidation of sewage and made them artificial, while at the same time made the industrial process natural, by connecting it to natural selection and the science of ecology. The industrialization of sewage treatment processes and their subsequent denaturing had helped make these natural factories possible.

Here, in the microbial cell factory, the contradictions of the industrial ecosystem extended to the cellular level. The contradiction between profit and purification was embodied in the microbe's internal systems for removing "misfolded and incompletely synthesized proteins," a kind of cellular "quality control system." While these quality control systems assured a high purity product, they also slowed down the production of proteins by the cell. "This poses a fascinating challenge for cell factory engineering," observed researchers, "since inactivation of quality control systems can improve protein production yields, but these improved yields might be at the expense of product quality."[31] Perhaps the concept of the biological machine reaches its greatest expression in the attempt to develop a "synthetic platform cell" that maximizes production of pharmaceuticals or other industrial products. To accomplish this, biotechnologists must overcome these contradictions of efficiency and accuracy, profit and purification.[32]

The focus on the techniques of genetic manipulation and the production of novel strains of microbes as the key component of the new biotechnology, however, obscured the continuing importance of an older aspect of the industrial ecosystem— the ability to grow the organisms. This focus on the refinement of the organisms, rather than the ability to grow them, has slowed the development of the biotechnology industry, according to some observers: "In contrast to the immense achievements in fundamental molecular biological sciences, the fermentation and downstream processing technologies used in industry have not developed at the same pace . . . A misbalance between new cellular systems and production technologies appeared."[33]

This conflict was present in Genentech, at the birth of the new biotechnology industry. Even before Genentech had developed any novel organisms, it bought a used fermenter from a pharmaceutical company and hired an engineer, a specialist in fermentation, as vice president of manufacturing. "You give me the bugs and I'll make the product," he was reported to say.[34] This engineer, however, was a representative of an older biotechnology industry, "the process historically used to make antibiotics," as Genentech's chief financial officer put it.[35] Genentech's molecular biologists were confused as to whether he was an engineer or simply a fermenter operator, and he had a difficult fit within Genentech's culture of DNA manipulation.[36] "He was supposed to be able to grow vats of bacteria—or yeast, I forget which microbe," recalled the CFO, as if the distinction were irrelevant.[37] But with no microbes yet, most scientists at Genentech couldn't figure out why he was there, and he became an object of ridicule.[38] In fact, Genentech did not possess the expertise to manufacture and commercialize its recombinant *E. coli*. That fell to Eli Lilly, whose engineers took the process from the 5- and 7-liter fermentation reactors of Genentech's laboratory to 38,000-liter reactors of the production plant.[39]

As the reception of the fermentation engineer at Genentech exposes, the importance of culturing—the understanding of the organism's place in the industrial ecosystem—was being overshadowed by the novel and exciting developments in genetic technology. The skills in managing the industrial ecosystem were being lost to an emphasis on genetics. But as the novel organisms were pressed into industrial service, the contradiction between craft and science reemerged, as did the emphasis on process control. Calling fermentation "an art from the past, a skill for the future," Brian McNeil and Linda M. Harvey argued that even though the quality of the "'new' fermentation products" was determined by the skill in monitoring and controlling fermentations, expertise in precisely this area was disappearing.[40] Referring to the microbes as "biocatalysts" rather than living organisms, researchers emphasized the importance of process control. "The biocatalyst is sensitive to environmental conditions. Even a small uncontrolled change in composition, pH, temperature, and pressure can alter cell metabolism and change radically process efficiency and productivity and can even render the process unprofitable."[41]

Just as in sewage treatment plants, this new emphasis centered on the operator. In phrases that could have been taken verbatim from the Water Environment Federation's *Operations Forum*, control engineers in biotechnology wrote that "the importance of effective operator control cannot be underestimated as the performance of a fermentation is very much dependant upon the ability to keep the system operating smoothly." These researchers emphasized the skill of the operator: "The operator uses his experience and knowledge of the fermentation process, together with information provided by supervisory control systems to detect potential problems and make modifications when necessary."[42]

Fermentation process engineers saw automation as a solution to the control problem. As in sewage treatment, the biotechnology industry lagged behind the chemical process industries in automation, and just as in the sewage treatment plant, it was the complexity of the biological processes in the fermentation reactor that made the penetration of automation difficult. "Difficulties with the application of conventional theory to complex and only qualitatively understood biotechnological systems are the reason why the control of fermentation processes is still fundamentally manual," stated an early review of fermentation process control.[43] Again, as in the sewage treatment plant, process control engineers concluded that they would

have to incorporate the experience, knowledge, and reasoning of the human operator into automatic control. Sewage treatment processes had actually been the first
biological processes to be modeled using fuzzy control and other knowledge-based
systems. These techniques were then applied to industrial fermentations.[44] Process
engineers studied the behavior of the human operator and tried to incorporate that
knowledge and reasoning into their control algorithms. "We have used the heuris-
tics and knowledge of an experienced bioprocess operator," wrote the developers of
one knowledge-based automation system. Interestingly, they called their program
"SUPERVISOR," leaving it ambiguous as to whether it was supervising operators or
the bacterial production process itself.[45]

The Expanding Industrial Ecosystem

Under the influence of the new genetic biotechnology industry, the industrial ecosystem is rapidly expanding. Genetic recombination has not so much transformed
the logic of the industrial ecosystem as it has dramatically increased its spread. The
techniques of genetic engineering have revived the industrial ecosystem for the production of many products that were earlier made uneconomical by improvements in
organic and inorganic chemistry. The fermentative production of butyl alcohol, for
instance, the subject of Weizmann's and Prescott's patents, had, by the 1950s, been
superseded by the purely chemical processes of the petrochemical refinery. With
increased interest in butyl alcohol as a biofuel though, research on *Clostridium* has
been revived, with the hope that the techniques of genetic engineering might make
the fermentation once again economical. To this end, researchers are investigating
ways of improving *Clostridium* by engineering increased alcohol production, sup-
pressing the production of acetone, increasing oxygen tolerance, allowing growth
on cellulosic materials, and increasing the bacterium's tolerance to the alcohol it
produces.[46]

Because of similar pressures for alternatives to petroleum, ethanol production
plants may now be the most rapidly increasing industrial ecosystem. Both the number and capacity of ethanol production plants have been expanding exponentially.
U.S. capacity in 1980 was 175 million gallons per year (mgy), and increased to
1,701 mgy in 1999. By 2010, it was 11,877 mgy.[47] So too, as with butyl alcohol,
researchers are seeking to improve the yeasts and other organisms involved in ethanol production using genetic engineering. Most observers consider the fermentation
of cellulosic feedstock essential to the future of biofuel production and genetic engineering of yeast critical to this effort. Genes for converting cellulose to sugar have
been inserted into yeast from various fungi. If these innovations prove to be technically and economically successful, ethanol fermentation plants will expand even
more.

But the reach of the industrial ecosystem is expanding even more as genetic engineering extends the contradictions of the industrial ecosystem to natural ecosystems more generally. Because the fundamental resource of the new biotechnology is the gene, scientists have begun to "mine" the natural world for genes that can be bred or engineered into other species. In searching for these novel genes that could confer useful properties to drug- and food-producing organisms, pharmaceutical and agrochemical firms have made the natural ecosystems of the rain forest, coral reef, and subtropical drylands an integral part of the industrial ecosystem of biotechnology-based manufacture. Even the quintessentially "wild" ecosystems of Yellowstone National Park have been incorporated into the industrial ecosystem, as genes from bacteria that live in the park's geothermal features have been spliced into various organisms used in genetic engineering.[48] "The business of the biotechnology-based industries is founded on the search for exploitable biology, or . . . genetic resources," wrote Alan Bull in the introduction to a text on "bioprospecting."[49] In this view, the earth's biota is "a biochemical laboratory unmatched for size and innovation." The natural laboratory of the ecosystem becomes incorporated into industrial production as the source of innovation to be screened and developed by the industrial laboratory.[50] In connecting the natural ecosystems supporting this diversity to the industrial ecosystem, scientists, industrial firms and governments have also profoundly affected the management and conservation of natural ecosystems far removed from the industrial laboratories and production plants.

The prospecting for biodiversity has implicated the key contradictions of the industrial ecosystem and shifted them from the industrial production plant to the "natural" environment. For many conservationists, bioprospecting represents a means of resolving the contradiction between profit and purification, or, in this context, conservation.[51] Conservationists have seen in the incorporation of natural ecosystems into the industrial ecosystem a means to value biodiversity in an industrial market. Forests, rather than being clear-cut for short-term profits, might have even greater economic value intact, as sources of raw inputs into the pharmaceutical and agrochemical industries. By drawing ecosystems into the market, conservationists would tie the management of forests and other habitats to the protection of biodiversity. Local communities would have incentives for protecting habitats, and habitats would be able to support sustainable development projects. The contradictory goals of profit and, in this case, conservation are aligned.

But crucial to this alignment is navigation of the contradiction between the public and private spaces of both the ecosystems themselves and the products derived from them. Up until the early 1990s, industrial and agricultural scientists in the developed countries had considered genetic diversity to belong to the international community at large, as the "common heritage of mankind." But once those genes were bred into, or sequenced and inserted into other organisms, they became the private intellectual property of the agricultural and pharmaceutical companies. As the economic value of genetic resources became clearly exposed by the burgeoning biotechnology industry, this inequitable intellectual property framework became politically untenable. It was the incorporation of wild biodiversity into the industrial ecosystem that first exposed the contradictions of public and private rights. Recognizing this inequity, countries like Costa Rica began to make bilateral agreements with pharmaceutical and biotechnology companies to provide access to natural diversity in return for cash payments and access to technology. Even the U.S. government entered into agreements for bioprospecting in its national parks. How to share the benefits of genetic resources among the local communities, indigenous groups, and governments of biologically rich countries and the biotechnology firms of the developed nations became the subject of heated debate. In 1992, the international community adopted the Convention on Biological Diversity to establish a framework for sharing the benefits of biodiversity. This convention, however, essentially affirmed existing intellectual property relations and made biodiversity just one more product subject to exchange in the market.[52]

The controversy over access to genetic diversity also centers on the conflict between craft and science, and how each is valued. The property rights scheme that recognized genetic diversity as common heritage, but products developed using that diversity as private property, singularly valued the scientific knowledge used to identify and isolate particular genes and devalued the stewardship and contribution made by the application of craft knowledge developed by the peoples who have been cultivating, selecting, breeding, and maintaining diversity of crop plants or ecosystems. Yet the new property rights scheme that recognizes a property right in biodiversity invests that right in the state, rather than the farmers, indigenous people, or local stewards, maintaining the structural devaluation of the craft knowledge of these peoples.[53]

The International Cooperative Biodiversity Groups (ICBG), a joint project of U.S. scientific and health agencies, tried to resolve these joint contradictions of profit, public, and craft. Cori Hayden, in a fascinating ethnography of bioprospecting in Mexico, shows how conservationists, biotechnology companies, and university researchers tried to negotiate the multiple contradictions posed by the extension of the industrial ecosystem into the natural world.[54] The ICBG project was an agreement among pharmaceutical and agricultural companies, researchers in the United States and Mexico, and the Mexican government, to search for plants in the arid areas of northern Mexico that might possess some previously unknown compounds useful in the manufacture of pharmaceuticals or agricultural chemicals. Project organizers recognized an ethical responsibility to benefit the communities or indigenous groups that provided both the plants as well as knowledge about their distribution and use. If any of the plants led to the development of a drug, royalties would be returned to these communities. In this way, project organizers hoped that there would be increased incentives among local people to be stewards of their ecosystems. As Hayden shows, however, the project met with mixed results and foundered on the very contradictions I have identified in the industrial ecosystem. Because plants were easier to obtain, but also because they came severed from their connections to local communities, ethnobotanists searched for potentially pharmacologically useful plants, not in wild ecosystems, but in the "public" spaces of the herbal medicine stands in urban markets. These herbs, already in a market system, did not come "bundled" with the claimants and owners. Once ethnobotanists bought herbs in the market, their relationship with the collectors and vendors was finished. Project organizers in the United States criticized the Mexican ethnobotanists for severing the intellectual property link between the "authors" of ecosystem knowledge and the pharmaceutical firms. Exchange on the market, project organizers feared, was being replaced with piracy—the theft of plants and ecosystem knowledge. For their part, Mexican researchers were using collecting in the marketplaces as a shortcut to identify potentially useful species and would later trace the plants from the markets back to the ecosystems and communities from which they came. Under the weight of these contradictions, however, the Mexican bioprospecting agreement was not renewed. Pharmaceutical firms wanted to shift the prospecting agreements from the plants, with all of their complex connections to local craft knowledge and communities that claimed ownership, to marine microorganisms that might be more easily separated from ethical connections with local communities. Bioprospecting has thus moved out into the deep ocean, where claims of ownership are less contested.[55] Marine scientists look to the organisms of the oceanic ecosystem using the same metaphors that had been used for bacteria in the sewage treatment plant. Blue-green algae (cyanobacteria) to these researchers are the "blue-collar workers," or "workhorses" of the ocean that can be exploited for society's needs.[56] The ocean becomes an industrial ecosystem in its own right, and the industrial ecosystem has expanded from the spaces of the industrial plant to the world at large.

Hybridizing Sewage, Biotechnology, and Nature

The industrial ecosystem has penetrated the natural world more than metaphorically. The genetic technology that has hybridized the natural and artificial in the industrial ecosystem is now extending itself into the wild as these genetic chimeras are literally hybridizing with wild species. Go out into the field and sample the DNA from wild plants, and you may find genes derived from the sewage treatment plant and patented by multinational life-sciences companies. The boundary between the industrial and the wild ecosystem is being dissolved.

In 1986, Terry Balthazor and Laurence Hallas, two scientists working for Monsanto, the giant agricultural and chemical firm, reported for the first time the isolation of microorganisms capable of degrading Monsanto's own herbicide, glyphosate, the active ingredient in their tradename herbicide Roundup. They had isolated these microbes from the activated sludge of the sewage treatment plants where Monsanto produced the herbicide. These organisms were critical, because the Environmental Protection Agency was stepping up its efforts to regulate the discharge of glyphosate into the environment. Every day, Monsanto's glyphosate manufacturing facility in Muscatine, Iowa, for instance, produced as much as 1,500 pounds of glyphosate waste and as much as 500,000 gallons of glyphosate-contaminated wastewater. This waste was all sent to a biological sewage treatment plant on site before it was discharged into the Mississippi River. The activated sludge process in Monsanto's waste treatment plant was capable of removing almost 100 percent of the glyphosate from the waste stream, but that efficiency often dropped to as little as half. Thus scientists began to study more closely the organisms thought responsible for the biodegradation of the glyphosate in order to comply with impending federal regulations.[57] They began to isolate the specific microbes responsible for metabolizing and degrading the herbicide, and they began to develop advanced, tertiary treatment systems for glyphosate degradation.

These experimental systems were conceptually identical to Schloesing and Muntz's glass columns filled with sand first used to study nitrification in sewage, or the intermittent sand filters of Frankland or the Lawrence Experiment Station, or even the design of acetic acid fermentation facilities. Monsanto scientists filled a column with diatomaceous earth, and inoculated it with activated sludge from the plant's sewage treatment facility. Over the following months, they followed the fate of glyphosate waste as it flowed past the particles in the column. Like Schloesing and Muntz's and Frankland's columns, there was little effect on the waste at first, as the bacteria grew from their small initial populations. But over the following weeks, as populations grew and bacteria attached to the particles, the amount of glyphosate in the effluent dropped precipitously. These columns were able to consistently remove over 98 percent of the herbicide from the wastewater. The EPA, however, never did require effluent limits on glyphosate from Monsanto's plants, and Monsanto continued to use its original activated sludge process, rather than these new tertiary treatment columns, to treat glyphosate waste.

The herbicidal properties of glyphosate, the active ingredient marketed by Monsanto in Roundup, were identified and patented by Monsanto scientists in 1971.[58] Glyphosate is a broad-spectrum herbicide, killing almost all plants by interfering with the organisms' production of amino acids, vitamins, and hormones. Glyphosate became the best-selling agricultural chemical ever and the most widely used herbicide in the United States as well as worldwide. By the 1990s, however, the global patents on glyphosate began to expire, culminating in the patent's U.S. expiration in 2001. Anticipating the patents' expiration and availability of cheaper generics,

Monsanto developed a corporate plan to maintain market share for its leading product.[59]

Glyphosate's major drawback was that it killed both crops and weeds indiscriminately, limiting its use during the crops' growing season. Using the increasingly powerful techniques of molecular biology and genetic engineering, Monsanto scientists began to look for ways to introduce tolerance to Roundup into the germ line of important crop species. If successful, farmers could spray herbicides on their growing crops, killing weeds but leaving the crops intact. Monsanto would have two products that reinforced each other, herbicide and herbicide-resistant seed, protecting market share despite the loss of patent protection for the herbicide itself.[60]

After several failed attempts using naturally occurring plant and *E. coli* mutants to produce crops that displayed commercially useful levels of tolerance to glyphosate, Monsanto scientists turned to the industrial ecosystem for the source of its genetic material: the sewage treatment plants where the constant selection for herbicide resistance had revealed several microorganisms that could grow in the presence of glyphosate.

One of these organisms was a strain of the bacteria *Agrobacterium* that possessed enzymes resistant to the effects of glyphosate while at the same time retained their metabolic efficiency.[61] From this strain, dubbed CP4, Monsanto isolated the gene for glyphosate resistance and transferred it to a wide variety of crop plants, including soybeans, canola, cotton, and corn (figure 6.1). In 1994, the firm filed for a patent on the C4 gene, "useful in producing transformed bacteria and plants which are tolerant to glyphosate herbicide."[62] Monsanto claimed the gene, CP4, that conferred tolerance. It also claimed the plants that were transformed themselves. That is, it claimed patents on the living organisms of soy, corn and other crops. Finally, it also claimed a patent on the method of using its invention: "a method for selectively controlling weeds in a crop field." Monsanto marketed these plants as "Roundup Ready® soybeans" and "Roundup Ready® canola"—plants that quickly transformed the agronomic landscape. Over 90 percent of the soybeans planted in the United States have the roundup-ready trait, and over 60 percent worldwide.[63] Through the combination of herbicide and herbicide-resistant crop, Monsanto was able to expand market share for its herbicide and seeds, while it also sells licenses for its resistance technology. The roundup-ready technology is the single most valuable genetic technology in agriculture, and it accounts for over one-half of Monsanto's revenues.[64]

With the expanding cultivation of genetically modified crops, however, the industrial ecosystem is now reaching into the wild, literally hybridizing the wild and industrial ecosystems. The CP4 gene, originally isolated from the industrial ecosystem of the sewage treatment plant, has now spread into the wild. The Scotts Company, under license from Monsanto, developed a variety of the turfgrass species, creeping

Figure 6.1
A test field for maize seed varieties in east-central Illinois, 2010. These varieties have "stacked" traits for resistance to multiple herbicides as well as for insect resistance, including Monsanto's Roundup Ready gene (CP4 EPSPS) for tolerance to glyphosate. The glyphosate tolerance gene was isolated from bacteria growing in the activated sludge sewage treatment plants of Monsanto's herbicide production facilities. Monsanto patented numerous crop plants modified with this gene and has licensed the Roundup Ready technology to other seed companies. In 2010, 70 percent of the maize and over 90 percent of the soybeans planted in the United States were genetically modified with herbicide resistance. In Canada, over 95 percent of the canola, or oil seed rape (*Brassica napus*), was Roundup Ready. Canola has hybridized with its wild relative, wild mustard (*Brassica rapa*). The gene, originally from an activated sludge bacterium, is now found as a stable part of the genome in wild populations of mustard. *Source*: Photo by author

bentgrass (*Agrostis stolonifera*), that incorporated the CP4 gene for glyphosate tolerance. Despite opposition from many turfgrass farmers concerned that pollen or seed from the genetically modified crop would contaminate their own crop, Scotts received permission from the U.S. Department of Agriculture to plant a test field of the herbicide-resistant grass on 400 acres in central Oregon in 2003. Creeping bentgrass is wind-pollinated and has extremely small seeds that are also easily carried by the wind. Within a year, scientists began to find escaped herbicide resistant plants outside the planting area and evidence of pollen movement up to 21 km away. Offspring of wild plants up to 14 km away contained the CP4 gene, including plants of another species of turfgrass, redtop (*Agrostis gigantea*). This experiment was widely regarded as a disaster. Scotts discontinued planting and undertook almost immediate efforts to remove genetically modified plants and prevent the spread of the gene. Nevertheless, even three years after planting stopped, researchers were discovering the herbicide-resistant plants at ever increasing distances from the plots and finding increasing evidence of hybridization with naturalized grasses.[65]

On the other side of the continent, roundup-ready canola had already been approved and dominated commercial production. Farmers had planted roundup-ready canola in two of their fields in Quebec in 2000 and 2001. By 2001, the canola (*Brassica napus*) had already hybridized extensively with a related wild species, wild mustard (*Brassica rapa*), carrying its bacterial gene with it. These hybrids have backcrossed into the wild population of *B. rapa*. By 2006, there were persistent populations of wild mustard that had incorporated the bacterial gene for herbicide resistance.[66] A gene derived from the bacteria living in an activated sludge sewage treatment plant has become established in wild plant species across the North American continent and perhaps the world. The industrial and the wild have been thoroughly hybridized by the industrial ecosystem.

7

Conclusion: The Living Machine®

In 1852, in a speech before a Belgian conference on hygiene, English sanitarian F. O. Ward invoked an explicitly organic metaphor to describe the circulatory system of water and waste that he and Edwin Chadwick were proposing for English cities. Ward described the water pipes and sewers as the arteries, veins, and capillaries of "an immense organism." Comparing the linked urban and rural areas to a living being, Ward was justifying the water carriage scheme by likening it to the organization of a natural body. By invoking this organic metaphor, the recycling of town waste on agricultural fields and the cleansing of sewage in soil were all justified as part of a natural plan. Yet, completely interspersed with these natural allusions, Ward also invoked the industrialization then spreading across England and the Continent. The system of circulation that he proposed was to be set in motion by huge pumps powered by steam engines, a system originally developed to drain water from England's coal mines. It was "nothing more than an adaptation of the great invention of Watt"; it was to be "*Hygiene by steam power.*"[1] In the middle of his arterial and venous pipes, Ward placed "a motive organ—a central heart so to say—in a word, a steam engine, which sets the whole system in motion." The steam engine, in service of sanitation, would transform cities as profoundly as it had altered "all the other branches of industrial art." Ward's "immense organism" was, in fact, a metaphorical hybrid, part nature, part industry. From the beginning of sewage treatment in England, promoters mobilized these powerful metaphors of both industry and nature to describe and justify their vision for water carriage sanitation in the nineteenth-century city.[2] Yet it remained just that: a metaphor. The pipes were not, in fact, veins and arteries; the steam-driven pump was not a heart.

With the development of biological sewage treatment, however, this metaphorical hybrid of Ward's became literal—the industrial ecosystem. From the sewage farms and irrigation of the mid-nineteenth century, to the septic tanks and bacterial filters of the 1890s to the activated sludge process of 1914, sanitary scientists and engineers created ecosystems where the microbial purification of a "natural" soil became progressively intensified and simplified—industrialized. This process complicated the categories "natural" and "artificial" and created something new that was at once neither and both.

Since the beginnings of the Industrial Revolution, observers as well as participants had been commenting on the unique hybrid created by the intrusion of the industrial into the natural world. Leo Marx called this landscape "a middle ground" that existed "somewhere 'between,' yet in a transcendent relation to, the opposing forces of civilization and nature."[3] Like Marx's "machine in the garden," the industrial ecosystem was neither natural nor artificial, but a transcendent hybrid of the two. Marx demonstrated the power of this cultural symbol in the literature of the nineteenth century. Among the many examples that he analyzes, he most forcefully identifies this theme in the writings of Melville: in *Moby Dick*, Ishmael finds the skeleton of a beached whale that inhabitants of a remote Pacific island have erected as a temple to nature. Exploring this temple, he discovers in the whale's center a tangled, verdant growth of vines, nature at its wildest. But the interweaving of the vines, the movement of sun, and the smoke from an altar flame reminds Ishmael, of all things, of a New England textile mill. For Marx, this scene from Melville is an expression of the power of this symbolic landscape, or metaphoric design: "Ishmael deliberately making his way to the center of primal nature only to find, when he arrives, a premonitory sign of industrial power. Art and nature are inextricably tangled at the center."[4]

Not long after Melville wrote *Moby Dick*, this tangle had become the industrial ecosystem. Built on the tropes of nature and industry that were commonplace in the nineteenth century, embodied in the sanitarians' organicist descriptions of urban processes, the industrial ecosystem made those metaphors real. While using the metaphorical power of nature to justify their work, some developers of the industrial ecosystem also understood that they were no longer speaking figuratively. Not only was the industrial ecosystem a metaphoric hybrid of the natural and artificial, it was a literal hybrid as well.

The inventors of the septic tank were explicit. They described the septic tank as the joint creation of both humans and nonhumans, built literally "partly by the mechanics . . . and partly by the anaerobes."[5] This hybrid was "a peculiar structure, produced in part by the hand of man, in part by the forces of nature. After the labors of the mason, the plumber, and the ironworker are over the microorganisms are set to work, and in the course of from six weeks to two months they add an inside floor and a roof of scum to the masonry structure, whereupon its temporary iron cover may be discarded." Only then, "for the first" do "we have the septic tank."[6] The scum, they claimed, "is as much a fixture of the apparatus as though built by carpenters or masons instead of by anaerobes."[7] Without a properly constructed tank, they argued, the anaerobic bacteria could not flourish. Similarly, without the scum on the surface that was created by anaerobic activity, the conditions in the tank could not be sustained. Neither alone was the "septic tank," only the "combination."

A century later, when John Todd asked whether there "can . . . be a synthesis" of ecology and technology, he was reviving this explicit hybrid of nature and industry.[8] His answer to this question was the "living machine," similarly a hybrid of the natural and technological. In fact, Todd's decision to call his sewage treatment systems "living machines" suggests that he had a more complex view of the ideas of "natural" and "artificial." Todd recognized that natural processes needed to be concentrated in smaller areas to efficiently treat sewage. Rather than relying on natural processes operating at natural rates, Todd sought to design treatment processes from scratch, using what he called ecological design principles. Living machines were not, therefore, natural ecosystems, but created ones. Living machines, he argued, took advantage of the "same design principles used by nature to build and regulate its great ecologies in forests, lakes, prairies, or estuaries." But, he emphasized, they were "totally new contained environments." In his patent for the solar aquatics systems, Todd had emphasized the artificial nature of the ecological community assembled in the tanks. The particular arrangement of the solar aquatic tank and its seeding with bacteria and other species, he asserted, "has spawned communities of organisms which have not been seen before either in such quality or such quantities."[10] The ecological community of the solar aquatic tank was new to nature. The builder of the living machine reassembled organisms "in unique ways for specific purposes. Their parts or living components can come from almost any region of the planet and be recombined in utterly new ways."[11]

Todd acknowledged that the term "machine" might trouble many in the environmental movement. "The idea of the living machine is going to be disturbing to some people," he said. "But basically what it's about is taking bits and pieces from nature, using the whole planet as a contributory, and reassembling them."[12] "So what's a living machine?" he asked. "How is it designed? What does it look like? It has engineering components. It has material components. It has living components. And they are all completely integrated."[13] Living machines would "do the work of society—produce food and fuel, treat waste, provide heating and cooling. All of the basic work of society could be done by living machines."[14] Todd's conception of the living machine was essentially that of the industrial ecosystem.

In his history of the Columbia River ecosystem, Richard White adopts the similar metaphor of the "organic machine." In the Columbia River, wrote White, "the human and the natural, the mechanical and the organic had merged so that the two could never be ultimately distinguished."[15] For Todd and White, their terms "living machine," and "organic machine" are meant not so much as a metaphor, but rather an actual description of the ecosystem. In a similar fashion, Donna Haraway means the term "cyborg" in a literal sense, a fusing of the natural and technoscientific."[16]

"What is real is the mixture," White says. Just as the septic tank was "a peculiar structure, produced in part by the hand of man, in part by the forces of nature," the Columbia River "is at once our creation and retains a life of its own beyond our control."[17] The natural and the human in ecosystems can no longer be separated; they are hybrids, living machines, cyborgs, industrial ecosystems.

Nevertheless, it has proved very difficult to accept the implications of the hybrid nature of the industrial ecosystem. Despite the references to hybrids, from Emerson and Melville to Cameron and Martin, sanitary scientists and engineers have been unable to understand the significance for sewage treatment of considering their plant as a true hybrid. The artificial and natural elements of biological sewage treatment were in constant tension as engineers sought to simultaneously naturalize and industrialize the treatment processes. As the history of activated sludge demonstrates, these industrial ecosystems could not be manipulated at will. Natural characteristics will always assert some agency, placing limits on what we can expect from the industrial ecosystem. As Arthur Martin wrestled with the difficulty of replacing the living bacterial floc of activated sludge with artificial materials, he acknowledged that "the whole subject bristled with conundrums."[18] The conundrum was caused by the hybrid nature of the industrial ecosystem. Was activated sludge natural or artificial? Was the "activity" of activated sludge due to its vital or material properties? Could an artificial sludge regenerate itself? As he discovered, it was impossible to separate the natural and artificial; activated sludge was both natural and artificial at the same time. Yet scientists, engineers, and the courts refused to acknowledge this hybrid character and repeatedly tried to place a line precisely where nature ended and artificiality began.

As the industrial ecosystem has impinged on the broader biosphere, this hybrid of industrial and wild in nature has proved equally difficult to navigate. As just one example, in 2009, the U.S. 9th Circuit Court of Appeals in San Francisco issued a ruling on how to distinguish between the "natural" and "artificial" when considering a species for protection under the Endangered Species Act. The central contradiction of the industrial ecosystem was reappearing in a controversy not over the management of a sewage treatment plant or chemical production process, but rather the application of one of the central laws in the United States for the protection of wild ecosystems.[19]

This controversy had its roots in the creation of White's "organic machine," the conversion of the Columbia River into an industrial ecosystem. In 1991, the National Marine Fisheries Service (NMFS) began an evaluation of the status of the coho salmon in the coastal rivers of Oregon to determine if the species was eligible for the protections of the Endangered Species Act. A fundamental part of its determination involved distinguishing between "naturally spawned" and "artificially propagated" fish.[20] Because of widespread overfishing, dam construction, logging in the headwaters, pollution, and diversion of water for irrigation, the salmon runs in the Pacific Northwest had been declining for decades. State, tribal, and federal agencies responded to the decline by producing thousands of fish in industrial-scale fish hatcheries and releasing them into rivers of the Pacific Northwest. Hatcheries, though, produced their own problems (figure 7.1). The environmental conditions of the hatchery artificially selected traits that allowed fish to survive in hatcheries rather than the river. Salmon had been domesticated. These artificially selected fish are genetically distinct from the naturally selected wild fish of the river even though they may have come from the same original broodstock. Fisheries scientists feared that if the hatchery fish bred with wild fish, the locally selected traits that fit the organism to its environment would be lost. In the early 1990s, the NMFS officially distinguished between "natural populations" of salmon and "artificially propagated" hatchery fish, and counted only natural fish to determine the size of a population for the purposes of the act. Because there were so few wild, naturally born fish, they listed the salmon populations of coastal Oregon rivers as "threatened."[21]

The distinction between natural and artificial fish became a political battle in 1999, when a small-town Oregon banker, Ronald Yechout, witnessed hatchery workers on Fall Creek, Oregon, killing returning hatchery fish. The workers were acting to prevent hatchery fish from spawning in the wild. Yechout, who had videotaped the workers clubbing the fish with baseball bats, took his video on the road, showing it to local business groups in Oregon and throughout the salmon territory. The video created a political firestorm. The Oregon State Legislature initiated hearings on the wild fish policy, and legislators expressed their confusion over the contradiction between natural and artificial fish. Suppose "I've got a hatchery fish that makes it out to the ocean, completes its life cycle and comes back up to spawn," speculated State Representative Jeff Kruse. "It is still not a natural fish." "But," he asked the NMFS, "the eggs would be, right?" "That's the point that confuses me," he continued. "Any fish that is strong enough to complete its life cycle and come up and make it up to spawn is a healthy, strong fish. . . . There are weaker strains, they don't make it. So it seems to me you've got a contradiction here." Why, he asked, would the NMFS want to "prevent hatchery fish from coming up and spawning naturally?"[22]

For this legislator and for Yechout, hatchery salmon were the same as wild salmon; there was no distinction between natural and industrial organisms. "We have raised the condor in the laboratory; it's considered wild," Yechout said. "You can raise children in a test tube; they have full constitutional rights. But we all of a sudden have a higher standard for fish. That doesn't make sense." Yechout thought that the government was differentiating between hatchery and wild fish as part of a government plot to limit property rights and put environmental restrictions into place. "It's heavy stuff," he argued. "But stop and think about it: Can you think of

Figure 7.1
The industrial ecosystem has intruded on the wild. Built in Oregon's Willamette National Forest in 1951, the Marion Forks Fish Hatchery appears strikingly similar to an activated sludge plant and carries with it the contradictions of the industrial ecosystem. This hatchery currently produces coho salmon for three river systems in Oregon. Natural selection in rivers has produced salmon adapted to the local conditions of their natal river. When artificially propagated hatchery salmon interbreed with these wild fish, the local adaptation can be lost, reducing the genetic variability and viability of the species in general. *Source*: Oregon State Archives, Secretary of State, OWR0076

better reason why? How better could they come out and control these waterways and control people's property rights?"[23]

A coalition of property rights interests, the Alsea Valley Alliance, took up the cause and, supported by a right-wing legal group, challenged the listing of Oregon coastal coho salmon. Their chief argument was that the NMFS's distinction between natural and artificial fish was "untenable." Hatchery fish that were added to the river in one year would be considered artificially propagated. But if, as in Kruse's example, they escaped capture when they returned from the ocean in following years and reproduced in the river, their offspring would be counted as natural. The U.S. District Court agreed with the Alsea Valley Alliance: "The NMFS listing decision created the unusual circumstance of two genetically identical coho salmon swimming side-by-side in the same stream, but only one receives ESA protection while the other does not. The distinction is arbitrary."[24] This decision raised the possibility that protection under the Endangered Species Act would be denied wild salmon because thousands of hatchery-produced fish would be counted in population estimates. Indeed, based on this decision, an assortment of irrigation, landowner, and business interests throughout Washington, Idaho, Oregon, and California sought to remove the protections of the Endangered Species Act from at least fifteen distinct populations of salmon.[25]

Environmental groups, government agencies, and business interests repeatedly contested this distinction in the courts.[26] Under guidance from the court decision, the NMFS removed the salmon from listing under the Endangered Species Act and developed new criteria for determining whether artificially spawned fish would be included with naturally spawned fish, based on the extent of their genetic differences.[27] Nevertheless, after applying these criteria, the NMFS concluded once again that coho on the Oregon coast were threatened. Property rights groups once again appealed the NMFS's policy, claiming that there should be no distinction between hatchery and wild fish. Illustrating the confusing nature of these cases, the same judge who had earlier decided that distinctions between hatchery and wild fish were arbitrary now ruled that the NMFS had properly made this distinction in their new policy.[28]

When the NMFS used these same criteria to reduce the protections for steelhead salmon in Washington, environmental groups sued the government for not sufficiently distinguishing between natural and artificially spawned fish. The NMFS policy had attempted to resolve the contradiction between hatchery and natural fish but instead got caught in the conflicting definitions created. In the steelhead case, the U.S. District Court noted the tension in the agency's policy, finding it to be "internally contradictory." Parts of the policy based decisions on the whole population, including both hatchery and wild individuals, while other parts of the policy considered only the naturally spawned fish. The district court rejected this and gave the "natural" primacy by deciding that the agency must use "the health and viability of natural populations as the benchmark."[29] This decision, however, was soon reversed by the appeals court[30] that ruled that the NMFS must determine the balance between the contradictory elements of natural and artificial in any population on a case-by-case basis. In a situation where the natural and artificial were confounded, there was no basis for automatically giving nature priority.

As the history of sewage treatment demonstrates, hybrids have proved nearly impossible to recognize. The patent examiner for the septic tank, for instance, rejected Cameron's claim to have invented a hybrid structure of tank and bacteria. The claim, he explained, was for "an aggregation" that included both "apparatus for treating Sewage," meaning the tank itself, as well as the "means for dissociating the solid matter contained," meaning the bacteria. The examiner categorically denied the existence of this aggregation: "There is no combination between the bacteria and the tank," stated the examiner, and denied the claim.[31] The appeals court in the *Cameron* case also rejected Cameron's suggestion that the septic tank itself was something new—a hybrid structure created by bacterial and human labor, both natural and artificial. Acknowledging the argument as at least "interesting," the court nevertheless viewed the masonry structure alone, separate from the bacteria. They declared the tank an old design lacking inventive novelty and denied a patent on the apparatus. The appeals court in the *Activated Sludge* case looked at a gradient of natural and artificial treatment processes and placed a rigid divide between them in order to classify activated sludge as artificial. It could not be both at once. So the NMFS recognized that, indeed, "natural populations represent a spectrum of influence from artificial propagation." But, just as in these earlier sewage cases, the NMFS insisted on being able to draw a line between the two categories, natural and artificial. In this case, fish were either "natural populations" or "hatchery stocks."[32] The NMFS explicitly refused to define salmon populations as both natural and artificial, as what they called "mixed." Even though it recognized these mixed population as hybrids—"a population in which hatchery-origin and natural-origin fish spawn naturally and interbreed, and/or natural-origin fish are regularly incorporated into the hatchery broodstock"—it could not recognize the hybrid itself as being real. Are salmon natural or artificial? By now they are both, and they are neither. They are hybrids of the two. This is the contradiction of the industrial ecosystem.

As White describes the idea of the organic machine, "What is real is the mixture and we seem unable to come to terms with this even though we created it."[33] In seeking to explain society's inability to come to terms with hybrids, Bruno Latour argues that modernity is conditioned by the twin processes of what he calls purification and mediation.[34] Purification is the division of the world into binary categories, between the natural and social, humans and nonhumans. Mediation is the simultaneous creation of hybrids that transcend these two separate poles. For Latour, the creation of these hybrids is one of the characteristics of the modern world behind the dynamism of the Industrial Revolution. But modern society, he argues, can create these hybrids only by refusing to acknowledge their hybrid character. The process of purification assures that we deny the existence of the very things we are busy creating. The industrial ecosystem is just such a hybrid, and in the sewage and salmon cases we can see society creating these hybrids and simultaneously denying their existence, forcing them into either of the twin poles of natural and artificial.

By combining the natural and artificial, the industrial ecosystem holds enormous creative and productive potential. The incorporation of the power of microorganisms into the purification of sewage provided improved public health and protected aquatic ecosystems from pollution. The biochemical diversity of microorganisms allows for the production of penicillin and other antibiotics as well as the production of chemicals and minerals in less environmentally damaging ways. Scientists are currently developing "microbial fuel cells" in which sewage treatment plant bacteria can generate electrical power directly as they decompose waste. But by denying their hybrid existence, we allow these hybrids to proliferate beyond our control, and we have failing sewage treatment plants, herbicide-resistant plants growing in the wild, and endangered salmon, not to mention "tank-bred tuna," "cyborg beetles," "robo-penguins," and "meat farms."[35] Bioscientists have patented a transgenic salmon that can grow over twice as quickly as a wild fish, one of the thousand or so animals patented in the wake of *Diamond v. Chakrabarty*. This fish is a genetically engineered Atlantic salmon that incorporates genes from two other species, chinook salmon and ocean pout, that overcome the normal seasonality of salmon growth and allow growth during the winter. A biotechnology company, AquaBounty Technologies, has licensed these patents and is seeking to commercialize production of these fish, trademarked "AquAdvantage." The fish is opposed by environmental activists, fishermen, fish farmers, and fisheries scientists over concerns that escaped transgenic individuals, like roundup-ready canola or bentgrass, will mate with or outcompete wild salmon.[36] Nevertheless, in May 2011, as I finish this book, the FDA seems poised to approve this animal for commercial production.[37]

Latour argues that the only way to prevent the uncontrolled proliferation of these "monsters" is, paradoxically, to explicitly recognize them. By rejecting the idea that society and nature are distinct, separate entities, and recognizing that humans and nonhumans both have agency, we can get beyond the contradictions of the industrial ecosystem. Only by recognizing them and representing them in a democratic process of decision making can we control which hybrids are created and which avoided.[38]

The industrial ecosystem was born when scientists and engineers enlisted natural organisms in the industrial production process. But as the experience of both salmon hatcheries and herbicide resistance shows, this movement between natural and industrial ecosystems goes both ways. Just as the natural was enlisted for the industrial ecosystem, so too has the industrial invaded the wild. As human activity dominates more and more of the earth, the "natural" ecosystems of river, forest, ocean, and lake become increasingly indistinguishable from an industrial ecosystem. And as the industrial ecosystem spreads, it becomes more and more urgent to face the contradictions inherent in this mode of production. To manage and protect the planet's ecosystems means confronting the inherent contradictions in the human use of those ecosystems, the contradictions of the industrial ecosystem that were so important in the history of the sewage treatment plant.

Notes

Introduction

1. U.S. EPA, "Clean Watersheds Needs Survey, 2004. Report to Congress," U.S. EPA, January 2008, <http://www.epa.gov/cwns/2004rtc/cwns2004rtc.pdf>, (accessed May 27, 2009); G. L. Van Houtven, S. B. Brunnermeier, and M. C. Buckley, "A Retrospective Assessment of the Costs of the Clean Water Act: 1972 to 1997," U.S. Environmental Protection Agency, Office of Water, 2000.

2. My treatment of the "industrial ecosystem" differs from the recently established field of "industrial ecology." In this field, the idea of the ecosystem as a bounded area with inputs and outputs of energy and materials has been used to analyze the industrial firm and describe its connections to suppliers and customers. Industrial ecology tries to close the loop between inputs and waste, devising ways for the waste of one firm to form the supply of another, and create a more environmentally sustainable economy. The idea of the ecosystem, though, is used primarily as a metaphor. In fact, in the statement announcing the creation of the *Journal of Industrial Ecology*, Reid Lifset explicitly identified the term industrial ecology as a metaphor, titling his introduction "A Metaphor, A Field, A Journal." Industrial ecology, he wrote, looks "to the natural world for models of highly efficient use of resources, energy, and wastes." In contrast, as an ecologist and historian, I use the term literally. Reid Lifset, "A Metaphor, a Field, and a Journal," *Journal of Industrial Ecology* 1 (1997): 1–3, on 1. See also Ralf Isenmann, "Industrial Ecology: Shedding More Light on Its Perspective of Understanding Nature as Model," *Sustainable Development* 11 (2003): 143–158. For other metaphorical uses of the term, see Harold L. Platt, *Shock Cities: The Environmental Transformation and Reform of Manchester and Chicago* (Chicago: University of Chicago Press, 2005), preface and chapter 6.

3. Edmund Russell, "The Garden in the Machine: Toward an Evolutionary History of Technology," in *Industrializing Organisms: Introducing Evolutionary History*, ed. Susan R. Schrepfer and Philip Scranton (New York: Routledge, 2004), 2.

4. Keith Vernon, "Pus, Sewage, Beer and Milk: Microbiology in Britain, 1870–1940," *History of Science* 28 (1990): 289–325; Keith Vernon, "Microbes at Work. Micro-organisms, the D.S.I.R. and Industry in Britain, 1900–1936," *Annals of Science* 51 (1994): 593–613.

5. A. Chaston Chapman, "The Employment of Micro-organisms in the Service of Industrial Chemistry. A Plea for a National Institute of Industrial Micro-biology," *JSCI* 38 (1919): 282–286, on 283.

6. Kristoff Glamann, "The Scientific Brewer: Founders and Successors during the Rise of the Modern Brewing Industry," in *Enterprise and History: Essays in Honor of Charles Wilson*, ed. D. C. Coleman and Peter Mathias (Cambridge, UK: Cambridge University Press, 2006), 186–198.

7. Testimony of Ralph McKee, *Record, Vol. II, Union Solvents vs. Guaranty Trust*, Equity no. 802, District Court, Delaware, RG 21, Box 272, NARA Mid-Atlantic, 945; Lloyd C. Cooley, "Acetone," *Industrial and Engineering Chemistry* 29 (1937): 1399–1407.

8. Tetsuo Oka, "Amino Acids, Production Processes," in *Encyclopedia of Bioprocess Technology: Fermentation, Biocatalysis, and Bioseparation*, ed. Michael C. Flickinger and Stephen W. Drew (New York: Wiley, 1999), 89–100; L. Eggeling, W. Pfefferle, and H. Sahm, "Amino Acids," in *Basic Biotechnology*, 3rd ed., ed. Colin Ratledge and Bjørn Kristiansen (Cambridge, UK: Cambridge University Press, 2006), 335–357. Executives of the agribusiness giant ADM and other firms were convicted of price fixing in the marketing of L-lysine, which is the subject of the recent Hollywood film *The Informant!* (2009).

9. On biomining, see D. E. Rawlings and B. D. Johnson, eds., *Biomining* (Berlin: Springer, 2007), and G. J. Olson, J. A. Brierley, and C. L. Brierley, "Bioleaching Review Part B: Progress in Bioleaching: Applications of Microbial Processes by the Minerals Industries," *Applied Microbiology and Biotechnology* 63 (2003): 249–257.

10. Shu-Jen D. Chiang and Jonathan Basch, "Cephalosporins," in *Encyclopedia of Bioprocess Technology*, ed. Flickinger and Drew, 560–570. These bacteria were first isolated from sewage-contaminated seawater. J. M. T. Hamilton-Miller, "Sir Edward Abraham's Contribution to the Development of the Cephalosporins: A Reassessment," *International Journal of Antimicrobial Agents* 15 (2000): 179–184.

11. José Manuel Otero, Gianni Panagiotou, and Lisbeth Olsson, "Fueling Industrial Biotechnology Growth with Bioethanol," *Advances in Biochemical Engineering/Biotechnology* 108 (2007): 1–40.

12. The brief overview of the history of sewage and sewage treatment that follows is drawn chiefly from the following works: Christopher Hamlin, *What Becomes of Pollution? Adversary Science and the Controversy on the Self-Purification of Rivers in Britain, 1850–1900* (New York: Garland Publishing Inc., 1987); Martin V. Melosi, *The Sanitary City: Urban Infrastructure in America from Colonial Times to the Present* (Baltimore: The Johns Hopkins University Press, 2000); Jamie Benidickson, *The Culture of Flushing: A Social and Legal History of Sewage* (Vancouver: University of British Columbia Press, 2007); and Joel A. Tarr, *The Search for the Ultimate Sink: Urban Pollution in Historical Perspective* (Akron: University of Akron Press, 1996); H. H. Stanbridge, *History of Sewage Treatment in Britain*, 12 vols. (Maidstone, UK: The Institute of Water Pollution Control, 1976, 1977). Christopher Hamlin has produced the most thorough understanding of sewage and sanitation in nineteenth-century England: Christopher Hamlin, *Public Health and Social Justice in the Age of Chadwick: Britain, 1800–1854* (Cambridge, UK: Cambridge University Press, 1998); Christopher Hamlin, "Providence and Putrefaction: Victorian Sanitarians and the Natural Theology of Health and Disease," *Victorian Studies* 28 (1985): 381–411; Christopher Hamlin, "William Dibdin and the Idea of Biological Sewage Treatment," *Technology and Culture* 29 (1988): 189–218; Christopher Hamlin, "Muddling in Bumbledom: On the Enormity of Large Sanitary Improvements in Four British Towns, 1855–1885," *Victorian Studies* 32 (1988): 55–83. See also Bill Luckin, *Pollution and Control: A Social History of the Thames in the Nineteenth Century* (Bristol: Adam Hilger, 1986), and Platt, *Shock Cities*. The role of sewage in polluting rivers, lakes, and

oceans, and particularly its impacts on public health, have been extensively treated, both in the historical and technical literature. On public health, see John Duffy, *The Sanitarians: A History of American Public Health* (Urbana: University of Illinois Press, 1990); Judith Walzer Leavitt, *The Healthiest City: Milwaukee and the Politics of Health Reform* (Princeton: Princeton University Press, 1982); Michael Worboys, *Spreading Germs: Disease, Theories, and Medical Practice in Britain, 1865–1900* (Cambridge: Cambridge University Press, 2000). On water pollution, see H. B. N. Hynes, *Biology of Polluted Waters* (Liverpool: Liverpool University Press, 1963), and Charles E. Warren, *Biology and Water Pollution Control* (Philadelphia: Saunders, 1971), Philip V. Scarpino, *Great River: An Environmental History of the Upper Mississippi, 1890–1950* (Columbia: University of Missouri Press, 1986).

13. Henry Austin, *Report on the Means of Deodorizing and Utilizing the Sewage of Towns: Addressed to the Rt. Hon. The President of the General Board of Health* (London: Eyre and Spottiswoode, 1857), [2262], 7.

14. Stephen Halliday, *The Great Stink of London: Sir Joseph Bazalgette and the Cleansing of the Victorian Metropolis* (Stroud: Sutton Publishing, 1999).

15. Royal Commission on Metropolitan Sewage Discharge, *First Report of the Commissioners*, 1884 [C. 3842], 253 (Baldwin Latham), 154 (Thomas Long), lii (harbormasters), 201 (Henry Jones). On Thames mud butter, see Frances Power Cobbe, Review of "*The Modern Householder: A Manual of Domestic Economy*," *The Academy* 7 (Apr. 3, 1875): 340–341, on 340. Rumors of butter made from Thames mud were so persistent that the *Sanitary Record* dispatched a team to investigate the rumors and assure Londoners that their butter was pure. "Thames Mud and Butter," *Boston Medical and Surgical Journal* 94 (1877): 150.

16. Upton Sinclair, *The Jungle* (New York: Doubleday, Page, and Company, 1906), 112.

17. C.-E. A. Winslow, "The Disposal of City Sewage," in *Municipal Chemistry*, ed. Charles Baskerville (New York: McGraw-Hill, 1911), 276–299, on 277.

18. Preliminary Report of the Commission Appointed to Inquire Into the Best Mode of Distributing the Sewage of Towns, and Applying it to Beneficial and Profitable Uses (London: Eyre and Spottiswoode, 1858), [2372], 6.

19. Hamlin, "Muddling in Bumbledom."

20. For Frankland's involvement in sewage and water issues, see Hamlin, *What Becomes of Pollution?*, 294–431.

21. *First Report of The Commissioners Appointed in 1868 to Inquire Into The Best Means of Preventing the Pollution of Rivers. (Mersey and Ribble Basins.) Vol. I. Report and Plans*, 1870 [C. 37], 62–70.

22. See Hamlin, *What Becomes of Pollution?*, 430, 522. See also Michelle Allen, *Cleansing the City: Sanitary Geographies in Victorian London* (Athens: Ohio University Press, 2008).

23. When the United States took over important German patents for chemical manufacture during the war, it also claimed Karl Imhoff's sewage treatment tank patent.

1 Natural vs. Artificial

1. *Sewage of Towns. Third Report and Appendices of the Commission Appointed to Inquire Into the Best Mode of Distributing the Sewage of Towns and Applying it to Beneficial and Profitable Uses*, 1865 [3472], 2.

2. George Vivian Poore, *The Earth in Relation to The Preservation and Destruction of Contagia, Being the Milroy Lectures Delivered at the Royal College of Physicians in 1899 Together With Other Papers on Sanitation* (London: Longmans, Green, and Co., 1902), 168.

3. Charles Dickens wrote in *Hard Times*: "It was a town of machinery and tall chimneys, out of which interminable serpents of smoke trailed themselves for ever and ever, and never got uncoiled. It had a black canal in it, and a river that ran purple with ill-smelling dye . . . where the piston of the steam-engine worked monotonously up and down." On English literature and its critique of industrialization, see Raymond Williams, *Country and City*, rev. ed. (Oxford: Oxford University Press, 1975); Jeremy Burchardt, *Paradise Lost: Rural Idyll and Social Change in England since 1800* (London: I. B. Tauris, 2002), 25–34, 67–76; and Michelle Allen, *Cleansing the City: Sanitary Geographies in Victorian London* (Athens: Ohio University Press, 2008). Leo Marx also identifies the conflict between the pastoral and industrial as the central theme of nineteenth-century American literature. Leo Marx, *The Machine in the Garden: Technology and the Pastoral Ideal in America* (Oxford: Oxford University Press, [1964] 2000).

4. "The Cremation of Sewage," *Times* (London), Sept. 15, 1898.

5. For biological treatment to be adopted by cities, biology also had to acquire a "constituency," a process described by Christopher Hamlin in *What Becomes of Pollution?* Naturalizing biology, I argue, was a key part of this process.

6. William Crookes and Ernst Röhrig, *A Practical Treatise on Metallurgy: Adapted from the Last German Edition of Prof. Kerl's Metallurgy, Vol. II, Copper, Iron* (London: Longman's, Green, and Co., 1869), 24–28, 358–363.

7. See Carroll Pursell, *Technology in Postwar America: A History* (New York: Columbia University Press, 2007); William Cronon, *Nature's Metropolis: Chicago and the Great West* (New York: W. W. Norton, 1991), particularly chapters 3 and 5; James C. Scott, *Seeing Like a State: How Certain Schemes to Improve the Human Condition Have Failed* (New Haven: Yale University Press, 1998), esp. 11–22; Anthony Giddens, *The Consequences of Modernity* (Stanford: Stanford University Press, 1990), 17–28; Wolfgang Schivelbusch, *The Railway Journey: The Industrialization of Time and Space in the Nineteenth Century* (Berkeley: University of California Press, 1986), 33–44.

8. Hamlin, *What Becomes of Pollution?*, 366–368, 392–395, 429–431, quotation on 430.

9. For agency of nonhumans, see Jim Johnson [Bruno Latour], "Mixing Humans and Nonhumans Together: The Sociology of a Door-closer," *Social Problems* 35 (1988): 298–310.

10. *Sewage of Towns. Second Report of the Commission Appointed to Inquire Into the Best Mode of Distributing the Sewage of Towns and Applying it to Beneficial and Profitable Uses*, 1861 [1962], 12.

11. *Sewage of Towns. Third Report*, 1865 [3472], 2.

12. George Shepherd, "The Sewage Experiments of Tottenham and Leicester," *Times* (London), Aug. 6, 1868.

13. Hamlin, "Providence and Putrefaction," 391.

14. *First Report of The Commissioners Appointed in 1868 to Inquire Into The Best Means of Preventing the Pollution of Rivers. (Mersey and Ribble Basins.) Vol. I. Report and Plans*, 1870 [C. 37], 71.

15. The connection between the rural and natural was supported by the writings of Gilbert White. See Donald Worster, *Nature's Economy: A History of Ecological Ideas*, 2nd ed. (Cambridge: Cambridge University Press, 1994), 3–25.

16. Hamlin, "Providence and Putrefaction." Quotation from Charles Kingsley on 403.

17. W. C. Sillar in discussion of Henry Robinson, "On 'Sewage Disposal,'" *Transactions of the Sanitary Institute of Great Britain* 6 (1884–1885): 216–222, 239–240, on 239.

18. Quoted in W. H. Corfield, *The Treatment and Utilization of Sewage* (London: Macmillan and Co.: 1887), 375.

19. Wm. Ham. Hall, *The Sewage Question in California. Report of the State Engineer, Wm. Ham. Hall, to the Board of Directors of the Stockton Insane Asylum on the Sewerage for the Institution in Their Charge* (Sacramento: State Office, 1883), 29, 49.

20. R. Warington, "On Nitrification: A Report of Experiments conducted in the Rothampsted Laboratory," *Chemical News* 36 (1877): 263–264; Robert Warington, "On Nitrification," *Journal of the Chemical Society* 33 (1878): 44–521; Robert Warington, "On Nitrification. (Part II)," *Journal of the Chemical Society* 35 (1879): 429–456; Hamlin, *What Becomes of Pollution?*, 429–430.

21. Henry Robinson, "Address," *Transactions of the Sanitary Institute of Great Britain* 4 (1882–1883): 139–156; Henry Robinson, "On 'Sewage Disposal,'" *Transactions of the Sanitary Institute of Great Britain* 6 (1884–1885): 216–222, 239–240; Henry Robinson, "On 'River Pollution,'" *Transactions of the Sanitary Institute of Great Britain* 8 (1886–1887): 175–184; On Robinson, see "Obituary," *Transactions of the Institution of Water Engineers* 20 (1915): 94.

22. Hamlin, *What Becomes of Pollution?*, 431, 512–513.

23. Wm. Ripley Nichols and George Derby, "Sewerage; Sewage; The Pollution of Streams; The Water-Supply of Towns. A Report to the State Board of Health of Massachusetts," Fourth Annual Report of the State Board of Health of Massachusetts, January 1873, 19–132.

24. Harrison P. Eddy, "Massachusetts—The Cradle of Public-Health Engineering," *SWJ* 2 (1930): 394–403.

25. On the Frankland experiments, see *First Report of The Commissioners Appointed in 1868*, 60–70.

26. On the establishment of the Lawrence experiment station, see "Water Supply and Sewerage," *Nineteenth Annual Report of the State Board of Health of Massachusetts* (Boston: Wright & Potter Printing Company, 1888), 1–96; Robert A. McCracken and Dennis Sebian, "Lawrence Experiment Station: Birthplace of Environmental Research in America," *The Diplomate* 24 (2) (1988): 12–18, WEF Archives, Record Group 2, Box 56.

27. William T. Sedgwick and Edmond B. Wilson, *General Biology* (New York: Henry Holt & Co., 1889), 104, 108, 161. In an 1895 revision of this textbook, Sedgwick added chapters on bacteria that demonstrated that he thought of bacteria in the same terms as animals and plants.

28. James Strick, "Evolution of Microbiology as Seen in the Textbooks of Edwin O. Jordan and William H. Park," *Yale Journal of Biology and Medicine* 72 (1999): 321–328.

29. William T. Sedgwick, "A Report of the Biological Work of the Lawrence Experiment Station," in *Experimental Investigations by the State Board of Health of Massachusetts, upon*

the Purification of Sewage by Filtration and by Chemical Precipitation and upon the Inter-mittent Filtration of Water. Made at Lawrence, Mass., 1888–1890, Part II of Report on Water Supply and Sewerage (Boston: Wright and Potter, 1890), 795–862, on 823. Part V of this report, "On Certain Species of Bacteria Observed in Sewage," 821–844, is separately authored by Edwin O. Jordan. On evolutionary theory and germ theory, see Nancy Tomes, *Gospel of Germs: Men, Women, and the Microbe in American Life* (Cambridge, MA: Harvard University Press, 1998), 33–34.

30. Sedgwick, "A Report," 823.

31. Robert M. Young, *Darwin's Metaphor: Nature's Place in Victorian Culture* (Cambridge, UK: Cambridge University Press, 1985), 81–124.

32. Gregory J. Cooper, *The Science of the Struggle for Existence: On the Foundations of Ecology* (Cambridge, UK: Cambridge University Press 2003), 1–12.

33. Mikulas Teich, "Fermentation Theory and Practice: the Beginnings of Pure Yeast Cultivation and English Brewing, 1883–1913," *History of Technology* 8 (1983): 117–133. See also E. M. Sigsworth, "Science and the Brewing Industry, 1850–1900," *The Economic History Review* 17 (1965): 536–550.

34. Emil Chr. Hansen, *Practical Studies in Fermentation, Being Contributions to the Life History of Micro-organisms*, trans. Alex K. Miller (London: E & FN Spon, 1896), iv, 155; emphasis in original.

35. John Simmons Ceccatti, "Science in the Brewery: Pure Yeast Culture and the Transformation of Brewing Practices in Germany at the End of the 19th Century," PhD diss., University of Chicago, 2001.

36. Emil Chr. Hansen, "Über "künstliche" und "natürliche" Hefenreinzucht," *Zeitschrift für das gesammte Brauwesen* 18 (1895): 113–115, translated and quoted in Ceccatti, "Science in the Brewery," 123.

37. Max Delbrück, "Die Carlsberger reingezüchtete Hefe, *Wochenschrift fur Brauerei*, 1885, 2 (1885):126–128, translated and quoted in Ceccatti, "Science in the Brewery," 106.

38. Sedgwick, "A Report," 845.

39. Ibid., 846.

40. Ibid., 859–860.

41. Ibid., 861.

42. H. H. Stanbridge, *History of Sewage Treatment in Britain, Vol. 6, Biological Filtration* (Maidstone, UK: The Institute of Water Pollution Control, 1976), 25.

43. W. D. Scott-Moncrieff, *Chadwick Lecutres, Session 1907–1908* (London: St. Bride's Press, 1909), 23.

44. W. D. Scott Moncrieff, "Recent Improvements in the Construction of Cultivation Tanks," *AMSDW*, 1915, 82–88, on 83.

45. "'Cultivation Filters' for Sewage Disposal," *The Engineer* 74 (1892): 330.

46. Donald Cameron, "The Septic Tank System of Sewage Treatment," *The Builder*, 1897, 27.

47. J. H. Garner, "Historical Development of Sewage Purification Processes," *Surveyor* 66 (1924): 213–216.

48. William Joseph Dibdin, "Sewage-Sludge and Its Disposal," *Minutes of Proceedings of the Institution of Civil Engineers* 88 (1887): 155–298, on 162. On Dibdin, see Hamlin, "William Dibdin and the Idea of Biological Sewage Treatment."

49. Henry C. H. Shenton, "Recent Practice in Sewage Disposal," *Transactions, Society of Engineers*, 1901: 217–266, on 219. For similar sentiments in the American literature, see A. C. Abbott, "The Utilization of Bacteria and Bacteriologic Methods in Sanitary Engineering," *Proceedings of the Engineers' Club of Philadelphia* 27 (1900): 47–63.

50. A. Bostock Hill, "Some Points in the Modern Treatment of Sewage," *AMSDW*, 1905, 58–65, on 60.

51. On Boyce, see C. S. Sherrington, "Boyce, Sir Rubert William (1863–1911)," rev. Claire E. J. Herrick, *Oxford Dictionary of National Biography* (Oxford: Oxford University Press, 2004).

52. Professor Boyce and Doctors Grünbaum, MacConkey, and Hill, "Investigation of the River Severn in the Shrewsbury District," *Second Report of the Commissioners Appointed in 1898 to Inquire and Report What Methods of Treating and Disposing of Sewage May Properly be Adopted*, 1902 [Cd. 1178], 105.

53. Ibid., 108.

54. Hamlin, "Muddling in Bumbledom."

55. Hamlin, *What Becomes of Pollution?*, 37.

56. Barwise in Arthur J. Martin, *The Sewage Problem: A Review of the Evidence Collected by the Royal Commission on Sewage Disposal* (London: The Sanitary Publishing Company, 1905), 32.

57. Alan Wilson, "Technology and Municipal Decision-Making: Sanitary Systems in Manchester 1868–1910," PhD thesis, University of Manchester, 1990.

58. "Royal Commission on Sewage Disposal," *Times* (London), Nov. 2, 1898, 4.

59. "How Shall Exeter Dispose of Its Sewage? A New System Proposed," *Sanitary Record* 20 (1897): 584–585.

60. "First Annual Dinner," *AMSDW*, 1903, 28–40, on 29.

61. Henry Kenwood and William Butler, "Some Observations on the Natural Purification of Sewage," *Journal of the Sanitary Institute* 19 (1898): 671–685.

62. H. Alfred Roechling, "The Present Status of Sewage Irrigation in Europe and America," *Journal of the Sanitary Institute* 27 (1896): 483–504, on 484.

63. Discussion of papers by W. Kaye Parry, W. J. Dibdin and G. Thudichum, Dr. Barwise, Prof. A. Bostock Hill, Donald Cameron and E. J. Silcock, *Journal of the Sanitary Institute* 18 (1897): 576–590, on 577.

64. J. Bailey-Denton, *Sewage Disposal. Ten Years' Experience (Now Fourteen Years) in Works of Intermittent Downward Filtration, Separately and in Combination with Surface Irrigation; With Notes on the Practice and Results of Sewage Farming* (London: E & FN Spon, 1885). Indeed, George Thudichum later attributed the failure of intermittent filtration to take hold to a lack of just such an understanding between bacteria and the environmental conditions in which they might thrive. Intermittent filtration, he said, "is practically that of the modern bacteria bed." "Its failure to make a permanent stand as the true system of purification," he continued, "is due almost entirely to the fact that its promoters, whilst recognising

its powers, failed to realise the true causes of its action, and therefore did not take care to supply the conditions which are now known to be essential. George Thudichum, *The Bacterial Treatment of Sewage. A Handbook for Councillors, Engineers, and Surveyors* (London: The Councillor and Guardian Offices, n.d.), 11–12.

65. "Obituary," *JRSI* 41 (1920): 44.

66. Discussion of papers by W. Kaye Parry, W. J. Dibdin and G. Thudichum, Dr. Barwise, Prof. A. Bostock Hill, Donald Cameron, and E. J. Silcock, *Journal of the Sanitary Institute* 18 (1897): 576–590, on 583.

67. Alfred S. Jones, "Bacterial Methods of Sewage Disposal," *Times* (London), May 2, 1900, 14.

68. "Discussion," *Journal of the Sanitary Institute* 19 (1898): 717–729, on 720.

69. Alfred S. Jones and H. Alfred Roechling, *Natural and Artificial Sewage Treatment* (London: E & FN Spon, 1902), 2.

70. Ibid., 37.

71. Martin, *The Sewage Problem*, 52.

72. Ibid.

73. Alfred S. Jones, "Sewage Disposal," *Times* (London), Dec. 27, 1907, 14.

74. Jones and Roechling, *Natural and Artificial*, 2, 19.

75. Ibid., 34, 37–38, 71.

76. George Vivian Poore, *Essays on Rural Hygiene* (London: Longmans, Green, and Co., 1893), 60–61

77. Martin, *The Sewage Problem*, 52.

78. Ibid., 64.

79. Samuel Rideal, *Sewage and the Bacterial Purification of Sewage* (New York: John Wiley, 1900).

80. "Annual General Meeting," *AMSDW*, 1902, 9–20, on 10, 12–13.

81. George Thudichum, *The Bacterial Treatment of Sewage. A Handbook for Councillors, Engineers, and Surveyors* (London: The Councillor and Guardian Offices, n.d.), 88–90, on 89–90.

82. Royal Commission on Sewage Disposal, *Interim Report of the Commissioners Appointed in 1898 to Inquire and Report What Methods of Treating and Disposing of Sewage May Properly be Adopted*, Vol. 1, 1901 [Cd. 685]. Speech by S. Rideal, "Annual General Meeting," *AMSDW*, 1902, 9–20.

83. Martin, *The Sewage Problem*, 284.

84. Discussion of Sidney H. Chambers, "Some Interpretations of Sewage Purification Phenomena," *Proceedings of the Incorporated Association of Municipal and County Engineers* 34 (1907): 30–73, on 66.

85. Alfred Stowell Jones and William Owen Travis, "On the Elimination of Suspended Solids and Colloidal Matters from Sewage," *Minutes of Proceedings of The Institution of Civil Engineers* 154 (1906): 68–94, on 77; discussion on 95–195.

86. Ibid., 79.

87. Ibid., 80, 85.

88. Discussion of Jones and Travis, "On the Elimination," 112 (Dibdin), 119 (Rideal), and 127, 128 (Scott-Moncrieff).

89. Sidney H. Chambers, "Some Interpretations of Sewage Purification Phenomena," *Proceedings of the Incorporated Association of Municipal and County Engineers* 34 (1907): 30–73, on 42.

90. W. Owen Travis, "The Hampton Doctrine in Relation to Sewage Purification," *Surveyor* 34 (1908): 63–65, on 63, 66; George Lewis Travis, "The Hampton Interpretation of the Operation of Sewage Purification," *British Medical Journal*, 1908 (II): 575–577.

91. Leonard P. Kinnicutt, C.-E. A. Winslow, and R. Winthrop Pratt, *Sewage Disposal* (New York: John Wiley and Sons, 1910), 156–159; George W. Fuller, *Sewage Disposal* (New York: McGraw-Hill, 1912), 593.

92. Lieut.-Col. Jones, "President's Address, Practice with Science In Cooperation," *AMSDW*, 1909, 14–22; W. Owen Travis, "Some Observations on the Principles of Sewage Purification," *AMSDW*, 1911, 175–196, "heresy" on 191, Travis on 196. The language used by sanitarians in this debate—"doctrine," "heresy," "immaculate," "preachers"—illustrate the characteristics of religious dogma applied to germ theory more broadly. See Tomes, *Gospel of Germs*.

93. Royal Commission on Sewage Disposal, *Fifth Report of the Commissioners Appointed to Inquire and Report What Methods of Treating and Disposing of Sewage (Including Any Liquid from Any Factory or Manufacturing Process) May Properly Be Adopted. Methods of Treating and Disposing of Sewage. Report*, 1908 [Cd. 4278], 9.

94. John D. Watson, in "Discussion on the Final Report of the Royal Commission on Sewage Disposal, opened by Samuel Rideal and J. D. Watson," *JRSI* 36 (1915): 452–465, on 457.

95. Discussion of John Manly, "Sewage Treatment.—Advantages of Land over Artificial Schemes," *JRSI* 34 (1913): 53–59, on 57.

96. A. S. Jones in discussion of S. Rideal, "Sewage Purification," *JRSI* 32 (1911): 429–438, on 438.

97. Arthur J. Martin, "The Evolution of Sewage Disposal," *JRSI* 34 (1913–1914): 469–474.

98. C. Meymott Tidy, *The Treatment of Sewage* (New York: D. Van Nostrand, 1887), 51 [abridged from the *Journal of the Society of Arts*].

99. Dibdin, *The Purification of Sewage and Water*, 3rd ed. (London: Sanitary Publishing Company, 1903), 112.

100. Quoted in Martin, *The Sewage Problem*, 272.

101. Frank Clowes, "The Experimental Bacterial Treatment of London Sewage," *Proceedings of the Incorporated Association of Municipal and County Engineers* 29 (1903): 265–279, on 267.

102. Thudichum, *The Bacterial Treatment of Sewage*, 70.

103. "Discussion on Modern Methods of Sewage Disposal," *JRSI* 28 (1908): 1–14, 2.

104. Frank Clowes and A. C. Houston, *The Experimental Bacterial Treatment of London Sewage, Being an Account of the Experiments Carried out by the London County Council Between the Years 1892 and 1903*, London County Council, No. 679, 1904, 126–129.

105. Quoted in Martin, *The Sewage Problem*, 269.

106. Discussion of papers by Dr. H. Kenwood and Dr. W. Butler, S. Rideal, W. D. Scott-Moncrieff, and W. E. Adeney, *JRSI* 19 (1898): 717–729, on 718.

107. *Cameron v. Saratoga Springs*, 159 F. 458.

108. W. Noble Twelvetrees, "Disinfection: Physical, Chemical, Mechanical," *Sanitary Record* 19 (1897): 483, as cited in Hamlin, *What Becomes of Pollution?*, 542.

109. A. S. Jones, "Remarks on the Eighth Report of the Royal Commission on Sewage Disposal," *AMSDW*, 1912, 121–122, on 122.

110. H. A. Roechling, in Discussion of Samuel Rideal, "The Purification of Sewage by Bacteria," *Journal of the Sanitary Institute* 18 (1897): 59–82, on 77.

111. Discussion of Arthur E. Collins, "Special Features of the Travis Hydrolytic System of Sewage Tanks Being Constructed at Norwich," *JRSI* 30 (1909): 261–288, on 284–285.

112. See William Cronon, *Nature's Metropolis: Chicago and the Great West* (New York: W. W. Norton & Co., 1991), 229–230.

113. John Manley, "Sewage Treatment–Advantages of Land over Artificial Schemes," *JRSI* 34 (1913): 53–59, on 53. On the Back to the Land Movement in England during this period, see Jan Marsh, *Back to the Land: The Pastoral Impulse in Victorian England from 1880 to 1914* (London: Quartet, 1983).

114. Jones and Roechling, *Natural and Artificial*, 15.

115. Jones, "President's Address," 18.

116. "Annual Meeting in London," *AMSDW*, 1916, 51.

117. Lloyd G. Stevenson, "Science Down the Drain: On the Hostility of Certain Sanitarians to Animal Experimentation, Bacteriology and Immunology," *Bulletin of the History of Medicine* 29 (1955): 1–26, on 14.

118. Jones and Roechling, *Natural and Artificial*, 19; Sir William Collins, "The Man versus the Microbe," *JRSI* 23 (1902): 335–356, on 339.

119. Jones and Roechling, *Natural and Artificial*, 3–4.

120. Collins, "The Man versus the Microbe," 344.

121. Ibid., 339.

122. George Wilson, *A Handbook of Hygiene and Sanitary Science*, 2nd ed. (Philadelphia: Lindsay and Blakiston, 1873).

123. George Wilson, *A Handbook of Hygiene and Sanitary Science*, 8th ed. (London: J&A Churchill, 1898).

124. "Obituary," *Public Health* 35 (1921): 32. For reviews, see *The Lancet*, 1898, pt. II: 1555; *New York Medical Journal* 69 (1899): 101; and *Public Health* 11 (1899): 371–373.

125. Ceccatti, "Science in the Brewery."

126. Max Delbrück and Franz Schönfeld, eds., *System der naturlichen Hefereinzucht* (Berlin: Paul Parey, 1903), translated and quoted in Ceccatti, "Science in the Brewery," 89; Max Delbrück, "Die natürliche Hefenreinzucht," *Wochenschrift fur Brauerei* 12 (1895): 66–67, translated and quoted in Ceccatti, "Science in the Brewery," 121.

127. Emil Chr. Hansen, "Über "künstliche" und "natürliche" Hefenreinzucht," *Zeitschrift für das gesammte Brauwesen* 18 (1895): 113–115, translated and quoted in Ceccatti, "Science in the Brewery," 123.

128. Ceccatti, "Science in the Brewery," 89.

129. Dibdin, "Sewage-Sludge and Its Disposal." His claims for priority were disputed by a number of writers who credited the Lawrence, Massachusetts, experiment station for the genesis of the idea. Others gave ultimate credit to Frankland, for his intermittent filtration experiments, even though Frankland did not recognize the role of bacteria until much later. Professor Dr. Dunbar, *Principles of Sewage Treatment*, trans. H. T. Calvert (London: Charles Griffin & Co., 1908), 188.

130. Dibdin, "Sewage-Sludge and Its Disposal," on 161–162. See also Dibdin, quoted in "Purification of Sewage by Microbes," *Engineering* 54 (1892): 453.

131. Christopher Hamlin discusses Dibdin's paper and its reception in Hamlin, "William Dibdin and the Idea of Biological Sewage Treatment," 206–208. Comments of Jones in Dibdin, "Sewage-Sludge and Its Disposal," 196. This laughter was frequently brought up again by supporters of bacterial treatment, even fifteen years later. See, for example, Thudichum, *The Bacterial Treatment of Sewage*, 14.

132. Comments of Dr. A. Angell in Dibdin, *Minutes of Proceedings of the Institution of Civil Engineers* 88 (1887): 204.

133. Charles E. Avery, "Manufacture of Lactates," U.S. Patent 330,815, Nov. 17, 1885. On Avery, see H. Benninga, *A History of Lactic Acid Making: A Chapter in the History of Biotechnology* (Dordrecht: Kluwer Academic Publishers, 1990), 86–97. On other early applications of industrial microbiology see, for instance, John C. Pennington, "Retting-Bath," U.S. Patent 509,396, Nov. 28, 1893; Carl Wehmer, "Process of Making Citric Acid," U.S. Patent 515,033, Feb 20, 1894; Joseph T. Wood, "Bate," U.S. Patent 638,828, Dec. 12, 1899. By the first decades of the twentieth century, these ideas had coalesced into an overall program for industrial microbiology. Chapman, "The Employment of Micro-organisms in the Service of Industrial Chemistry," 283T. See Keith Vernon, "Pus, Sewage, Beer and Milk" and "Microbes at Work."

134. Deposition of Gilbert Fowler, Plaintiffs' Depositions, GSD-Dep, 469–470; Ernest Moore Mumford, "A New Iron Bacterium," *Journal of the Chemical Society Transactions* 103 (Pt. 1) (1913): 645–650.

135. Depositions of Stephen DeM. Gage, 418–419, and Harry W. Clark, 11–12, Defendant's Depostions, Vol. 1, ASM-Dep.

136. Deposition of Gilbert Fowler, Plaintiffs' Depositions, 464–470, M.7 quotation on 470, GSD-Dep.

137. Gilbert J. Fowler and E. Moore Mumford, "Preliminary Note on the Bacterial Clarification of Sewage," *JRSI* 34 (1913): 497–500, on 500, "Discussion," *JRSI* 34 (1913): 501–510.

138. Deposition of Edward Ardern, Plaintiffs' Depositions, 512, GSD-Dep.

139. Deposition of William Lockett, Defendants' Depositions, 488–490, GSD-Dep.

140. Deposition of William Lockett, Defendants' Depositions, 512–514, GSD-Dep.

141. Deposition of William Lockett, Defendants' Depositions, 901, GSD-Dep.

142. Edward Ardern and William T. Lockett, "Experiments on the Oxidation of Sewage without the Aid of Filters," *JSCI* 33 (1914): 523–539.

143. Deposition of Edward Ardern, Plaintiffs' Depositions, 520, GSD-Dep.

144. Epoch-making quotation from Ardern and Lockett, "Experiments on the Oxidation of Sewage," 536. For the political and environmental context of the Manchester experiments, see Wilson, *Technology and Municipal Decision-Making*, and Harold L. Platt, *Shock Cities: The Environmental Transformation and Reform of Manchester and Chicago* (Chicago: University of Chicago Press: 2005), 408–441.

145. F. W. Mohlman, "A Census of Sewerage Systems and Sewage Treatment Plants," *SWJ* 14 (1942): 731–732. In the United States, a population of 22,143,000 was served by secondary treatment plants, although Mohlman considered this figure to be high because chemical precipitation plants were undoubtedly overestimated. The true figure was closer to 18,000,000. Of these, there were 302 Activated Sludge plants, serving a population of 10,480,000 and 1,486 Trickling Filters, serving 8,425,000. Using the 18,000,000 figure, activated sludge treated 58 percent of the population served by secondary treatment.

146. Deposition of William Lockett, Defendants' Depositions, 900, GSD-Dep.

147. This point is made by Hamlin, "William Dibdin and the Idea of Biological Sewage Treatment," 212–213, from which these quotations are taken.

148. H. W. Clark and Stephen DeM. Gage, *A Review of Twenty-one years' Experiments upon the Purification of Sewage at the Lawrence Experiment Station*, reprinted from the 40th Annual Report of the State Board of Health (Boston: Wright & Potter Printing Co., State Printers, 1909), 114.

149. Deposition of Gilbert Fowler, Plaintiffs' Depositions, 472–473, GSD-Dep.

150. Deposition of Gilbert Fowler, Plaintiffs' Depositions, 473, GSD-Dep.

151. Comments of H. Alfred Roechling in discussion of G. J. Fowler, "The Activated Sludge Process of Sewage Purification," reprinted from the *Journal of the Institute of Sanitary Engineers*, March and April 1916, 13, BP.

152. Arthur J. Martin, *Sewage and Sewage Disposal* (London: MacDonald and Evans, 1930), 44.

153. Comments of John D. Watson on J. P. Wakeford, "Brief Notes of Experiments in Sewage Purification by Forced Aeration," *Surveyor* 48 (1915): 132–146, on 144.

154. J. Grossmann in discussion of S. E. Melling, "The Purification of Salford Sewage Along the Lines of the Manchester Experiments," *JSCI* 33 (1914): 1124–1130, on 1128; Ardern, in discussion of Melling, "Purification," 1129; Eddy, "Lights and Shadows," 260.

155. Fowler to Bartow, Mar. 31, 1915, BP, Box 3.

156. F. W. Harris, T. Cockburn, and T. Anderson, "Observations on Biological and Physical Properties of Activated Sludge and the Principles of its Application," *AMSDW*, 1926, 52–68, on 52.

157. Deposition of Edward Ardern, GSD-Dep., 504. See also M. N. Baker, "Activated Sludge in America: An Editorial Survey," *Engineering News* 74 (1915): 164–171.

158. Ardern and Lockett, "Experiments on the Oxidation of Sewage," 535; Edward Ardern and William T. Lockett, "The Oxidation of Sewage without the Aid of Filters. Part II," *JSCI* 33 (1914): 1122–1124.

159. Robbins Russel and Edward Bartow, "Bacteriological Study of Sewage Purification by Aeration," n.p., n.d., 348–358, BP, Box 3.

160. J. W. Haigh Johnson, "A Contribution to the Biology of Sewage Disposal. Pt. II," *Journal of Economic Biology* 9 (1914): 127–164.

161. A. M. Buswell and H. L. Long, "Microbiology and Theory of Activated Sludge," *Journal of the American Water Works Association* 10 (1923): 309–321, on 317.

162. Gilbert J. Fowler, "Aeration of Activated Sludge," *Surveyor* 67 (1925): 305.

163. William B. Fuller, "Sewage Disposal by the Activated Sludge Process: An Epoch-Making Discovery," *The American City* 14 (1916): 78–81.

164. Buswell and Long, "Microbiology," 317.

165. "Exhibition of Photo-Micrographs," *AMSDW*, 1923, 21–24, on 22.

166. W. E. Speight, in discussion of F. R. O'Shaughnessy, "Sewage Disposal and the Community," *AMSDW*, 1924, 68–83, on 78.

167. Thresh, "The Sewage Works Manager of the Future," *AMSDW*, 1909, 18.

168. Minutes of the Purification Works Subcommittee of the Manchester Rivers Committee, No. 32, Mar. 20, 1922, on 138, and May 22, 1932, on 149–150, Manchester Archives, Manchester Central Library, Manchester, UK.

169. Editorial, *Surveyor* 43 (1923): 525.

170. Fritz Simmer, "Process for Biological Purification of Waste Water," U.S. Patent 1,751,459, Mar. 18, 1930. Simmer also filed for patents on this invention in Austria, Germany, Czechoslovakia, Poland and Great Britain, and assigned it to a Danish fermentation company, Aktieselskabet Dansk Gaerings Industri.

171. John Van Nostrand Dorr, "The Conversion of Batch into Continuous Processes," *Industrial and Engineering Chemistry* 21 (1929): 465–471. Dorr made equipment for sewage treatment plants.

172. See Michael Nuwer, "From Batch to Flow: Production Technology and Work-Force Skills in the Steel Industry, 1880–1920," *Technology and Culture* 29 (1988): 808–838.

173. See Ardern on obviousness of continuous process, Ardern deposition, 520, GSD-Dep. Leslie C. Frank and Calvin W. Hendrick, "Co-operation Sought in Conducting Activated Sludge Experiments at Baltimore," *ER* 71 (1915): 521–522; Edward Ardern, "A Recent Development of the Sewage Problem," *AMSDW*, 1914, 61–71, on 69; H. C. H. Shenton, "The Activated Sludge Process," *Surveyor* 59 (1921): 323–324.

174. J. A. Reddie, in discussion of F. R. O'Shaughnessy, "The Physical Aspect of Sewage Disposal," *JSCI* 42 (1923): 359T–370T, on 368T.

175. F. R. O'Shaughnessy, "The Physical Aspect of Sewage Disposal," *JSCI* 42 (1923): 359T–370T.

176. Arthur J. Martin, *The Activated Sludge Process* (London: MacDonald and Evans, 1927), 314.

177. H. T. Calvert, "Sanitation and Water Purification," *Reports of the Progress of Applied Chemistry* 7 (1922): 460–487, on 464.

178. Arthur John Martin, "The Bio-Aeration of Sewage," *Minutes of Proceedings of the Institution of Civil Engineers* 217 (1924): 96–205, on 166.

179. E. Williams, "The Purification of Sewage by the Activated Sludge Process: An Historical Survey and Some Suggestions," *Sewage Purification* 2 (1940): 125–133; Paul Zigerli, U.S. Patent 2,158,954, "Purification of Sewage or Waste Liquors," May 16, 1939.

180. Ralph A. Stevenson, "Chemical Purification of Sewage," *Western Construction News and Highways Builder* 7 (1932): 527–528, on 527; Ralph A. Stevenson, "Chemical Sewage Purification at Palo Alto: Regeneration of Spent Coagulent Effects Complete Sewage Treatment," *SWJ* 5 (1933): 53–60.

181. "Studies of Sewage Purification III. The Clarification of Sewage—A Review," *U.S. Public Health Reports* 50 (1935):1581–1595, on 1594.

182. Emery J. Theriault, "Activated Sludge as a Biozeolite," *Industrial and Engineering Chemistry* 27 (1935): 683–686, on 686.

183. Emery J. Theriault, "A Biozeolitic Theory of Sewage Purification," *Industrial and Engineering Chemistry* 28 (1936): 83–86; "Biological Methods of Sewage Treatment Identified with Water Softening," *U.S. Public Health Reports* 50 (1935): 143–144.

184. F. C. Vokes, "Development and Application of the Partial Activated Sludge Method of Treatment at Birmingham. (a) Development," *Institute of Sewage Purification. Journal and Proceedings*, 1954 (Part 3): 220–223.

185. Harvey O. Banks, "The Palo Alto Sewage Treatment Plant," *SWJ* 8 (1936): 68–80.

186. C. E. Keefer, "Sewage Sludge as a Biozeolite," *SWJ* 19 (1947): 980–988, 987–988.

187. Arthur M. Buswell, *The Chemistry of Water and Sewage Treatment* (New York: The Chemical Catalog Company, 1928), 318.

188. Chapman, "The Employment of Micro-organisms in the Service of Industrial Chemistry," 283T.

2 Public vs. Private

1. "Statement of the Association for the Defense of Septic Process Suits," *EN* 60 (7) (1908): 181.

2. "Septic Tank Patent Settlement in Sight," *ENR* 82 (22) (1919): 1041.

3. *Cameron Septic Tank Co. v. Village of Saratoga Springs et al.*, 151 F. 242 (1907), 261.

4. R. S. Eisenberg, "Public Research and Private Development: Patents and Technology Transfer in Government-Sponsored Research," *Virginia Law Review*, 82 (1996): 1663–1727. Henry Etzkowitz and Loet Leydesdorff, *Universities and the Global Knowledge Economy: A Triple Helix of University-Industry-Government Relations* (London: Pinter, 1997). Sally Smith Hughes, "Making Dollars Out of DNA: The First Major Patent in Biotechnology and the Commercialization of Molecular Biology, 1974–1980," *Isis* 92 (2001): 541–575. Daniel J. Kevles, "*Diamond v. Chakrabarty* and Beyond: The Political Economy of Patenting Life," in *Private Science: Biotechnology and the Rise of the Molecular Sciences*, ed. Arnold Thackrey (Philadelphia: University of Pennsylvania Press, 1998), 65–79. Daniel L. Kleinman, *Impure Cultures: University Biology and the World of Commerce* (Madison: University of Wisconsin Press, 2003). Harold Varmus, "Perspectives from Different Sectors: Government," in National Research Council, *Intellectual Property Rights and Research Tools in Molecular Biology: Summary of a Workshop Held at the National Academy of Sciences, February 15–16, 1996*. A. Webster and K. Packer, "When Worlds Collide: Patents in Public-Sector Research," in Etzkowitz and Leydesdorff, *Universities and the Global Knowledge Economy*, 48–59.

5. Creager and a number of other researchers have excavated much earlier controversies over the patenting of scientific research supported by universities and the government. Angela

N. H. Creager, "Biotechnology and Blood: Edwin Cohn's Plasma Fractionation Project, 1940–1953," in Thackrey, *Private Science*, 39–62; Rima D. Apple, "Patenting University Research: Harry Steenbock and the Wisconsin Alumni Research Foundation," *Isis* 80 (1989): 375–394; C. Weiner, "Patenting and Academic Research: Historical Case Studies," in *Owning Scientific and Technical Information: Values and Ethical Issues*, ed. Vivian Weil and John W. Snapper (New Brunswick, NJ: Rutgers University Press, 1989), 87–109; John W. Servos, "Engineers, Businessmen, and the Academy: The Beginnings of Sponsored Research at the University of Michigan," *Technology and Culture* 37 (1996): 721–762.

6. Debate in the literature on the engineering profession, its accommodation to capital, and attempts at reform largely ignore the role of publicly employed, municipal engineers. With interests very different from privately employed engineers, they strongly promoted the public interest and were the ones involved in the patent controversies. David F. Noble, *America by Design: Science, Technology and the Rise of Corporate Capitalism* (New York: Oxford University Press, 1979), 84–109. See Peter Meiksins, "The 'Revolt of the Engineers' Reconsidered," *Technology and Culture* 29 (1988): 219–246, for a reexamination of Noble's argument.

7. Brad Sherman and Lionel Bently, *The Making of Modern Intellectual Property Law: The British Experience, 1760–1911* (Cambridge, UK: Cambridge University Press, 1999). This transition was by no means smooth. Adrian Johns describes the powerful midcentury attacks on patents by important industrialists in Great Britain that was almost successful in repealing the whole patent system. Nonetheless, despite the developing critique, the patent system emerged even stronger, and by the end of the nineteenth century, intellectual property rights systems were further reinforced through national laws and international conventions. Adrian Johns, *Piracy: The Intellectual Property Wars from Gutenberg to Gates* (Chicago: University of Chicago Press, 2009), 258–289.

8. Donald Cameron and Frederick James Commin, "Improvements in the Treatment of Sewage, and in Apparatus therefor," British Patent 21,142, Apr. 25, 1896. Donald Cameron, "The Septic Tank System of Sewage Treatment," *The Builder* 74 (July 10, 1897): 27; "The Exeter Septic Tank System," *Engineering Record* 39 (1899): 379–380; Testimony of Mr. Donald Cameron, Oct. 18–19, 1898, *Royal Commission on Sewage Disposal, Vol. II, Evidence*, 1902 [Cd. 686]. Donald Cameron, Frederick J. Commin, and Arthur J. Martin, "Process of and Apparatus for Treating Sewage," U.S. Patent 634,423, Oct. 3, 1899.

9. A. Marston, "Septic Tank Patents," *Municipal Engineering* 35 (1908): 288–312.

10. Leonard Metcalf, "The Antecedents of the Septic Tank," *Proceedings of the American Society of Civil Engineers* 27 (1901): 506–521. See also "Review of the History of the Septic Tank," *EN* 56 (5) (1906): 111–112.

11. Frank Herbert Snow, "Discussion—Antecedents of the Septic Tank," *Transactions, American Society of Civil Engineers* 46 (1901): 472–481, on 472.

12. Testimony of John W. Alvord, *Cameron Septic Tank Co. v. Village of Saratoga Springs et al.*, 1908, *Record*, vol. II, 852–855, quotation on 852, Box 1100, Record Group 276, U.S. Court of Appeals, 2nd Circuit, National Archives and Records Administration, Northeast Region, New York, NY (hereafter cited as Transcript of *Cameron v. Saratoga Springs*).

13. "Patent Claims. Plainfield's Case," *Report of the State Sewerage Commission of the Legislature, Session of 1905*, 13–15.

14. *Cameron Septic Tank Co. v. Village of Saratoga Springs et al.*, 151 F. 242 (1907), 248–249. The decision was also reprinted in the engineering press. "Decision in the Cameron Septic Tank Suit against Saratoga Springs, N.Y.," *EN* 57 (1907): 347–351.

15. *Cameron Septic Tank Co. v. Village of Saratoga Springs et al.*, 159 F. 453 (1908), 458.

16. "Septic Tank Patents Sustained," *Municipal Engineering* 34 (2) (1908): 94.

17. Marston, "Septic Tank Patents," 294.

18. "Statement of the Association for the Defense of Septic Process Suits," *EN* 60 (7) (1908): 181.

19. Marston, "Septic Tank Patents"; "The Association for the Defense of Septic," *EN* 61 (1) (1909): 26–27, on 26. The appeals court noted that the septic tank differed from previous tanks in that there was no accumulation of sludge, so the tank never needed cleaning. Engineers argued that by cleaning tanks periodically, they would no longer infringe on the patent. A. Marston to E. W. Stanton, May 27, 1909, Anson Marston Papers, University Archives, Iowa State University, Ames, IA.

20. "The Septic Tank in California," *Municipal Journal and Engineer* 28 (1910): 119.

21. "End of Septic Tank Patent," *EN* 74 (14) (1916): 663.

22. "League of Iowa Municipalities Calls Conference to Contest Claims for Payment of Royalty for Use of Septic Process of Sewage Treatment," *Engineering and Contracting* 45 (1916): 216; "National Septic Process Protective League," *EN* 75 (12) (1916): 578; "The National Septic Process Protective League," *EN* 76 (3) (1916): 141; "A Septic Tank Infringement Suit," *EN* 76 (18) (1916): 866.

23. "Septic Tank Patent Settlement in Sight," *ENR* 82 (22) (1919): 1041.

24. Bartow to Fowler, Feb. 12, 1915, BP, Box 3.

25. Fowler to Bartow, Mar. 3, 1915, BP, Box 3.

26. The following history of negotiations among Jones & Attwood, Fowler and Manchester is reconstructed from the Deposition of Gilbert Fowler, Plaintiffs' Depositions, GSD-Dep, letters of the parties and minutes of the Rivers Commission contained in the Deposition of George F. Walter, Defendants' Depositions, Vol. II, 1094–1335, GSD-Dep, and a self-published history of Activated Sludge Ltd. in the British Library, E. P. Coombs, *Activated Sludge Ltd.—The Early Years* (Bournemouth: C. Roy Coombs, 1992), 22–28.

27. Fowler to Walter Jones, Dec. 3, 1914, Defendants' Depositions, Vol. II, 1178, GSD-Dep.

28. Fowler to O. Wethered, Oct. 13, 1914, in Coombs, *Activated Sludge Ltd.*, 23.

29. Fowler to Walter Jones, Dec. 3, 1914, Defendants' Depositions, Vol. II, 1178, GSD-Dep.

30. There were a number of patents on both process and apparatus. Walter Jones and Jones & Attwood Ltd., "Improvements in and Connected with the Purification of Sewage and Analogous Liquids," British Patents 22,736 and 22,737, both issued Dec. 20, 1915, were the key British patents whose priority allowed the American patents to proceed. Walter Jones, "Purification of Sewage and Analogous Liquids," U.S. Patents 1,247,540, 1,247,542, and 1,247,543 all issued Nov. 20, 1917, and 1,286,017, Nov. 26, 1918, were the key process patents. Apparatus patents included Walter Jones, "Apparatus for the Purification of Sewage and Analogous Liquids," U.S. Patent 1,282,587, Oct. 22, 1918, and "Purification of Sewage and Other Liquids," U.S. Patent 1,341,561, May 25, 1920.

31. Fowler to Chairman of the Rivers Committee, Defendants' Depositions, Vol. II, 1129, GSD-Dep.

32. Fowler to Walter Jones, Dec. 3, 1914, Defendants' Depositions, Vol. II, 1178, GSD-Dep.

33. Minutes, Rivers Committee, Vol. 21, Sept. 1914–May 1916, Mar. 8, 1915, 180–181. Manchester Archives, Manchester Central Library, Manchester, England.

34. Deposition of John Haworth, Defendants' Depositions, Vol. I, 466, GSD-Dep.

35. See David F. Noble, *America By Design: Science, Technology and the Rise of Corporate Capitalism* (New York: Oxford University Press, 1979), 84–109, and Leonard S. Reich, *The Making of American Industrial Research: Science and Business at GE and Bell, 1876–1926* (Cambridge: Cambridge University Press, 1985), 81–82, 235–238.

36. Deposition of John Haworth, Defendants' Depositions, Vol. I, 468–470, GSD-Dep.

37. T. Chalkely Hatton, "Detail Report of Inspection of English Sewage Disposal Plants, as made from July 12th to 17th, 1924," Box 9171, MMSD.

38. Discussion of John Haworth, "Sheffield Sewage Disposal Works and the Experimental Work," *AMSDW*, 1916, 31–39, on 38.

39. See Judith Walzer Leavitt, *The Healthiest City: Milwaukee and the Politics of Health Reform* (Princeton: Princeton University Press, 1982), esp. 59–61, for pressures to build a sewage treatment plant in Milwaukee and for the relation between municipal politics and sanitation in general.

40. "British and American Patents on Activated Sludge," *EN* 75 (4) (1916): 189–190, on 189.

41. Bartow to Fowler, Mar. 18, 1915, BP, Box 3.

42. "Appoint Committee on Activated-Sludge Process," *Engineering Record* 74 (3) (1916): 91. The English engineering press also noted the formation of the committee: "The Activated Sludge Process. An American Committee," *Surveyor* 50 (1916): 247.

43. Deposition of Harry W. Clark, Defendants' Depositions, Vol. 1, 387, ASM-Dep. Engineers also made "careful inquiry" into the patent situation in England to ascertain among sanitary engineers and city officials "their attitude towards the patent rights." *Eleventh Annual Report of the Sewerage Commission of the City of Milwaukee Wisconsin for the Year 1924*, 41. While the situation in England is less clear, it appears that many cities there also refused to pay royalties to Activated Sludge. Like with the septic tank, smaller cities with fewer resources often chose to pay royalties rather than fund legal battles, while larger cities, whose royalties would be higher to begin with refused. Hatton, "Detail Report."

44. Hatton was trained as a civil engineer and began his career as a railway engineer for the St. Paul, Minneapolis, and Manitoba Railway in 1878. He was assistant and then chief engineer for Wilmington's streets and sewer department from 1883 to 1899. In 1900 he went into business as a consulting engineer. *Who Was Who in American History—Science and Technology* (Chicago: Marquis Who's Who, 1976), 258; "Statement of the Association for the Defense of Septic Process Suits," *EN* 60 (7) (1908): 181.

45. Hatton to Fowler, Aug. 10, 1914, Fowler to Hatton, Jan. 8, 1915, DHP, Box 31, File 777.

46. Hatton to Fowler, June 25, 1913, DHP, Box 31, File 777.

47. Hatton to Fowler, July 31, 1915, in Brief for the City of Milwaukee, Defendant-Appellant, *Milwaukee vs. Activated Sludge Inc.*, Box 8303, MMSD, 325.

48. See correspondence in DHP, Box 31, File 777. Jones & Attwood spun off the activated sludge portion of their business into a separate company, Activated Sludge Ltd., in 1919.

49. Deposition of Glenn West, Defendants' Depositions, Vol. II, 989–1004, ASM-Dep. "California Sewage Works Association Holds 7th Spring Conference at Fresno," *Water Works and Sewerage* 82 (1935): 226.

50. "Sewage Plant System Stolen, Geiger Rules," *MS*, Feb. 8, 1933.

51. USPO, Patent #1,247,540.

52. *Guthard et al. v. Sanitary Dist. of Chicago*, 8 F. Supp. 329 (1934). MMSD Attorney Thomas J. Crawford argues that the judge in the case, Ferdinand Geiger, was ardently pro-business and particularly antagonistic to Milwaukee's Socialist administration, accounting in large part for the ruling against the Sewerage District. "The Intellectual Property Legacy of Milwaukee's Sewer Socialist," Presented to the Government Lawyers Division, Wisconsin State Bar, Feb. 13, 2008. I thank Thomas Crawford for sharing his research and analysis with me.

53. *City of Milwaukee v. Activated Sludge Inc.*, 69 F. 2d 577 (1934), *Sanitary District of Chicago v. Activated Sludge Inc.*, 90 F. 2d 727 (1937). "Milwaukee Wins Cut In Sewage Disposal Award," *CDT*, July 24, 1938, "Sanitary Board Sues to Reopen $7,500,000 Case," *CDT*, Oct. 22, 1939. "Sludge Firm Wins $950,000 Judgment Suit," *CDT*, Jan. 29, 1946. USPO, Patent #1,247,540.

54. For the record of denials and appeals of the sewage patents see USPO, Patent #634,423, #1,247,540, #1,247,542, and #1,247,543.

55. On *Cameron*, for example, see Stephen A. Bent, Richard L. Schwaab, David G. Conlin, and Donald D. Jeffery, *Intellectual Property Rights in Biotechnology Worldwide* (New York: Stockton Press, 1987), 108.

56. Paul Lucier, *Scientists and Swindlers: Consulting on Coal and Oil in America, 1820–1890* (Baltimore: The Johns Hopkins University Press, 2008), 162–185; Jackson quotation on 181.

57. Snow, "Discussion—Antecedents of the Septic Tank." "Obituary, James Owen," *Good Roads* 61 (3) (1921): 40. For an account of Owens's dealings with Cameron, see "Septic Tank Patents," *Report of the State Sewerage Commission of the Legislature, Session of 1905*, 42–43.

58. "Killed With Own Revolver," *CDT*, Jan. 18, 1907, 1.

59. Archibald R. Eldridge, "Is It Unprofessional for an Engineer to Be a Patentee?," *Transactions, American Society of Civil Engineers* 48 (1902): 314–326, on 316. Curiously, given the vehemence of his arguments, Eldridge had no patents.

60. On engineers and professionalism, see Edwin Layton, *The Revolt of the Engineers: Social Responsibility and the American Engineering Profession* (Cleveland: The Press of Case Western Reserve University, 1971); Samuel Haber, *The Quest for Authority and Honor in the American Professions, 1750–1900* (Chicago: University of Chicago Press, 1991), 294–318.

61. Comments of S. Whinery and James Owens in Eldridge, "Is It Unprofessional?," quotation on 324–235; "Engineering Patents. An Informal Discussion at the Annual Convention, June 27th, 1906," *Transactions, American Society of Civil Engineers* 57 (1906): 83–90.

62. Testimony of Mr. William Joseph Dibdin, Oct. 19, 1898, *Royal Commission on Sewage Disposal, Vol. II, Evidence*, 1902 [Cd. 686].

63. Fowler conducted extensive correspondence on the "distorting element of commercialism" in sewage research with George Hart, Leeds's sewerage engineer. Hart discussed the unethical patenting practices of "manufacturing Engineers . . . whose actions are prostituting scientific knowledge and investigation." Fowler to Hart, Feb. 6, 1914, and Hart to Fowler, Feb. 7, 1914, Deposition of George A. Hart, Defendants' Depositions, Vol. I, 386, 388, GSD-Dep.

64. "Engineering Patents. An Informal Discussion," 88. On divisions within the profession based on work conditions and the relation to professional ethics, see Meiksins, "The 'Revolt of the Engineers' Reconsidered," and Bruce Sinclair, "Local History and National Culture: Notions on Engineering Professionalism in America," *Technology and Culture* 27 (1986): 683–693.

65. Deposition of James A. Coombs, Plaintiffs' Depositions, 159, GSD-Dep; Deposition of Harry E. Partridge, Defendants' Depositions, Vol. II, 705–711, GSD-Dep. For similar trends among American Engineers, see Haber, *The Quest for Authority*, 317. While the monetary rewards were most important, it is interesting that in addition to royalties or bonuses, engineer employees also struggled with their firms for acknowledgment of their intellectual contributions to the patents. Jones & Attwood's policy was to file patents developed by employees under the firm's name, obscuring the work of the actual inventors. Harry Partridge fought to have his contribution credited and this conflict led to his resignation.

66. Deposition of James A. Coombs, Plaintiffs' Depositions, 153–166, GSD-Dep.

67. Eldridge, "Is It Unprofessional?"; "Engineering Patents. An Informal Discussion," 90. Samuel Whinery had a troubled history with monopoly that no doubt conditioned his attitude toward the patent question in engineering. A paving engineer, he was president of the Warren-Scharf Asphalt Company when that company was acquired by the American Asphalt Company, or the "Asphalt Trust." Whinery and all the officers were fired. He then became a consulting engineer. "Samuel Whinery Dead," *NYT*, Jan. 16, 1925, 17. He did have several patents related to paving and roadways. See, for example, Archibald Eldridge, "Roadway," U.S. Patent 675,694, June 4, 1901.

68. Discussion of Donald Cameron, "Notes on Bacterial Treatment of Sewage," *Proceedings of the Incorporated Association of Municipal and County Engineers* 24 (1898): 289–305, on 297.

69. Arthur J. Martin wrote: "Nearly thirty years ago. . . I was associated with a patented process. I soon realized that, in view of the feeling which existed both among engineers and on the part of local authorities, it was not wise to press the question of patents." Martin supported the right of the national government to buy the rights to patents of great public interest, like sewage patents. Arthur J. Martin, "Some Aspects of Sewage Treatment," *The Canadian Engineer* 50 (1926): 517–518, on 518.

70. "Sewage Disposal Patents," *Municipal Journal and Engineer* 30 (1911): 53.

71. Hatton to Fowler, July 31, 1915, DHP, Box 35, File 777.

72. See discussion in Layton, *Revolt*, 154–178. See also Bruce Sinclair, *A Centennial History of The American Society of Mechanical Engineers. 1880–1980* (Toronto: University of Toronto Press, 1980), esp. 96–112.

73. Angell, "Inaugural Address," 2, 25.

74. Sinclair, *A Centennial History*, 98–99.

75. "Information concerning the American Society of Municipal Improvements: Its Aims and Accomplishments, Its Consitution, Officers, List of Members, and Papers Published," n.d., n.p. Interestingly, among the "Associate Members," mostly firms doing municipal business, was H. D. Wyllie, general manager of the Cameron Septic Tank Company.

76. Morris Llewellyn Cooke, "Some Factors In Municipal Engineering," *Transactions, American Society of Mechanical Engineers* 36 (1915): 605–629, on 605.

77. Ibid.

78. Morris Llewellyn Cooke, "The Public Interest as the Bed Rock of Professional Practice," *Transactions, American Society of Mechanical Engineers* 40 (1918): 85–100, on 93–94.

79. Morris Llewellyn Cooke, *Snapping Cords: Comments on the Changing Attitude of American Cities Toward the Utility Problem* (n.p., 1915), 11. For another critique of municipal utility patents, see Morris Llewellyn Cooke, *Our Cities Awake: Notes on Municipal Activities and Administration* (New York: Doubleday, Page & Company, 1918), 72.

80. "Big Asphalt Trust Formed," *NYT*, June 30, 1899, 3; "Denounce Asphalt Trust," *NYT*, June 14, 1900, 11; "Asphalt Trust Spreads," *NYT*, Jan. 31, 1901, 6; "Brief Sketch of the Development of the Asphalt Trust," DHP, Box 14, File 345.

81. Daniel T. Rodgers, "In Search of Progressivism," *Reviews in American History* 10 (1982): 113–132, on 124; David P. Thelan, *The New Citizenship: Origins of Progressivism in Wisconsin, 1885–1900* (Columbia: University of Missouri Press, 1972), 223–289; Scott L. Bottles, *Los Angeles and the Automobile: The Making of the Modern City* (Berkeley: University of California Press, 1987), 22–51.

82. Noble, *America By Design,* 84–109.

83. Daniel B. Luten, "How Shall Patented Materials or Processes on Public Works be Handled?," *ER* 72 (1915): 546–548, on 546. These patents were later found invalid in a series of scathing decisions: "While Mr. Luten has made no invention, he has made a great *discovery*, namely that not more than one city or county attorney or Attorney General in 10 knows anything about patent law." *Luten v. Wilson Reinforced Concrete Co.*, 263 F. 983 (1920), 985. See also *Luten v. Marsh et al.*, 254 F. 701 (1919).

84. "The Septic Tank Litigation," *Pacific Municipalities and Counties* 32 (1918): 70–71, 71.

85. Report of a Committee Appointed by the President of the Local Government Board to Inquire into the Several Modes of Treating Town Sewage, 1876 [C. 1410], 105–115.

86. J. W. Slater, *Sewage Treatment, Purification, and Utilization* (London: Whittaker & Co., 1888), 117.

87. Henry Letheby, *The Sewage Question* (London: Baillière, Tindall and Cox, 1872).

88. Joel A. Tarr discusses the controversy between Waring's separate system and the "rational" system of combined sewerage in *The Search for the Ultimate Sink: Urban Pollution in Historical Perspective* (Akron: University of Akron Press, 1966), 131–158. For the patent controversy, see George E. Waring Jr., "Sewering and Draining Cities," U.S. Patent 236,740, Jan. 18, 1881; C. W. Chancellor, "Sewage Disposal. A Rejoinder to Col. George E. Waring (In Two Parts)," *The Sanitarian* 184 (1885): 193–208; Rudolph Hering, "The Separate-System of Sewerage and the 'Waring Patent'" *The Sanitary Engineer* 9 (1884): 546; "The Waring Patents and the Separate-System of Sewerage," *The Sanitary Engineer* 9 (1884): 572.

89. "Proceedings of the Suffolk District Medical Society," *Boston Medical and Surgical Journal* 140 (1884): 81–86, on 83.

90. Comments of J. Lemon, on Donald Cameron, "Notes on Bacterial Treatment of Sewage," *Proceedings of the Incorporated Association of Municipal and County Engineers* 24 (1898): 289–305, on 296–297.

91. T. Chalkley Hatton, "Sewerage Disposal as Applied to Small Cities and Towns," *Proceedings of the Eleventh Annual Convention of the American Society of Municipal Improvements*, 1904, 149–159.

92. "The Milwaukee Sewerage Problem and the Sewage Treatment Testing Station," *Engineering and Contracting* 42 (1914): 367–369, on 368.

93. Joseph L. Tropea, "Rational Capitalism and Municipal Government: The Progressive Era," *Social Science History* 13 (1989): 137–158.

94. J. Simpson, "Use of Patented Articles. Court Decisions in the Several States as to Conditions under which Cities May Contract for Patented Pavements and Other Articles," *Municipal Journal* 38 (1) (1915): 13–15. See also John Simpson, "Municipal Use of Patented Articles," *Municipal Journal and Engineer* 22 (1907): 543–545, and "Practice Relating to Patented Pavements in American Municipalities," *Engineering and Contracting* 44 (1915): 103–108.

95. S. Whinery, "State Should Deal Directly with Patentees of Processes and Materials in Public Work," *ER* 72 (3) (1915): 73–74.

96. Snow, "Discussion—Antecedents of the Septic Tank," 475.

97. A. B., "The Commission for the Disposal of Sewage," *Times* (London), Nov. 1, 1898, 11.

98. "Obituary. John Haworth," *JRSI, Supplement*, Mar. 1934, 54: 163–164.

99. Defendants' Depositions, Vol. II, 1164, GSD-Dep.

100. Soon after Fowler was subject to the investigation by the Manchester Rivers Committee and his role in the activated sludge patents became known, he quit his post and moved to India to become professor of applied chemistry at the Indian Institute of Science in Bangalore. He spent the rest of his career there. He was also Activated Sludge's "technical representative" for India and the East. There is no indication of why Fowler chose this path. On the movement of British public health doctors and scientists to Indian universities, see Helen Power, "The Calcutta School of Tropical Medicine: Institutionalizing Medical Research in the Periphery," *Medical History* 40 (1996): 197–214.

101. In assessing damages for infringement in the Chicago case, the court took the public interest literally into account, *increasing* the damages because of the value of the activated sludge process in avoiding "noxious odors, putrid streams, pestilential flies." *Activated Sludge, Inc., et al. v. Sanitary Dist. of Chicago*, 64 F. Supp. 25 (1946), 32. In just one aspect of the case did the courts act to protect the public interest. The trial judge in the Milwaukee case, in ruling against the city, had enjoined it from using the activated sludge process altogether, which would have required the city to close its sewage treatment plant and dump raw sewage into Lake Michigan. On appeal, the court ruled that "the health and the lives of more than half a million people are involved" and overturned the injunction, while letting the rest of the ruling stand. *City of Milwaukee v. Activated Sludge Inc.*, 69 F. 2d 577 (1934), 593. This decision is frequently cited as a useful precedent in requiring pharmaceutical and other manufacturers to license their patents when the public health is at stake. See, for instance,

Maureen A. O'Rourke, "Toward a Doctrine of Fair Use in Patent Law," *Columbia Law Review* 100 (2000): 1177–1250; Barbara M. McGarey and Annette C. Levey "Patents, Products, and Public Health: An Analysis of the CellPro March-in Petition," *Berkeley Technology Law Journal* 14 (1999): 1095–1116.

102. Patent Act of 1790, Ch. 7, 1 Stat. 109–112 (Apr. 10, 1790).

103. A. D. Thompson, "A Municipal Testing Laboratory," *Paving and Municipal Engineering* 10 (1896): 1–11, and "Cement Testing in Municipal Laboratories," *Engineering Record* 46 (1902): 27–29. J. O. Preston, "The Organization of a Standard Municipal Testing Laboratory," *The Canadian Engineer* 36 (1919): 520–522. On concrete testing, see Amy E. Slaton, *Reinforced Concrete and the Modernization of American Building, 1900–1930* (Baltimore: The Johns Hopkins University Press, 2001), 20–61. Christopher Hamlin examines the role of municipal chemists in urban environmental history in "The City as a Chemical System? The Chemist as Urban Environmental Professional in France and Britain, 1780–1880," *Journal of Urban History* 33 (2007): 702–728, but the extent of basic and applied research performed by city governments has not been fully examined in the literature on science and the city. See, for example, Sven Dierig, Jens Lachmund, and J. Andrew Mendelsohn, eds., *Science and the City, Osiris 18* (Chicago: University of Chicago Press, 2003), 18.

104. George C. Whipple, "Municipal Water-works Laboratories," *Popular Science Monthly* 58 (Dec. 1900): 172–182, on 172.

105. "The Baltimore Sewage Testing Station," *Engineering Record* 54 (1906): 550–552, on 550.

106. Robert Cramer, "Future Possibilities of the Activated Sludge Process," reprinted from *Bulletin of the Associated State Engineering Societies*, July 1929, MMSD Box 10244. Milwaukee spent $781,006.75 in their experimental account, which included "cost of constructing and operating the experimental plant," 12.1. Langdon Pearse, in discussion of William T. Lockett, "Activated Sludge Process: Aeration and Circulation," *Association of Managers of Sewage Disposal Works. List of Members and Associates and Proceedings*, 1924, 92–110. In comparison, General Electric's industrial research laboratory had an annual budget between $105,000 and $553,000 during roughly the same period. Reich, *The Making of American Industrial Research*, 80–92. Municipal laboratories continued to dominate sewage research in the United States through the 1930s. Harold E. Babbitt, "Research in Sewage Treatment at Educational Institutions in the United States," in *Modern Sewage Disposal: Anniversary Book of the Federation of Sewage Works Associations*, ed. Langdon Pearse (New York: Federation of Sewage Works Associations, 1938), 237–247.

107. Deposition of John D. Watson, Plaintiff's Depositions, 290–291, GSD-Dep.

108. Francis Richard O'Shaughnessy, "Improvements in the Treatment of Sewage Liquor and like Impure Liquids," British Patent #218,399, June 27, 1924.

109. *Fourth Annual Report of the Sewerage Commission of the City of Milwaukee, Wisconsin*, 1918, 20; C. H. Nordell, "Aerating Apparatus," U.S. Patent 1,208,821, Dec. 19, 1916. Copeland to Hatton, n.d., "Testing Station," MMSD Box 9172. Nordell moved into industry, patenting over sixty inventions related to waste treatment. The commission later changed its policy, requiring employees to assign any patents to the commission. James Brower, *Report: Resume of Twenty Five Years of Plant Operation in Connection with Drying Sewage Sludge*, MMSD, files of Tom Brennan, 27.

110. Leslie Frank, "Process of Purifying Sewage or Other Wastes and Apparatus Therefor," Patent 1,139,024, May 11, 1915. Deposition of Earle B. Phelps, Defendants' Depositions, Vol. II, 877, ASM-Dep. USPO, Patent #1,247,542 and #1,247,543.

111. Deposition of Edward Ardern, Plaintiffs' Depositions, 496, GSD-Dep.

112. Langdon Pearse, Discussion of Lockett, "Activated Sludge Process," 108.

113. Langdon Pearse, "Sewage Disposal in the United States: A General Review," *Water Works and Sewerage* 82 (1935): 33–41, on 41.

114. Deposition of Harry H. Tomson, Defendants' Depositions, Vol. II, 818, GSD-Dep.

115. "Discussion—Antecedents of the Septic Tank," *Transactions, American Society of Civil Engineers* 46 (1901): 472–481, on 480.

116. Fowler to Clark, Mar. 30, 1914, in H. W. Clark, "Origin of the Activated Sludge Process," *Surveyor* 57 (1920): 308.

117. Calculated from J. Edward Porter, *The Activated Sludge Process of Sewage Treatment: A Bibliography of the Subject*, 2nd ed. (Rochester, NY: General Filtration Company, 1921).

118. Fowler to Weizmann, January 21, 1916, Scientific Material, Box No. 1A, Years: 1912–1914, Weizmann Archives, Rehovot, Israel. I thank Ms. Merav Siegel, Director, for providing this document, which was cited in Jehuda Reinharz, *Chaim Weizmann: The Making of a Zionist Leader* (Oxford: Oxford University Press, 1985), 423.

119. Fowler to George A. Hart, in Deposition of George A. Hart, Defendants' Depositions, Vol. I, 381–382, GSD-Dep.

120. Coombs, *Activated Sludge Ltd.*, 24.

121. See discussion by Fowler and Duckworth in Edward Ardern, "A Recent Development of the Sewage Problem," *AMSDW*, 1914, 61–71, on 69.

122. Edward Ardern and William T. Lockett, "Experiments on the Oxidation of Sewage without the Aid of Filters," *JSCI* 33 (1914): 523–539.

123. The patent was only approved when Jones provided evidence that his first British patent had been applied for before Ardern and Lockett's paper was read. USPO, Patent #1,247,540. Defendants' Depositions, Vol. II, 1138–1139, GSD-Dep.

124. Metcalf, "Antecedents."

125. *The Battle of the Microbes: Nature's Fight for Pure Water*, The Merchants' Association of New York, 1908, 6.

126. USPO Patent #634,423.

127. Cameron, "The Septic Tank System of Sewage Treatment," 27. Nevertheless, when Cameron applied for his patent in the U.S. he emphasized in his specification that he had invented an "artificial method . . . for the purification of sewage." U.S. Patent 634,423.

128. Gilbert J. Fowler, "Aeration of Activated Sludge," *Surveyor* 67 (1925): 305.

129. *Cameron Septic Tank Co. V. Village of Saratoga Springs et al.*, 151 F 242 (1907), 261.

130. With the judges split, the case was not decided either way. *Boulton v. Bull*, 2 H Blackstone 463 (1795), 494. The patent was affirmed unanimously in a later case, *Hornblower v. Boulton* 8 Term Reports 95 (1799).

131. *Boulton v. Bull*, 2 H Blackstone 463 (1795), 494–495.

132. Ibid., 492, 494.

133. Sherman and Bently, *The Making of Modern Intellectual Property Law*, 108.

134. *Cochrane v. Deener*, 94 U.S. 780 (1877), 788.

135. *Le Roy v. Tatham*, 55 U.S. 156 (1853), 175.

136. George Tickner Curtis, *A Treatise on the Law of Patents for Useful Inventions in the United States of America* (Boston: Charles C. Little and James Brown, 1849). See also *O'Reilly v. Morse*, 56 U.S. 62 (1854).

137. *Le Roy v. Tatham*, 55 U.S. 156 (1853), 175.

138. Sherman and Bently, *The Making of Modern Intellectual Property Law*, 46.

139. For the United States, see William C. Robinson, *The Law of Patents for Useful Inventions, Vol. I* (Boston: Little, Brown, 1890), 230–259; for Great Britain, see Robert James, *The Grant and Validity of British Patents for Inventions* (London: John Murray, 1903).

140. For confusion of patent examiners on sewage process patents, see USPO, Patent #770,490. Quotation from E. D. Sewall, "The Status of Process Inventions," *Machinery* 17 (1910): 10–14, on 10.

141. Snow, "Discussion—Antecedents of the Septic Tank," 474.

142. Hatton to Fowler, June 25, 1915, DHP, Box 31, File 777.

143. "Institution of Municipal and County Engineers. Annual General Meeting in London," *Surveyor* 48 (1915):143–146, on 144. This complaint anticipates current objections to patents on fundamental biotechnology methods.

144. Karl Imhoff, "Sewage Treatment Apparatus," Patent 924,664, June 15, 1909. Leonard P. Kinnicutt, C.-E.A. Winslow, and R. Winthrop Pratt, *Sewage Disposal* (New York: Wiley, 1919), 174–175. "Cameron Septic Process and the Imhoff Tank," *ER* 66 (1912): 504. W. T. Waters, "Sewage Disposal Plans of Atlanta, Ga.," *Municipal Engineering* 40 (1911): 1–5. For a town with a population of 100, the license on an Imhoff tank was $10. For a city of 100,000, the fee was $2,550. "War Status of Imhoff and Other German Patents," *ENR* 78 (1917): 565. The primary opposition to Imhoff's tank came from other inventors. Imhoff was (unsuccessfully) sued in Germany for infringement on Travis's patent for the hydrolytic tank, and Cameron considered Imhoff's tank to infringe on the septic process and threatened suit, but these suits never materialized.

145. "Remodeling of Septic Tanks into Imhoff Tanks Eliminates Odors from Land Irrigation," *ER* 71 (1915): 747–748.

146. "Statement of the Association for the Defense of Septic Process Suits," *EN* 60 (7) (1908): 181.

147. Comments by Earle B. Phelps in George T. Hammond, *Sewage Treatment by Aeration and Activation*, reprinted from the *Proceedings of the American Society of Municipal Improvements*, 1916, 81–82.

148. Deposition of Earle B. Phelps, Defendants' Depositions, Vol. II, 880, 888, ASM-Dep.

149. U.S. Constitution, Art. 1, sec. 8.

150. Testimony of William Main, *Record*, Vol. I, 272, Transcript of *Cameron v. Saratoga Springs*.

151. *Cameron Septic Tank Co. V. Village of Saratoga Springs et al.*, 151 F 242 (1907), 259–260.

152. *Cameron Septic Tank Co. v. Village of Saratoga Springs et al.*, 159 F. 453 (1908), 462.

153. *City of Milwaukee v. Activated Sludge Inc.*, 69 F. 2d 577 (1934), 581.

154. *City of Milwaukee v. Activated Sludge Inc.*, 69 F. 2d 577 (1934), 580–581, 583–583.

155. Henry Field Smyth and Walter Lord Obold, *Industrial Microbiology: The Utilization of Bacteria, Yeasts and Molds in Industrial Processes* (Baltimore: The Williams and Wilkins Co., 1930), Chapter 37, "The Relation of Patent Law to Biological Processes," written by J. Howard Flint, 285.

156. See Reinharz, *Chaim Weizmann*, 423n25. "Obituary: William T. Lockett," *Institute of Sewage Purification, Journal and Proceedings,* 1960 (Part 3), 349–351.

157. Reinharz, *Chaim Weizmann*, 45–46.

158. See Reinharz, *Chaim Weizmann*, for the most detailed account of the development of this fermentation process.

159. The decision is in *Guaranty Trust Co. Of New York et al. v. Union Solvents Corporation*, 54 F. 2d 400 (1931); *Union Solvents Corporation vs. Guaranty Trust Company of New York, The Butacet Corporation, and Commercial Solvents Corporation*, Brief for Appellant, 74, Vol. 1725, NARA-Mid Atlantic, RG 276, U.S. Court of Appeals, 3rd Circuit, Case #4866.

160. Comment of Court, 53, Opening for Defendant, 63, Testimony of Clark, 53, *Record*, Vol. I, *Union Solvents vs. Guaranty Trust*, Equity no. 802, District Court, Delaware. NARA-Mid Atlantic RG 21, Box 272.

161. Brief for Appellant, 74–75, Vol. 1725, NARA-Mid Atlantic, RG 276, U.S. Court of Appeals, 3rd Circuit, Case #4866.

162. Testimony of Edwin O. Jordan, *Record*, Vol. II, 1317–1318, *Union Solvents vs. Guaranty Trust*, Equity no. 802, District Court, Delaware. NARA-Mid Atlantic RG 21, Box 272. On Jordan, see Carolyn G. Shapiro-Shapin, "'A Really Excellent Scientific Contribution': Scientific Creativity, Scientific Professionalism, and the Chicago Drainage Case, 1900–1906," *Bulletin of the History of Medicine* 71 (1997): 385–411.

163. *Union Solvents vs. Guaranty Trust*, Equity no. 802, District Court, Delaware. *Record*, Vol. II, 1524–1525, NARA-Mid Atlantic RG 21, Box 272.

164. Arguments on Final Hearing, 195. *Union Solvents vs. Guaranty Trust*, Equity no. 802, District Court, Delaware. NARA-Mid Atlantic RG 21, Box 272.

165. Ibid., 131.

166. K. Morikawa, "A New Butyl and Isopropyl Alcohols Fermentation," *Bulletin of the Agricultural Chemical Society of Japan* 3 (1927): 28–32.

167. Samuel A. Goldblith, *Pioneers in Food Science Volume I. Samuel Cate Prescott: M.I.T. Dean and Pioneer Food Technologist* (Trumbull, CT: Food & Nutrition Press, 1993). S. C. Prescott, "A Brief Sketch of the History of the 'Bug Club' (Boston Bacteriological Club)," appendix E in Goldblith, *Prescott*.

168. "Specification," 13; "Paper No. 2," 2, USPO, Patent #1,933,683.

169. "Paper No. 3a, Mar. 16, 1928," USPO, Patent #1,908,361.

170. See, for example, the patent application of Alexander Izsak and Forest J. Funk for a process of producing butyl alcohol, acetone, and isopropyl alcohol using *Clostridium saccharobutylicum-gamma*. The claims "are for no more than the inherent activity of the

organism named, or, in other words, for a principle of nature," stated Richard. USPO, Patent #1,908,361.

171. "Amended Feb. 26, 1931," July 30, 1931. USPO, Patent #1,933,683.

172. "In response to amendment filed Mar. 15, 1928," Jan. 4, 1929. USPO, Patent #1,933,683.

173. "Responsive to amendment of June 26 and letter of July 11, 1929," Feb. 1, 1930. USPO, Patent #1,933,683.

174. "Examiner's Statement," Apr. 20, 1932, 14. USPO, Patent #1,933,683.

175. "Brief for Applicants," Paper 16, 5, 8. USPO, Patent #1,933,683.

176. Ibid., 13.

177. Despite the repudiation of his arguments in this case, Richard was not a rogue examiner. He was a leading member of the Patent Office Society, was principal examiner in the Chemical Division of the Patent Office, and was later appointed examiner-in-chief of the Patent Office and member of the Patent Board of Appeals by President Roosevelt. "New Members of the Board of Appeals," *Journal of the Patent Office Society* 24 (1942): 736.

178. "U.S. Court Rules Sewage Microbe Patents Valid," *CT*, Mar. 3, 1934, 5.

179. Edward B. Mallory, "Oxidized Sludge Process and the 'Equilibrium Index,'" *Water Works & Sewerage* 88 (1941): 333–344, on 344. E. B. Mallory, "Process of Controlling the Purification of Sewage," U.S. Patent 2,154,132, Apr. 11, 1939.

180. This period saw the rise of industrial research laboratories generally. Reich, *The Making of American Industrial Research*, 252–256.

181. Mallory, "Oxidized Sludge Process," 344.

182. "Report by Hon. Examination and Research Secretary," *AMSDW*, 1929, 24.

183. D. H. A. Price, in discussion of "The Evolution and Development of the Activated Sludge Process of Sewage Purification in Great Britain. A Symposium," *Institute of Sewage Purification, Journal and Proceedings,* 1954 (Part 3), 174–272, on 257. J. Bolton, on the other hand, thought this trend necessary: "There was a limit to the amount of money which could be spent by public authorities on research work . . . Failing national assistance, it was necessary to turn to commercial firms" (269).

184. "Evaluation of Professional Objectives in the Design of Sanitary Engineering Works: Report of a Committee of the Sanitary Engineering Division," *Proceedings of the American Society of Civil Engineers* 72 (1946): 69–82, on 69, 74, 80.

3   Craft vs. Science

1. Paul W. Reed, "The Sewage Treatment Plant Operator: A Keystone of the Stream Pollution Abatement Program," *Sewage Gas*, 10 (1) (1947): 2–4, 3–4.

2. "The Activated Sludge Process—Part I," *The Digester*, February 1936, no. 1: 2–5; "The Activated Sludge Process—Part II," *The Digester*, May 1936, no. 2: 8–12; "The Activated Sludge Process—Part III," *The Digester*, August 1936, no. 3: 18–24.

3. "Real Estate Men See Sewage Plant," *MS*, Oct. 20, 1915.

4. Arthur M. Buswell, *The Chemistry of Water and Sewage Treatment* (New York: Chemical Catalog Company, 1928), 7.

5. John E. Farmer, "Sewage Disposal and Works Management," *The Sanitary Record and Municipal Engineering* 53 (1914): 287.

6. George Perazich, Herbert Schimmel, and Benjamin Rosenberg, "Industrial Instruments and Changing Technology, Works Progress Administration, National Research Project Report No. M-1," in *Research and Technology*, ed. I. Bernard Cohen (New York: Arno Press, 1980), 88.

7. Harry Braverman, *Labor and Monopoly Capital: The Degradation of Work in the Twentieth Century* (New York: Monthly Review Press, 25th anniversary ed., [1974)] 1998). David F. Noble, *Forces of Production: A Social History of Industrial Automation* (Oxford: Oxford University Press, 1984); Stuart Bennett, "'The Industrial Instrument–Master of Industry, Servant of Management': Automatic Control in the Process Industries, 1900–1940," *Technology & Culture* 32 (1991): 69–81; Stuart Bennett, "The Use of Measuring and Controlling Instruments in the Chemical Industry in Great Britain and the USA during the Period 1900–1939," in *Determinants in the Evolution of the European Chemical Industry, 1900–1939*, ed. A. S. Travist, Harm G. Schroter, Ernst Homburg, and Peter J. T. Morris (Dordrecht: Kluwer Academic Publishers, 1998), 215–237; John Simmons Ceccatti, "Science in the Brewery: Pure Yeast Culture and the Transformation of Brewing Practices in Germany at the End of the 19th Century," Ph.D. diss., University of Chicago, 2001. Dr. W. J. Clark, "The Automatic Control of Chemical Processes," *Proceedings of the Chemical Engineering Group* (Society of Chemical Industry) 18 (1936): 125, cited in Bennett, "The Use of Measuring." James Donnelly, "Consultants, Managers, Testing Slaves: Changing Roles for Chemists in the British Alkali Industry, 1850–1920," *Technology and Culture* 35 (1994): 100–128, on 109.

8. Martha Moore Trescott, *The Rise of the American Electrochemicals Industry, 1880–1910* (Westport, CT: Greenwood Press, 1981), 139–140; Donnelly, "Consultants."

9. John Lomas, *A Manual of the Alkali Trade* (London: Crosby Lockwood and Co. 1880), 46–58.

10. James Mactear, "On the Controlling of the Escapes of Sulphur Gases in the Manufacture of Sulphuric Acid," *Chemical News* 36 (1877): 49–51, on 51.

11. "President's Address," *JSCI* 7 (1882): 249–252, on 251, 252.

12. Karl Marx, *Economic & Philosophical Manuscripts of 1844*, as quoted in John Bellamy Foster, *Marx's Ecology: Materialism and Nature* (New York: Monthly Review Press, 2000).

13. On alienation and denaturing, see Foster, *Marx's Ecology*, and Peter Dickens, *Reconstructing Nature: Alienation, Emancipation and the Division of Labour* (London: Routledge, 1996).

14. Frederick Winslow Taylor, *Principles of Scientific Management* (New York: Harper and Brothers, 1911). For only a few of the sources in the long debate about Taylor and "scientific management," see Peter F. Meiksins, "Scientific Management and Class Relations: A Dissenting View," *Theory and Society* 13 (1984): 177–209; David F. Noble, *America by Design: Science, Technology and the Rise of Corporate Capitalism* (New York: Oxford University Press, 1979); Daniel Nelson, ed., *Scientific Management Since Taylor* (Columbus: Ohio State University Press, 1992); Kevin Whitston, "The Reception of Scientific Management by British Engineers, 1890–1914," *The Business History Review* 71 (1997): 207–229.

15. The "science of shoveling" was one of Taylor's common tropes. For shovel-leaners, see Taylor, *Principles of Scientific Management*, 76.

16. Testimony of F. W. Taylor, *The Taylor and Other Systems of Shop Management. Hearings Before Committee of the House of Representatives to Investigate the Taylor and Other Systems of Shop Management under Authority of H Res 90*, Vol. 3, 1912, 1393–1394.

17. Braverman, *Labor*, 113.

18. Eustace Thomas, "The Management of Engineering Workshops," *Journal of the Institution of Electrical Engineers* 41 (1908): 741–758, on 742, 751.

19. Kevin Whitston emphasizes this point in British engineering practice. Whitston, "The Reception of Scientific Management," 215.

20. For boiler control, see G. A. H. Binz, "Scientific Boiler Control," *The Canadian Engineer* 16 (1909): 292–296, Henry E. Armstron, "The Scientific Control of Fuel Consumption," *The Colliery Guardian* 96 (1908): 657–658. For sugar manufacture, see Guilford L. Spencer, "Report of Experiments in the Manufacture of Sugar By Diffusion, at Magnolia Station, Lawrence, LA, Season of 1888–1889," U.S. Department of Agriculture, Division of Chemistry, Bulletin No. 21, 1889; Ed. W. Knox, "1.–On An Application of Chemical Control to a Manufacturing Business," *Report of the Second Meeting of the Australasian Association for the Advancement of Science*, 1890, 372–379; and *Methods of Analysis and Laboratory Control of the Great Western Sugar Company* (Denver: Great Western Sugar Company, 1920). For machine tool making, see O. M. Becker, "Temperature Determination and Control for High-Speed Steel Treatment," *Engineering Magazine* 39 (1909): 174–185, 373–383. For steelmaking, see Alfred Harrison and Richar Vernon Sheeler, "The Chemical Control of the Basic Open-hearth Process," *The Iron and Coal Trades Review* 77 (1908): 1443–1445.

21. Foster, *Marx's Ecology*, 72.

22. J. W. Slater, *Sewage Treatment, Purification, and Utilization* (London: Whittaker & Co., 1888), 1, 5–6.

23. Slater, *Sewage Treatment*, 114.

24. Testimony of John W. Alvord, Transcript of Record, Vol. II, 1282, *The Cameron Septic Tank Company of Chicago vs. The Village of Saratoga Springs and the Sewer, Water and Streets Commission of Saratoga Springs*, National Archives and Records Administration, Northeast Region, RG 276 United States Court of Appeals for the Second Circuit Case Files, Case No. 2971, Box 1100.

25. Arthur J. Martin, "The Management of Sewage Disposal Works," *JRSI* 25 (1904): 660–667, on 662.

26. Gilbert J. Fowler, "Recent Experience in the Treatment of Manchester Sewage," *JRSI* 25 (1904): 620–642, on 627.

27. Donald Cameron, "Notes on Bacterial Treatment of Sewage," *Proceedings of the Incorporated Association of Municipal and County Engineers* 24 (1898): 289–305, on 300.

28. Martin, "The Management," 665–666.

29. Ibid., 665.

30. Ibid., 661.

31. Discussion by Martin in Fowler, "Recent Experience," on 640.

32. W. J. Dibdin, *The Purification of Sewage and Water* (London: The Sanitary Publishing Company Ltd., 1897), 80–82. This diatribe against the unscientific management of sewage treatment plants was removed from later editions of Dibdin's book.

33. George Ferme, *Local Board Sewage Farming: A Letter to Clare Sewell Read, Esq., M.P.* (London: Daldy, Isbister & Co., 1876), 4–5, 13.

34. Alfred S. Jones, "Bacterial Methods of Sewage Disposal," *Times* (London), May 2, 1900, 14.

35. "Association of Managers of Sewage Disposal Works, Summer Meeting at Leicester," *Surveyor* 34 (1908): 55–56, on 56.

36. Dibdin, *Purification*, 81.

37. Ibid.

38. Whitston, "The Reception of Scientific Management."

39. C. E. Kenneth Mees, "The Organization of Industrial Scientific Research," *Science* 43 (1118) (1916): 763–773, on 765.

40. "Activated-Sludge Sewage Treatment," *ER* 73 (1) (1916): 5.

41. Edward Ardern, "A Recent Development of the Sewage Problem," *AMSDW*, 1914, 61–71, on 68.

42. 3rd Annual Report of the Sewerage Commission of the City of Milwaukee, Dec. 31, 1916, 42.

43. "Activated-Sludge Sewage Treatment," *ER* 73 (1) (1916): 5.

44. John Haworth, "Sheffield Sewage Disposal Works and the Experimental Work," *AMSDW*, 1916, 31–39, on 37. See also Harrison P. Eddy, "Lights and Shadows of the Activated Sludge Process for the Treatment of Sewage and Industrial Wastes," *Journal of the Western Society of Engineers* 26 (7) (1921): 272.

45. "Activated-Sludge Sewage Treatment," *ER* 73 (1) (1916): 5.

46. E. Ardern and W. T. Lockett, "The Oxidation of Sewage Without the Aid of Filters. Part III," *JSCI* 34 (18) (1915): 937–943, on 937.

47. William T. Lockett, "Activated Sludge Process: Aeration and Circulation," *AMSDW*, 1924, 92–104, on 101.

48. Ralph A. Stevenson, "Chemical Purification of Sewage," *Western Construction News and Highways Builder* 7 (1932): 527–528, on 527; Ralph A. Stevenson, "Chemical Sewage Purification at Palo Alto: Regeneration of Spent Coagulent Effects Complete Sewage Treatment," *SWJ* 5 (1933): 53–60.

49. W. E. Speight, "Exhibition of Photo-Micrographs," *AMSDW*, 1923, 21–24, on 23; Joshua Bolton, "Aeration Experiments," *AMSDW*, 1922, 35–38, on 36.

50. Joshua Bolton, "Activated Sludge Experiments at Bury," *AMSDW*, 1921, 33–47, 46.

51. "13 Plead Guilty in Melrose Park Rum Ring Trial," *CDT*, Jan. 16, 1929, 14; "Rum Indictment Hits Village President, 100 in Melrose Park," *CDT*, June 2, 1928, 5; Agents Wreck 5 Stills in Melrose Park," *CDT*, Dec. 4, 1927, 7; "Sweep Rummers from West Suburb," *CDT*, July 24, 1927, 1.

52. For instance, N. Swaminadhan, "Filamentous Bacteria in Activated Sludge," in *Some Studies in Biochemistry by Some Students of Dr. Gilbert J. Fowler* (Bangalore: Phoenix Printing House, 1924), 27–31; "The Operation of Activated Sludge Plants," *AMSDW*, 1927, 33–37; Walter Scott, "Bulking of Activated Sludge: An Investigation as to its Cause," *AMSDW*, 1928, 45–52; G. M. Ridenour, "Observations on Bulking in a Surface Aera-

tion Activated Sludge Plant," *SWJ* 5 (1933): 74–82; E. J. Reese, *A Report on the Causes of Bulking of Activated Sludge*, thesis, University of California–Berkeley, 1932.

53. Gilbert J. Fowler, "Aeration of Activated Sludge," *Surveyor* 67 (1925): 305.

54. "Biology and Sewage Purification: Midland Branch Discussion," *Sewage Purification* 1 (1939): 312–314, on 313.

55. H. Heukelekian, "Activated Sludge Bulking," *SWJ* 13 (1941): 39–42, on 41, 42.

56. C. George Anderson, "Activated Sludge Control at Rockville Centre and the Prevention of Bulking," *SWJ* 8 (1936): 784–792, on 786.

57. Chapman, "The Employment of Micro-Organisms in the Service of Industrial Chemistry," 283T.

58. A. A. Backhaus, "Ethyl Alcohol," *Industrial and Engineering Chemistry* 22 (1930): 1151–1153; J. F. Garrett, "Lactic Acid," *Industrial and Engineering Chemistry* 22 (1930): 1153–1154.

59. "Control of Operation of Activated-Sludge Plants," *SWJ* 4 (1932): 338–340, on 338, 340.

60. See, among others, F. W. Harris, T. Cockburn, and T. Anderson, "Observations on Biological and Physical Properties of Activated Sludge and the Principles of its Application," *AMSDW*, 1926, 52–68; "Appendix 1. Activated Sludge Process. Microbiological Studies," City of Manchester. Rivers Department, *Annual Report* 1928, 40–46; H. Heukelekian, "Sewage Plant Operation by pH Control," *SWJ* 3 (1931): 428–429; G. M. Ridenour and C. N. Henderson, "Stale Return Sludge as a Factor in the Activated Sludge Process," *SWJ* 6 (1934):36–41; F. W. Mohlman, "The Sludge Index," *SWJ* 6 (1934): 119–122; H. Heukelekian, "Some Biochemical Indices of the Condition of Activated Sludge," *Proceedings of the Nineteenth Annual Meeting, New Jersey Sewage Works Association*, 1934, 50–55; George Anderson, "Running an Activated Sludge Plant on D.O. Determinations," *Water Works and Sewerage* 83 (1936): 418–419; Lewis H. Kessler, "The 'Odeeometer'—Its Place in the Control of Activated Sludge Plants," *Water Works and Sewerage* 83 (1936): 13–19; Milton Speigel, Stanley E. Kappe, and Gilbert M. Smith, "Control of the Activated Sludge Process," *Water Works and Sewerage* 84 (1937): 167–169; Edward B. Mallory, "Oxidized Sludge Process and the 'Equilibrium Index,'" *Water Works and Sewerage* 88 (1941): 333–344. Process control strategies were reviewed in "The Operation and Control of Activated Sludge Sewage Treatment Works: Report of Committee on Sewage Disposal, American Public Health Association, Public Health Engineering Section," *SWJ* 14 (1942): 3–69.

61. "The Activated Sludge Process—Part I," *The Digester*, February 1936, no. 1: 2–5, on 5.

62. Edmund B. Besselievre, "Sewage Treatment in America. Co-operation Between Engineers and Managers of Works," *Institute of Sewage Purification, Proceedings*, 1931, 177–185.

63. Quoted in Peter F. Meiksins, "Scientific Management and Class Relations," *Theory and Society* 13 (1984): 177–209, on 188.

64. Wm. P. Ellis, "Proposed Association of Sewage Works Managers," *The Sanitary Record* 28 (1901): 306.

65. J. Fieldhouse, "Proposed Association of Sewage Works Managers," *The Sanitary Record* 28 (1901): 193.

66. J. H. Barford, "The Sewage Works Manager–His Position and Responsibilities," *AMSDW*, 1903, 44–48, on 44; "Sewage Farming: An Association of Sewage Works Managers Suggested," *The Sanitary Record* 28 (1901), 159.

67. J. H. Barford, "The Sewage Works Manager—His Position and Responsibilities," *AMSDW*, 1903, 44–48, on 44.

68. Andrew Abbott, *The System of Professions: An Essay on the Division of Expert Labor* (Chicago: University of Chicago Press, 1988); Ibo Van de Poel, "The Bugs Eat the Waste: What Else is There to Know? Changing Professional Hegemony in the Design of Sewage Treatment Plants," *Social Studies of Science* 38 (2008): 605–634, examines the contest between bacteriologists and engineers in the design of sewage treatment plants in similar terms.

69. Edward Lawes, *The Act for Promoting the Public Health: With Notes, an Analytical Index, and (By Way of Appendix) the Nuisances Removal and Diseases Prevention Act, 1848* (London: Shaw and Sons, 1849).

70. See, for example, T. W. Stainthorpe, "The Status of the Local Government Board Surveyor," *Proceedings of the Association of Municipal and Sanitary Engineers and Surveyors* 9 (1883): 69–74.

71. Lewis Angell to C. B. Adderly, Mar. 9, 1871, reprinted in Appendix, *Proceedings of the Association of Municipal and Sanitary Engineers and Surveyors* 1 (1873–1874): 231.

72. Lewis Angell, "Inaugural Address," *Proceedings of the Association of Municipal and Sanitary Engineers and Surveyors* 1 (1873–1874): 17–28.

73. R. A. Buchanan, "Institutional Proliferation in the British Engineering Profession, 1847–1914," *The Economic History Review* 38 (1985): 42–60.

74. For the rise of sanitary engineering, see Martin V. Melosi, *The Sanitary City: Urban Infrastructure in America from Colonial Times to the Present* (Baltimore: The Johns Hopkins University Press, 2000), 104–116; Joel A. Tarr, *The Search for the Ultimate Sink: Urban Pollution in Historical Perspective* (Akron, OH: University of Akron Press, 1996), 195–199; "The Sanitary Engineering Section of the American Public Health Association," *American Journal of Public Health* 2 (1912): 276–277, on 276.

75. Charles Terry, "How to Advance the Interests of the Association," *AMSDW*, 1903, 71–79, on 72.

76. "Annual Dinner," *AMSDW*, 1905, 105–110.

77. Address of S. Rideal, in "Annual General Meeting," *AMSDW*, 1902, 9–20; comments of S. Rideal, "First Annual Dinner," *AMSDW*, 1902, 29, 30.

78. "New Jersey Sewerage Organization," *Municipal Journal* 40 (1916): 242.

79. Clyde Potts, "Twenty-five Years of Sanitary Progress in New Jersey," *Proceedings of the 25th Annual Meeting, New Jersey Sewage Works Association*, 1940, 153–166, on 155.

80. William J. Orchard, "The Need for a Strong National Sewage Works Association," *Proceedings of the 25th Annual Meeting, New Jersey Sewage Works Association*, 1940, 115–125.

81. C. A. Emerson Jr., "Forward," in *Modern Sewage Disposal: Anniversary Book of the Federation of Sewage Works Associations*, ed. Langdon Pearse (New York: Federation of Sewage Works Associations, 1938), vii–viii.

82. Orchard, "The Need," 121.

83. Adolph Kanneberg and L. F. Warrick, "Aims of the Central States Sewage Works Association," *SWJ* 1 (1928): 5–8.

84. "Twenty-five Years of Progress with the Central States Sewage and Industrial Wastes Association. Silver Anniversary Meeting, Madison, Wisconsin, June 25–27, 1952," WEF Archives, Record Group 4, Box 3.

85. Abbott, *The System of Professions*, 8, 102.

86. "Twenty-five Years of Progress."

87. Donald M. Pierce, "History of the Michigan Water Pollution Control Association 1925–1975," n.p., WEF Archives, Record Group 4, Box 3.

88. F. Herbert Snow, "The Skilled Supervision of Sewage Purification Works," *American Journal of Public Health* 3 (1913): 535–546.

89. C. T. Mudgett, "Certification," *SWJ* 12 (1940): 613–616.

90. J. H. Garner, "The Aims and Methods of the Sewage Works Manager," *Institute of Sewage Purification, Journal and Proceedings*, 1946 (Part 1), 29–38; "The Institute—Past, Present and Future," *Institute of Sewage Purification, Journal and Proceedings*, 1949 (Part 2), 166–169.

91. H. H. Wagenhals, E. J. Theriault, and H. B. Hommon, "Sewage Treatment in the United States: Report on the Study of Fifteen Representative Sewage Treatment Plants," *United States Public Health Service, Public Health Bulletin No. 132*, 1923, 104, 112.

92. H. Maclean Wilson, "Presidential Address," *AMSDW*, 1915, 15–21, on 15.

93. Dr. Thresh, "The Sewage Works Manager of the Future," *AMSDW*, 1909, 16–19, on 19.

94. Stephen Flinn, "Report on the Examination Scheme," *AMSDW*, 1911, 19–22.

95. Ibid.

96. Warren J. Scott, "Regulations Pertaining to Qualifications of Sewage Works Operators," *SWJ* 6 (1934): 90–99, on 99.

97. Comments of Paul Hansen in discussion of Snow, "The Skilled Supervision," 546.

98. Scott, "Regulations Pertaining to Qualifications of Sewage Works Operators"; Charles C. Agar, "Licensing of Sewage Plant Operators," *SWJ* 13 (1941): 89–100; Mudgett, "Certification," 616.

99. Discussion of J. B. Baty, "The Operator's Licensing Act," *Proceedings of the Seventeenth Annual Meeting, New Jersey Sewage Works Association*, 1932, 61–72, on 69.

100. Mudgett, "Certification"; "Report of Committee on Certification of Operators," WEF Archives, Record Group 2, Box 71.

101. "Certification of Sewage Treatment Operators," *The Digester*, November 1938, no. 11: 25–27, on 25.

102. Illinois, Service Rating Score Sheet, WEF Archives, Record Group 2, Box 70.

103. Dr. Thresh, "The Sewage Works Manager of the Future," *AMSDW*, 1909, 18; *AMSDW*, 1924, 20–21.

104. Associate Membership Examination, *Institute of Sewage Purification, Journal and Proceedings*, 1945 (Part 2), 191–194; John Hurley, "Presidential Address," *Institute of Sewage Purification, Journal and Proceedings*, 1945 (Part 2), 7–9.

105. "The Associate Class of the Institute," *Sewage Purification, Land Drainage, Water and River Engineering* 1 (1939): 195. "The prestige of the examination has steadily increased, so that it now ranks as a recognised technical qualification."

106. Ibid.

107. A. H. Cross, "The Associate Class," *Sewage Purification* 1 (1939): 286.

108. "Operator Certification Programs for Water and Wastewater Personnel," *Journal of the Water Pollution Control Federation* 44 (1972): 2218–2228.

109. Anonymous, "Short Courses for Sewage Plant Operators," *Proceedings of the Fourteenth Annual Meeting, New Jersey Sewage Works Association*, 1929, 11–12.

110. Professor Lendall, "The Rutgers Short Course," *Proceedings of the Fifteenth Annual Meeting, New Jersey Sewage Works Association*, 1930, 58–61.

111. "Engineering Section Project, American Public Health Association, Progress Report, Sewage Works Short Course Training, 1947" and "Federation of Sewage and Industrial Wastes Associations. Listing of Short Schools, April 1951," WEF Archives, Record Group 2, Box 70.

112. "Improving the Education Side of the Association," *AMSDW*, 1926, 91–93.

113. Program, Short Course for Sewage Treatment Works Operators to be held at University of Illinois, 1941, University of Illinois Archives, Short Course, Conference and Institute Programs and Proceedings, 1910–1966, Record Series 11/5/830, Box 1, Sewage Works Short Course.

114. Professor Lendall, "The Rutgers Short Course," *Proceedings of the Fifteenth Annual Meeting, New Jersey Sewage Works Association,* 1930, 58–61.

115. Twelfth Water and Sewage Short Course, July 20–24, 1936, Oklahoma A. & M. College.

116. New York University, Special Course for Sewage Treatment Plant Operators, Grade II, WEF Archives, Record Group 2, Box 69.

117. Whether scientific education for the working class represents social control or opportunities for upward mobility is debated in the literature. See Steven Shapin and Barry Barnes, "Science, Nature and Control: Interpreting Mechanics' Institutes," *Social Studies of Science* 7 (1977): 31–74, and John Laurent, "Science, Society and Politics in Late Nineteenth-Century England: A Further Look at Mechanics' Institutes," *Social Studies of Science* 14 (1984): 585–619. At least in later periods, technical training was probably instrumental in allowing some sewage treatment plant operators to move into the ranks of engineers. *Operations Forum* 2 (5) (1985): 24.

118. "The Activated Sludge Process—Part II," 8.

119. See Gilbert J. Fowler, *Sewage Works Analyses* (New York: Wiley, 1902); Leonard Metcalf and Harrison P. Eddy, *American Sewerage Practice, vol. III, Disposal of Sewage*, 2nd ed. (New York: McGraw-Hill, 1916), 44–77; Frank Bachmann, "Simple Tests for Sewage Plant Control," *Proceedings of the Eighth Texas Water Works Short School*, 1926, Bulletin No. 1: 151–155; Henry Weiner, "The Laboratory of the Sewage Treatment Plant," *SWJ* 1 (1929): 358–338; G. F. Catlett, "Laboratory Control of Sewage Treatment Works," *SWJ* 2 (1930): 48–54; "Symposium: Standard Methods for the Examination of Sewage and Sewage Sludge," *SWJ* 2 (1930): 347–386; W. D. Hatfield, "The Works Laboratory," in Pearse, ed.,

Modern Sewage Disposal, 126–131; W. A. Hardenbergh, *Operation of Sewage-Treatment Plants* (Scranton: International Textbook Company, 1939), 41–90.

120. "Sewage Works Operating and Cost Records," *SWJ*, 1931, 4:3–25, 3.

121. "The Operation and Control of Activated Sludge Sewage Treatment Works. Report of Committee on Sewage Disposal, American Public Health Association, Public Health Engineering Section," *SWJ* 14 (1942): 3–69.

122. "Sewage Works Operating and Cost Records," *SWJ* 4 (1931): 3–25, on 3.

123. H. H. Wagenhals, "Some Conclusions Drawn From a Recent Survey of Sewage Treatment Plants," *Journal of the Western Society of Engineers* 27 (1922): 239–252, on 247.

124. C. H. Nordell, "Sewage Disposal," U.S. Patent 2,225,437, Dec. 17, 1940.

125. Heukelekian, "Some Biochemical Indices of the Condition of Activated Sludge," 50.

126. "Mechanism of the Activated Sludge Process," *SWJ* 2 (1920): 146–147, on 147.

127. John Finch and Harold Ives, "Settleability Indexes for Activated Sludge," *Sewage and Industrial Wastes* 22 (1950): 833–839, on 833.

128. Heukelekian, "Some Biochemical Indices on the Condition of Activated Sludge," 50, 54.

129. J. A. Reddie, "Notes on the Bradford Experimental Activated Sludge Tanks," *AMSDW*, 1922, 121–124, on 122.

130. Joshua Bolton, "Activated Sludge Experiments at Bury," *AMSDW*, 1921, 33–47, on 47.

131. Eddy, "Lights and Shadows," 265.

132. Discussion on "The Operation of Activated Sludge Plants," *AMSDW*, 1927, 33–45, on 36.

133. Wagenhals, Theriault, and Hommon, "Sewage Treatment in the United States," 156.

134. "Activated Sludge Sewage Treatment," *Eleventh Annual Report, Ohio Conference on Sewage Treatment*, 1937, 82–83. See also C. N. Sawyer, "Activated Sludge Oxidations. V. The Influence of Nutrition in Determining Activated Sludge Characteristics," *SWJ* 12 (1940): 3–17, on 15.

135. W. W. Mathews, "Bark from the Daily Log," *SWJ* 18 (1946): 1198–1202, on 1199.

136. W. W. Mathews, "Bark from the Daily Log," *SWJ* 19 (1947): 1060–1065, on 1063.

137. W. W. Mathews, "Bark from the Daily Log," *SWJ* 20 (1948): 324–328.

138. G. G. Pointdexter, "Chemical Control," *SWJ* 11 (1939): 1025–1029, on 1028.

139. James Scott, *Seeing Like a State: How Certain Schemes to Improve the Human Condition Have Failed* (New Haven: Yale University Press, 1998).

140. H. C. H. Shenton, "The Activated Sludge Process," *Surveyor* 59 (1921): 289–290, on 289.

141. Dibdin, *Purification*, 79.

142. Sawyer, "Activated Sludge Oxidations. V," 15.

143. Heukelekian, "Some Biochemical Indices of the Condition of Activated Sludge," 50.

144. "Control of Operation of Activated-Sludge Plants," *SWJ* 4 (1932): 338–340.

145. John R. Palmer and F. W. Mohlman, "Laboratories of the North Side Sewage Treatment Works of the Sanitary District of Chicago," *SWJ* 2 (1930): 40–47. Sanitary District of Chicago, *Formal Opening, The North Side Sewage Treatment Project*, 1928.

146. Dickens, *Reconstructing Nature*, 129.

147. Chapman, "The Employment of Micro-organisms in the Service of Industrial Chemistry," 284T. For the persistence of craft in brewing in Chicago see Platt, *Shock Cities*, 129.

148. R. J. Spode, "The Operation of Activated Sludge Plants Practically Considered," *Institute of Sewage Purification, Journal and Proceedings*, 1933 (Part 1), 49–58, on 55.

149. William Joseph Dibdin, "Sewage-Sludge and Its Disposal," *Minutes of Proceedings of the Institution of Civil Engineers* 88 (1887): 155–298, on 155.

150. Discussion of E. Hannaford Richards and G. C. Sawyer, "Further Experiments with Activated Sludge," *JSCI* 41 (1922): 62T–72T, on 72T.

151. J. A. Reddie, "Series of Photo-Micrographs of Activated and Precipitated Sewage Sludges," *AMSDW*, 1922, 109–124; C. C. Ruchhoft and J. H. Watkins, "Bacteriological Isolation and Study of the Filamentous Organisms in the Activated Sludge of the Des Plaines River Sewage Treatment Works," *SWJ* 1 (1928): 52–58; H. Taylor, "Some Biological Notes on Sewage Disposal Processes," abstracted in *SWJ*, 1930, 2 (1930): 637–639; A. M. Buswell, "The Biology of Activated Sludge—An Historical Review," *SWJ* 3 (1931): 362–368; C. T. Butterfield, "Studies of Sewage Purification: A Zooglea-Forming Bacterium Isolated from Activated Sludge," *Public Works*, May 1935, 16–18, and June 1935, 23–26.

152. C. N. Sawyer and M. Starr Nichols, "Activated Sludge Oxidations: II. A Comparison of Oxygen Utilization by Activated Sludges Obtained from Four Wisconsin Municipalities," *SWJ*, 11 (1939): 462–471, on 462.

153. Discussion on "The Operation of Activated Sludge Plants," *AMSDW*, 1927, 33–45, on 43.

154. "The Way I See It," *Sewage Gas: The Operators' Journal* 15 (1952): 31–32, 32.

155. J. R. Palmer, "Activated Sludge Control," *Thirteenth Annual Report, Ohio Conference on Sewage Treatment*, 1939, 11–25, on 21.

156. Comments of H. Heukelekian to Sherwood Vermilye, "Activated Sludge—Some Notes and Comments," *Proceedings of the Nineteenth Annual Meeting, New Jersey Sewage Works Association*, 1934, 44–49, on 49.

157. Discussion of J. B. Baty, "The Operator's Licensing Act," *Proceedings of the Seventeenth Annual Meeting, New Jersey Sewage Works Association*, 1932, 61–72, on 67–68.

158. Paul Molitor Jr., "Operation of the Activated Sludge Treatment Plant at Morristown, N.J.," *New Jersey Sewage Works Association, Proceedings, Twenty-third Annual Meeting*, 1938, 39–46, on 45.

159. Comments of H. Heukelekian to Vermilye, "Activated Sludge—Some Notes and Comments," 49.

160. Earle B. Phelps, "Stream Pollution from the Operator's Point of Vew," *SWJ* 2 (1930): 555–560, on 560.

161. Researchers have noted the importance of these kinds of qualitative indicators of ecosystem state in the management of other ecosystems. These include the monitoring of fat content in Caribou as a measure of population growth and an indicator for adjusting the size of a hunt, or the presence or size of indicator species for assessing soil fertility or timing planting

in agricultural systems. Fikret Berkes, *Sacred Ecology: Traditional Ecological Knowledge and Resource Management* (Philadelphia: Taylor & Francis, 1999), 105–109; M. Tengö and K. Belfage, "Local Management Practices for Dealing with Change and Uncertainty: A Cross-Scale Comparison of Cases in Sweden and Tanzania," *Ecology and Society* 9 (3) (2004): 4, <http://www.ecologyandsociety.org/vol9/iss3/art4>; Scott, *Seeing Like a State*, 311–312. As Scott points out, craft knowledge is characterized generally by the use of measures of ecosystem state or function that are often qualitative rather than quantitative; are integrative, accumulating many partly redundant signals; and are economical, incorporating only those measures that suffice and dispensing with unnecessary measurement—all characteristics of craft management of sewage treatment plants.

162. Clarence MacCallum, "Training Operators of Sewage Treatment Plants," *SWJ* 12 (1940): 987–991, on 988, 991.

163. Phelps, "Stream Pollution," 560.

164. Reed, "The Sewage Treatment Plant Operator," 3–4.

165. Abel Wolman, "The Small Plant Operator as Scientist," *Journal of the American Water Works Association* 8 (1921): 359–361.

166. Eustace Thomas, "The Management of Engineering Workshops," *Journal of the Institution of Electrical Engineers* 41 (1908): 741–758, on 742, 751.

4   Profit vs. Purification

1. George Thudichum, *The Bacterial Treatment of Sewage. A Handbook for Councillors, Engineers, and Surveyors* (London: The Councillor and Guardian Offices, n.d.), 91.

2. How Fertilizers Help to Establish Dense, Luxuriant Turf, Apr. 18, 1927, N/MC.

3. Ibid. A. B. Granville, *The Great London Question of the Day; or, Can Thames Sewage Be Converted into Gold?* (London: Edward Stanford, 1865).

4. Quoted in Royal Commission on Metropolitan Sewage Discharge, *Second and Final Report of the Commissioners*, 1884 [C. 4253], xxxiii.

5. "Commercial Possibilities with Activated-Sewage Sludge," *ER* 74 (1916): 428.

6. Nicholas Goddard, "'A Mine of Wealth'? The Victorians and the Agricultural Value of Sewage," *Journal of Historical Geography* 22 (1996): 274–290. Richard P. Aulie, "The Mineral Theory," *Agricultural History* 48 (1974): 369–382.

7. Goddard, "'A Mine of Wealth'?," 276.

8. John Louis William Thudichum, "On an improved mode of collecting Human Voidings, with a view to their application for the benefit of Agriculture and the reduction of Local Taxation," Report from the Select Committee on Sewage (Metropolis); Together with the Proceedings of the Committee, Minutes of Evidence, Appendix and Index, July, 14, 1864 [487], 299–314, on 312, 313.

9. J. B. Lawes, "On the Sewage of London. Being a paper read at the thirteenth ordinary meeting of the Society of Arts, Wednesday, March 7, 1855," *The Rothamsted Memoirs on Agricultural Chemistry and Physiology, Vol. II* (London: William Clowes and Sons Ltd., 1898), 14.

10. Report from the Select Committee on Sewage (Metropolis), 591–592.

11. Corfield, *Treatment and Utilization of Sewage* (London: MacMillan and Co., 1887), 267–273; Goddard, "'A Mine of Wealth'?," 281.

12. Quoted in Robert Scott Burns, *Outlines of Modern Farming, Vol. V. Utilization of Town Sewage.—Irrigation—Reclamation of Waste Land,* 2nd ed. (London: Virtue Brothers, & Co., 1865), 15.

13. Burns, *Outlines,* 21–22. Lawes, "On the Sewage of London," 6–7.

14. Corfield, *Treatment,* 369–373.

15. Lawes, "On the Sewage of London," 32.

16. C. H. Cooper, "Activated Sludge," *Surveyor* 49 (1916): 55–56. Characterizing sewage as an "El Dorado" was also in E. J., "The Metropolitan Sewage Difficulty," *Times* (London*),* Aug. 22, 1885, 12, and Dr. Letheby, *The Sewage Question* (London: Bailliere, Tindall and Cox, 1872), vii.

17. Testimony of W. C. Mylne, *Report from the Select Committee on Metropolitan Sewage Manure,* 1846 [474], 27–28.

18. Royal Commission on Metropolitan Sewage Discharge, *First Report of the Commissioners,* 1884 [C. 3842], xl.

19. Report from the Select Committee on Sewage (Metropolis), 291.

20. For the history of Napier and Hope's proposal and the Metropolis Sewage & Essex Reclamation Company, I follow P. L. Cottrell, "Resolving the Sewage Question: Metropolis Sewage & Essex Reclamation Company, 1865–81," in *Cities of Ideas: Civil Society and Urban Governance in Britain 1800–2000,* ed. Robert Colls and Richard Rodger (Aldershot: Ashgate, 2004), 67–95.

21. Compare this with the average return on investment, from 1870 to 1880, including dividends and appreciation, of 3.8 percent on British government issues, 11.8 percent on London bank stock, and 9 percent on English railway stock. Among municipal infrastructure investments, waterworks had an average return of 10.1 percent, gas companies, 8.9 percent, and tramways, 6.2 percent. Robert Lucas Nash, *A Short Inquiry into the Profitable Nature of Our Investments* (London: Effingham Wilson, Royal Exchange, 1880). In contrast, Nash warned that sewage companies were among the gravest of risks during this period.

22. William H. Brock, *Justus von Liebig: The Chemical Gatekeeper* (Cambridge: Cambridge University Press, 1997), 268.

23. "Railway and Other Companies," *Times* (London), Apr. 6, 1871, 6.

24. W. Hope, "Can Sewage Be Utilised as Well as Purified," *Journal of the Society of Arts* 24 (1876): 624–628, on 626. See also W. Hope, *Food Manufacture versus River Pollution. A Letter Addressed to the Newspaper Press of England* (London: Edward Stanford, 1875).

25. *Times* (London), Aug. 22, 1873, 8; *Times* (London), Nov. 25, 1904, 4.

26. Hope, *Food Manufacture,* 34.

27. W. Hope, "The Treatment of Sewage," *Times* (London), Oct. 23, 1907, 4.

28. *Report of the Judges Appointed by the Royal Agricultural Society of England to Adjudicate the Prizes in the Sewage Farm Competition, 1879* (London: William Clowes, 1880).

29. These firms had names like the Sewage Phosphate Company and the General Sewage Manure Company. P. L Cottrell, "Resolving the Sewage Question," 92.

30. Royal Commission on Metropolitan Sewage Discharge, *Minutes of Evidence Taken Before the Commission, From May 1884 to October 1884 Together with a Selection from the Appendices, and a Digest of the Evidence, Vol. II*, 1885 [C. 4253-I], 25. There was much confusion as to what the "A" stood for. Most thought it stood for alum but at times stood for animal charcoal.

31. "The Native Guano Company," *Times* (London), Oct. 14, 1869, 3.

32. "Money-Market & City Intelligence," *Times* (London), Oct. 14, 1869, 5.

33. "Metropolitan Board of Works," *Times* (London), Dec. 24, 1869, 5.

34. Royal Commission on Metropolitan Sewage Discharge, *Minutes of Evidence Taken Before the Commission*, 31. I thank Larry Neal for helping me understand the workings of the London Stock Exchange during this period.

35. Testimony of R. Giffen. London Stock Exchange Commission, *Report of the Commissioners*, 1878 [C. 2157], 303.

36. Rivers Pollution Commission (1868), Second Report of the Commissioners Appointed in 1868 to Inquire into The Best Means of Preventing the Pollution of Rivers. The A.B.C. Process of Treating Sewage, 1870 [C. 180], 18.

37. J. Bailey Denton, "The Sewage Question," *Times* (London), Sept. 27, 1871, 4.

38. "Metropolitan Board of Works," *Times* (London), Jan. 28, 1871, 8; "Metropolitan Board of Works," *Times* (London), Apr. 7, 1871, 9.

39. "Money-Market and City Intelligence," *Times* (London), Sept. 23, 1871, 7.

40. "Metropolitan Board of Works," *Times* (London), Feb. 1, 1873, 7.

41. "Money-Market & City Intelligence," *Times* (London), Sept. 28, 1871, 5.

42. "Money-Market & City Intelligence," *Times* (London), Apr. 10, 1872, 10.

43. For Crookes's involvement in Native Guano, see William H. Brock, *William Crookes (1832–1919) and the Commercialization of Science* (Aldershot: Ashgate, 2008), 103, 286–293.

44. "Railway and Other Companies," *Times* (London), Jan. 9, 1873, 6.

45. Testimony of W. Sillar, Royal Commission on Metropolitan Sewage Discharge, 29. "Native Guano Company," *Journal of the Society of Arts* 21 (1873): 234.

46. "Metropolitan Board of Works," *Times* (London), Feb. 1, 1873, 7.

47. "Metropolitan Board of Works," *Times* (London), Feb. 1, 1873; "Metropolitan Board of Works," *Times* (London), Feb. 15, 1873, 12.

48. Testimony of William C. Sillar, *Report of Sir John Hawkshaw, The Commissioner Appointed to Inquire as to the Purification of the River Clyde. Together with Appendix and Minutes of Evidence*, 1876 [C. 1464], 14. Testimony of R. Giffen. London Stock Exchange Commission. *Report of the Commissioners*, 300–303.

49. Corfield, *Treatment*, 339–340.

50. Incorporated Council of Law Reporting for England & Wales, *The Weekly Notes, Pt. II*, Aug. 5, 1876, no. 32, 349.

51. "The Sewage Disinfecting and Manure Company," *Times* (London), Feb. 23, 1872. Incorporated Council of Law Reporting for England & Wales, *The Weekly Notes, Pt. II*, Jan. 11, 1873, no. 2, 26.

52. "Railway and Other Companies," *Times* (London), Sept. 15, 1875; "Money-Market and City Intelligence," *Times* (London), May 18, 1876, 4.

53. "The Utilization of Sewage," *Times* (London), July 27, 1880, 12.

54. "Railway and Other Companies," *Times* (London), Nov. 1, 1881, 7.

55. *Times* (London), July 20, 1883, 11.

56. M. Lefeldt quoted in William Crookes, "The Profitable Disposal of Sewage: A Paper Read Before the Congrès International d'hygiène et de sauvetage, Held at Brussels, October 4, 1876" (London: E. J. Davey, 1876)

57. Letheby, *The Sewage Question*, 166.

58. C. Norman Bazalgette, "The Sewage Question," *Minutes of Proceedings of the Institution of Civil Engineers*, 48 (1876–1877): 105–251, 113–114.

59. "The London Sewage Disposal, No. 1," *The Sanitary Engineer* 11 (1884): 350.

60. Robert Vawser, "The Purification and Disposal of Sewage," *The Sanitary Engineer* 12 (1885): 211.

61. C. Meymott Tidy, *The Treatment of Sewage* (New York: D. Van Nostrand, 1887), 203.

62. Thudichum, *The Bacterial Treatment*, 91.

63. Noel G. Coley, "Medical Chemists and the Origins of Clinical Chemistry in Britain (circa 1750–1850)," *Clinical Chemistry* 50 (2004): 961–972.

64. In discussion of William Joseph Dibdin, "Sewage-Sludge and Its Disposal," *Minutes of Proceedings of the Institution of Civil Engineers* 88 (1887): 155–298, on 214.

65. Royal Commission on Metropolitan Sewage Discharge, *Second and Final Report of the Commissioners*, 1884 [C. 4253], xxxvii.

66. Other coastal cities, including Dublin, Glasgow, and Manchester, used similar ocean dumping schemes. Royal Commission on Sewage Disposal, *Fifth Report of the Commissioners Appointed to Inquire and Report What Methods of Treating and Disposing of Sewage (Including Any Liquid from Any Factory or Manufacturing Process) May Properly Be Adopted*, 1908 [Cd. 4278], 162–167.

67. *Times* (London), Jan. 23, 1885, 11.

68. "Railway and Other Companies," *Times* (London), April 13, 1893, 12.

69. "Railway and Other Companies," *Times* (London), Jan. 28, 1896, 3.

70. Royal Commission on Sewage Disposal, *Fifth Report of the Commissioners*, 32, 230.

71. Daily Report of the Manager to the Chairman of the Rivers Committee, May 18, 1912, 97; Davidson & Morris to Town Clerk, June 5, 1912, 101, Minutes of the Davyhulme Subcommittee, Manchester Rivers Committee, No. 16, Manchester Archives, Manchester Central Library, Manchester, England.

72. See, for instance, their ads in *Times* (London), "Best and Cheapest Manure for Farm and Garden—Native Guano." Curiously, they were still marketing the manure for £3 10s per ton, the exact price they were charging in 1869, almost forty years earlier. *Times* (London), Apr. 16, 1906, 11.

73. Bazalgette, "The Sewage Question," 111.

74. Joel A. Tarr, *The Search for the Ultimate Sink* (Akron: University of Akron Press, 1996), 301. For an overly optimistic view of the potential for sewage irrigation, see George W.

Rafter, "Sewage Irrigation," *Water-Supply and Irrigation Papers of the United States Geological Survey No. 3*, 1897, and George W. Rafter, "Sewage Irrigation, Part II," *Water-Supply and Irrigation Papers of the United States Geological Survey No. 22*, 1899.

75. Discussion of Henry J. Barnes, "Sewage Systems and the Epuration of Sewage by Irrigation and Agriculture," *Boston Medical and Surgical Journal* 110 (1884): 609–611, on 609, 610.

76. "Sewage Peril Here, Sanitarians Find," *NYT*, April 21, 1913.

77. F. R. O'Shaughnessy, "The Physical Aspect of Sewage Disposal," 362T.

78. Fowler to Bartow, Dec. 14, 1915, BP.

79. Edward Bartow, "The Treatment of Sewage by Aeration in the Presence of Activated Sludge," *The Transactions of the American Institute of Chemical Engineers* 8 (1915): 119–131.

80. Fowler to Bartow, Mar. 31, 1915, BP.

81. Fowler to Bartow, May 12, 1915, BP.

82. Gilbert J. Fowler, *An Introduction to the Biochemistry of Nitrogen Conservation* (London: E. Arnold & Co., 1934). He also tried to commercialize what he called the "Activated Manure" process, which involved mixing sewage sludge with rubbish to produce compost. J. A. Coombs did the engineering work on this process. Ironically, perhaps, Coombs argued that composting was "nature's doing and thus the discovery is unpatented and unpatentable, as is the method of composting itself for the same reason." Coombs, discussion of Sir Albert Howard, "The Preservation of Domestic Wastes for Use on the Land," *Journal of the Institution of Sanitary Engineers* 43 (1939): 173–196, on 192.

83. Hamlin, "Muddling in Bumbledom"; Tarr, *Searching*, 163–170.

84. Gail Radford, "From Municipal Socialism to Public Authorities: Institutional Factors in the Shaping of American Public Enterprise," *The Journal of American History* 90 (2003): 863–890; J. R. Kellett, "Municipal Socialism, Enterprise and Trading in the Victorian City," *Urban History* 5 (1978): 36–45, Chamberlain quotation on 42; Malcolm Falkusa, "The Development of Municipal Trading in the Nineteenth Century," *Business History* 19 (1977): 134–161. For the broad exchange of progressive ideas and practices, see Daniel T. Rodgers, *Atlantic Crossings: Social Politics in a Progressive Age* (Cambridge, MA: Harvard University Press, 1998).

85. Kellett, "Municipal Socialism," Chamberlain quotation on 42. On debates over profit, see Edwin Cannan, "The Principle of Municipal Trading," *Independent Review* 7 (1905): 287–301, and Robert Donald, "Municipal Trading and Profits," *The Economic Journal* 9 (35) (1899): 378–383. On technical considerations of profit, see J. Row-Fogo, "The Statistics of Municipal Trading," *The Economic Journal* 11 (41) (1901): 12–22, and Hoan to Carl Thompson, August 20, 1914, DHP Box 3, File 81.

86. For this calculus in Milwaukee's system, see Kate Foss-Mollan, *Hard Water: Politics and Water Supply in Milwaukee, 1870–1995* (West Lafayette, IN: Purdue University Press, 2001).

87. David P. Thelen, *The New Citizenship: Origins of Progressivism in Wisconsin, 1885–1900* (Columbia: University of Missouri Press, 1972), 223–249, 256–287; Judith Walzer Leavitt, *The Healthiest City: Milwaukee and the Politics of Health Reform* (Princeton: Princeton University Press, 1982), 122–150.

88. *Annual Report of the Board of Public Works* (Milwaukee, 1892), vi–vii; *Annual Report of the Board of Public Works* (Milwaukee, 1893), 7; *Annual Report of the Board of Public Works* (Milwaukee, 1900); Foss-Mollan, *Hard Water*, 71–72.

89. Foss-Mollan, *Hard Water*, 86–90; First Annual Report of the Sewerage Commission of the City of Milwaukee, 1915, 3–20; Leavitt, *The Healthiest City*, 57–61. At the same time, the typhoid outbreak led to the overwhelming passage of a bond referendum to build a sewage treatment plant.

90. Tarr, *The Search*, 122–128, 191–194.

91. Hoan to Corcoran, Feb. 21, 1927, DHP, Box 31, File 776.

92. Through Hatton's experience at Wilmington, he was not convinced that filtration prevented all cases of typhoid, and that even with filtration, if it wasn't managed properly, typhoid would still be a threat. T. Chalkley Hatton, "Some Relations between Filtered Water and Typhoid Fever as Shown in the Two Years' Service of the Wilmington Filters," *American Journal of Public Health* 3 (1913): 742–745.

93. "The Truth About Filtration," DHP Box 15, File 371; Ellms to Hoan, Feb. 26, 1921, DHP Box 15, File 371; Hatton to Priebe, Feb. 5, 1916, DHP Box 31, File 775; Hoan to Eddy, Jan. 6, 1931, DHP Box 35, File 888; "Reasons Why City Engineer Should Be on the Sewerage Commission," DHP, Box 31, File 776. "Shoots Holes into Ruhland's Water Record," n.p., 1918, and "$4,500,000 for Water Filter Up to Council," n.p., 1926, newspaper clippings collection, MMSD Box 4823.

94. One reason Hoan wanted to take control of the sewage plant from the Milwaukee Sewerage Commission was its coal-fired generator used to power the compressed air pumps. Hoan saw municipal ownership of this generator as a way for the city to enter and build a market for municipal power production, a longtime goal of the Socialists. Another goal was to use sewerage to facilitate the annexation of outlying areas to the city. "Suggestions for a Power Policy for the City of Milwaukee," DHP, Box 6, File 156; Hatton to Hoan, June 13, 1925, DHP, Box 15, File 360.

95. "Report of Special Committee Appointed to Investigate the Sewage Disposal Plant, Milwaukee, Wis.," Oct. 25, 1926, DHP, Box 35, File 886. John Boettiger, "Engineer Quits in Milwaukee Row on Sewage," *CDT*, Jan. 4, 1927. Hatton continued as a paid consultant and was briefly rehired as acting chief engineer in 1932 before dying in an automobile accident in 1933.

96. T. Chalkley Hatton, "Municipal Ownership of Water Supplies," *Proceedings of the Eleventh Annual Convention of the American Society of Municipal Improvements*, 1904, 137–144.

97. "Nitrogen Recovery from Sewage Sludge Reaches Commercially Practicable Stage," *ER* 74(15) (1916): 444–445, on 445.

98. Eleventh Annual Report of the Sewerage Commission of the City of Milwaukee for the Year 1924, Jan. 1, 1925, 80.

99. "Nitrogen Recovery from Sewage Sludge," 445.

100. Eighth Annual Report of the Sewerage Commission of the City of Milwaukee, Jan. 1, 1922, 29.

101. Eleventh Annual Report of the Sewerage Commission of the City of Milwaukee for the Year 1924, Jan. 1, 1925, 79.

102. O. J. Noer, "Annual Report of Milwaukee Sewerage Commission Fellowship, for Year Ending February 1, 1926," N/MC.

103. Twelfth Annual Report of the Sewerage Commission of the City of Milwaukee for the Year 1925, Jan. 1, 1926, 9.

104. Alexander M. McIver & Son to V. H. Kadish, Dec. 9, 1926, N/MC.

105. O. J. Noer. "The Fertilizer Value of Activated Sludge with Particular Reference to Golf Courses. Progress Report for Seasons 1923 and 1924," N/MC; "Tentative Program of Field Work for 1924 in Connection with the commercial Development of Milwaukee's Activated Sludge, January 10th 1924," N/MC.

106. Twenty-third Annual Report of the Sewerage Commission of the City of Milwaukee for the Year 1946, Jan. 2, 1927, 17.

107. O. J. Noer, *The ABC of Turf Culture* (Cleveland, OH: The National Greenkeeper, 1928).

108. Hamlin, "The City as a Chemical System?"

109. V. H. Kadish to R. A. Oakley, Feb. 2, 1925; Kadish to Oakley, Mar. 13, 1925; Oakley to Kadish, Mar. 25, 1925, "Sludge Analyses," N/MC; O. B. Fitts to O. J. Noer, Nov. 12, 1925, "Market Dev Position," N/MC. "Questions and Answers," *The Bulletin of the United States Golf Association Green Section* 6 (1) (1926): 16–20.

110. Fertilizer Practice Simplified, Service Bureau, Milwaukee Sewerage Commission, July 1, 1931, "Market Dev. Position," N/MC. This material came from a publication of the Service Bureau of the Sewerage Commission. But it appeared verbatim under the authorship of a golf course greenkeeper. Whether the commission tried to plant the material under the imprimatur of a supposedly objective source, or whether the greenkeeper simply plagiarized the report, is unclear. Arthur Stephen, "My Experience with Fertilizers," *The National Greenkeeper* 6 (9) (1932): 5–6. Stephen was greenkeeper at Erie Down's Golf and Country Club Ltd.

111. Advertisement, *The National Greenkeeper* 5 (4) (1931): 28.

112. Advertisement, *The National Greenkeeper* 6 (5) (1932): 15. Advertisement, *The National Greenkeeper* 6 (9) (1932): 4. Advertisement, *The National Greenkeeper* 6 (12) (1932), inside back cover.

113. Nineteenth Annual Report of the Sewerage Commission of the City of Milwaukee for the Year 1932, Jan. 2, 1933, 163–179, on 172.

114. National Gardening Association, *National Gardening Survey, 1995–1996* (Burlington, VT: National Gardening Association, 1996), 87, and National Gardening Association, *National Gardening Survey, 1990–1991* (Burlington, VT: National Gardening Association, 1991), 86.

115. Weed control Plot. Milarsenite. Westmoor Country Club. 10-23-41. Chemical Weed Control. Plots. North Shore CC. Milwaukee, "Weed Control 1941," N/MC.

116. "Milarsenite," *The Greenkeepers' Reporter* 8 (6) (1940): 36. The advertisement "Milarsenite Proves Its Worth," is on the inside front cover.

117. How to Use MILARSENITE for Clover and Weed Control on Golf Greens, Milarsenite Leaflet # 1, "Weed Control Data + Correspondence," N/MC.

118. O. J. Noer to W. W. Allen, Dow Chemical Company, Apr. 23, 1945, "Weed Control Data + Correspondence," N/MC; Paul C. Marth, "New Developments in Weed Control with

2,4-D," *The Greenkeepers' Reporter* 14 (5) (1946): 9–12; "Timely Turf Tips, Weed Control," *The Greenkeepers' Reporter* 13 (5) (1945): inside front cover; "Timely Turf Tips, Common Chickweed in Watered Fairways," *The Greenkeepers' Reporter* 14 (6) (1946): inside front cover.

119. B. W. Bellinger to Victor Kadish, Mar. 21, 1945, "Milorganite Trials," N/MC.

120. Willaim E. Zimmerman, American Cyanamid, to O. J. Noer, June 14, 1951, "Milorganite Trials," N/MC.

121. For the history of 2,4-D during this period, see Gale E. Peterson, "The Discovery and Development of 2,4-D," *Agricultural History* 41 (1947): 243–254, and Nicolas Rasmussen, "Plant Hormones in War and Peace: Science, Industry, and Government in the Development of Herbicides in 1940's America," *Isis* 92 (2001): 291–316.

122. Bentley-Milorganite Company to V. H. Kadish, Sept. 6, 1947, "Milorganite on Other Crops, " N/MC.

123. Royal Commission on Metropolitan Sewage Discharge, *Minutes of Evidence Taken Before the Commission, From May 1884 to October 1884 Together with a Selection from the Appendices, and a Digest of the Evidence, Vol. II*, 1885 [C. 4253-I], 26–33, on 26, 27.

124. J. A. Voelcker, "Some Considerations Affecting the Agricultural Use and Value of Sewage," *AMSDW*, 1910, 71–81, on 80–81.

125. Fowler to Bartow, Apr. 21, 1915, BP.

126. Willem Rudolfs, "Fertilizer and Fertility Value of Sewage Sludge," *Water Works and Sewerage* 87 (1941): 575–578, on 575.

127. A. M Rawn, "Concerning Those Plus Fertility Values of Sewage Sludges," *Water Works and Sewerage* 88 (1941): 186–188.

128. Editor's comment, Willem Rudolfs, "Fertilizer and Fertility Value of Sewage Sludge," *Water Works and Sewerage* 87 (1941): 575–578, on 575.

129. Report 44–2 to The Sewerage Commission of the City of Milwaukee for the Period March 1 to May 1, 1944, MMSD Box 923.

130. *Research—A National Resource II.—Industrial Research. December 1940* (Washington, DC: National Resources Planning Board, 1941), 73–74.

131. Report 47–12 to The Sewerage Commission of the City of Milwaukee for the Period November 1, 1947 to January 1, 1948, MMSD Box 923.

132. Carl Shelly Miner Jr. and Bernard Wolnak, "Fermentation Activation," U.S. Patent 2,544,273, Mar. 6, 1951.

133. Tentative letter to R. F. Kneeland, Food and Drug Administration, May 1, 1952, MMSD Box 923.

134. C. W. Crawford to O. J. Noer, June 16, 1951, MMSD Box 923.

135. "Report on the Attitude of Three Large Companies in the Feed Industry Toward the Acceptability of Milorganite and Its Extracts as Vitamin B-12 Supplements in Animal Feeds," Dec. 17, 1952, MMSD Box 923.

136. "Engineering Study and Estimate of Costs for Proposed Milorganite Vitamin B-12 Extraction Plant. The Sewerage Commission of the City of Milwaukee, March 3, 1952," MMSD Box 3424; Carl Miner to Ray D. Leary, June 10, 1952, MMSD Box 929. Thomas R.

Wood, Murray Hill and David Hendlin, "Process for Production of Vitamin B$_{12}$," U.S. Patent 2,595,499, May 6, 1952.

137. "Vitamin B-12 Test Project Is Contracted," unidentified newspaper clipping, June 14, 1953, Milwaukee Sewerage Commission 1933–1960, xeroxed clippings, Milwaukee Public Library.

138. "B-12 Vitamin Production from Activiated [sic] Sludge Under Way at Milwaukee Sewage Plant," *Water & Sewage Works* 103 (10) (1956): 477.

139. Forty-third Annual Report of the Sewerage Commission of the City of Milwaukee For the Year 1956, 33; Forty-fifth Annual Report of the Sewerage Commission of the City of Milwaukee For the Year 1958, 31; Forty-seventh Annual Report of the Sewerage Commission of the City of Milwaukee For the Year 1960, 32–33.

140. "Benefits Seen in Milorganite," unidentified newspaper clipping, Mar. 31, 1952, Milwaukee Sewerage Commission 1933–1960, xeroxed clippings, Milwaukee Public Library.

141. *Utilization of Sewage Sludge as Fertilizer*, Manual of Practice No. 2. Federation of Sewage Works Associations, 1946, 58.

142. Annual Budget Report to the Board of Trustees, Making Recommendations for 1940, The Sanitary District of Chicago, 1940.

143. Report to the Board of Trustees, in Support of Recommendations for 1949 Budget, The Sanitary District of Chicago, 1949.

144. Advertisements for McIver & Son, *Commercial Fertilizer and Plant Food Industry* 91(3A) (1955): 188; *Commercial Fertilizer* 89 (3A) (1954): 152.

145. Pasadena, Los Angeles County, Inventory of the California Bureau of Sanitary Engineering Papers, 1911–1963, MS 80/3, Box 35, WRCA.

146. P. N. Daniels, "Increasing the Salability of Sludge," *Water Works and Sewerage* 81 (1934): 277–279, on 277, 279.

147. Daniels, "Increasing the Salability," 279.

148. C. K. Calvert, Almon L. Fales, C. G. Gillespie, C. E. Keefer, T. J. Lafreniere, F. W. Mohlman, Willem Rudolfs, F. M.Veatch, P. J. A. Zeller, and Langdon Pearse, "Utilization of Sewage Sludge as Fertilizer, Report of the Committee on Sewage Disposal, American Public Health Association, Public Health Engineering Section," *SWJ* 9 (1937): 861–912, on 903; Gilbeart H. Collings, *Commercial Fertilizers: Their Sources and Use*, 5th ed. (New York: McGraw-Hill, 1955), 131.

149. A. M. Rawn, "Concerning Those Plus Fertility Values of Sewage Sludges," *Water Works and Sewerage* 88 (1941):186–188, on 187.

150. Forty-first Annual Report of the Sewerage Commission of the City of Milwaukee For the Year 1954, 20; Fortieth Annual Report of the Sewerage Commission of the City of Milwaukee For the Year 1953, 28; Forty-second Annual Report of the Sewerage Commission of the City of Milwaukee For the Year 1955, 13–14. The commission prevented the use of "Daorganite" (Miami) and "Muskegonite" (Muskegon) as well. Alan K. Nees to Bill Parker, Feb. 20, 1992, MMSD Box 9000.

151. Except where noted, the history of Rapidgro is reconstructed from various newspaper clippings contained in a scrapbook from the Grand Rapids treatment plant records. Most of the article clippings in the scrapbook, however, have neither a source nor date, although they

appear to be chiefly from the 1930s. Series 16–20, Environmental Protection Department, Wastewater Treatment Plant, Misc. Historical Records, Grand Rapids City Archives.

152. Calvert et al., "Utilization of Sewage Sludge."

153. Arthur D. Caster, "Cities Report Varying Success in Drying Sludge for Sale," *Wastes Engineering* 26 (11) (1955): 610–611.

154. "Olis Proposes Package Sale of Fertilizer," *CDT*, Apr. 23, 1954.

155. "Sanitary Unit Can't Sell All Its Fertilizer," *CDT*, Aug. 10, 1957.

156. "Sanitary District Will Demolish Fertilizer Manufacturing Plant," *CT*, Dec. 1, 1968.

157. *Utilization of Municipal Wastewater Sludge*, Manual of Practice No. 2, Water Pollution Control Federation, 1971, 40.

158. James Brower, *Report: Resume of Twenty Five Years of Plant Operation in Connection With Drying Sewage Sludge*, 37–38, files of Tom Brennan, Milwaukee Metropolitan Sewerage District. I thank Tom Brennan for sharing his files of historical materials.

159. Radford, "From Municipal Socialism," 889.

160. *Utilization of Sewage Sludge as Fertilizer*, 98.

161. "Milorganite Moves from Profit to Loss," *Milwaukee Journal*, Nov. 13, 1960, Milwaukee Sewerage Commission 1933–1960, xeroxed clippings, Milwaukee Public Library.

162. "Fertilizer Plant's Sales Might Top Two Million," unidentified newspaper clipping, Sept. 13, 1951, Milwaukee Sewerage Commission 1933–1960, xeroxed clippings, Milwaukee Public Library.

163. "S. Shore Sewage Unit Urged Here," unidentified newspaper clipping, March 21, 1956, Milwaukee Sewerage Commission 1933–1960, xeroxed clippings, Milwaukee Public Library.

5   The Contradictions Continue

1. See, for instance, Harold B. Gotaas, Bert W. Johnson, F. H. Doe Jr., and Albert E. Berry, "Frontiers in Wastewater Management: Technological, Financial, Administrative," *Journal of the Water Pollution Control Federation* 38 (1996): 745–773.

2. Martin V. Melosi, *The Sanitary City: Urban Infrastructure in America from Colonial Times to the Present* (Baltimore: The Johns Hopkins University Press, 2000), 320–337.

3. Milorganite Advisory Council Meeting, Nov. 19, 1972, Toronto, Canada, MMSD Box 3673.

4. K. C. Flynn, "Sludge Marketing: The Quiet Revolution," *Journal of the Water Pollution Control Federation* 54 (1982): 1267–1269.

5. Fulton Rockwell, "New York City Moves Ahead with Biosolids Management," *Biocycle* 34 (10) (1993): 55–60; Tom Barron, "Can Cities Turn Profits from Sludge?," *Environment Today* 3 (4) (1992): 3–4.

6. Sandra Steingraber describes this dilemma well for waste incineration in *Living Downstream: An Ecologist Looks at Cancer and the Environment* (Reading, MA: Addison-Wesley, 1997), 215–217.

7. Vinton W. Bacon to Robert J. Borchardt, July 6, 1977, "Tech Bulletin 88," Noer/Milorganite® MMSD Collection, Turfgrass Information Center, Michigan State University Library.

8. Robert J. Borchardt to David G. Berger, Sept. 2, 1977, "Tech Bulletin 88," Noer/Milorganite® MMSD Collection, Turfgrass Information Center, Michigan State University Library.

9. Milorganite Advisory Council Meeting, Oct. 17, 1976, Buena Vista, Florida, MMSD Box 3673.

10. "Milorganite Gardening—A Health Hazard," a CBE comment, Mar. 1978, Citizens for a Better Environment, MMSD Box 14128. CBE also protested Chicago's sludge, called "Nu-earth," and forced the city to place warning signs where the city gave the sludge away to the public. "Plan Alert Signs at Nu-Earth Sites," *CT*, May 18, 1978.

11. Swanson to Hawkins, "Milorganite Retail Bag—NOTICE," Sept. 20, 1989, Cadmium Content Warning 1980–89, MMSD Box 7011.

12. Hawkins to McCabe & Graef, Sept. 27, 1989, "Milorganite Retail Bag," MMSD Box 7011.

13. Swanson to Hawkins, "Milorganite Retail Bag—NOTICE." A recent report suggests that brain trauma like that suffered by football players may mimic the symptoms of and be misdiagnosed as ALS. This report was based on studies of the neuropathology of yet another football player for the 49'ers initially diagnosed with ALS. Alan Schwarz, "Study Says Brain Trauma Can Mimic ALS," *NYT*, Aug. 17, 2010.

14. Michael Zahn, "Safer Milorganite to Be Sold in Snazzier Bags," *MJ*, June 3, 1982.

15. Michael Zahn, "Maryland Milorganite Ban Fought," *MJ*, Aug. 15, 1982.

16. "Resolution Authorizing Approval to Modify the 'Notice' on the Milorganite® 40 lb. Retail Bag," Sept. 27, 1993, 93–165–9(02), Item 10, MMSD Box 9000.

17. There remains great controversy surrounding the safety of sewage sludge, even that considered "exceptional quality." See John Stauber and Sheldon Rampton, *Toxic Sludge Is Good For You!* (Monroe, ME: Common Courage Press, 1995), 99–122, for a critique of sludge safety.

18. "Resolution Authorizing Approval to Modify the 'Notice' on the Milorganite® 40 lb. Retail Bag," Sept. 27, 1993, 93–165–9(02), Item 10, MMSD Box 9000.

19. Market Probe, Inc., "Milorganite Market Positioning Study, External Interviews," Marketing Study 1994/1995, MMSD Box 9000.

20. Charles G. Wilson to William Katz, July 25, 1977, Milorganite "C" "F" "X," MMSD Box 3673.

21. "Toward an Overall MMSD Marketing Approach: A Proposal for a Product Development Function," submitted to Patrick Marchese and Stephen P. Graef, Nov. 18, 1985, prepared by Rosemary Murphy, Daniel Smith. Solids Study— "Marketing" Plans, MMSD Box 5708, 3, 5.

22. "Report on the Milwaukee Metropolitan Sewerage District Solids Utilization Department," by Pamela J. Falvey and Geoffrey Hurtado, n.d., 14, 27, MMSD Box 5708.

23. Terry Ward, "Marketing Task Force," Feb. 25, 1991, MMSD Box 9000.

24. "Milorganite Market Positioning Study, Summary of Findings and Recommendations," Marketing Study 1994/1995, MMSD Box 9000.

25. Tom Barron, "Can Cities Turn Profits from Sludge?"

26. Michael Specter, "Ultimate Alchemy: Sludge to Gold," *NYT*, Jan. 25, 1993.

27. Terry Ward, "Inputs from Marketing Staff for Review," Nov. 27, 1990, MMSD Box 7011.

28. Stauber and Rampton, *Toxic Sludge*, 99–122.

29. Hill to Mees, "Milorganite Pricing Proposal," July 28, 1992, MMSD Box 9000. Data are for 1991.

30. "Milorganite Market Positioning Study, Summary of Findings and Recommendations," Marketing Study 1994/1995, MMSD Box 9000.

31. Jim Spindler to Alan Nees, "Bio Gro Product Dumping," April 14, 1994, MMSD Box 3673, "Milorganite Market Positioning Study, Summary of Findings and Recommendations," Marketing Study 1994/1995, MMSD Box 9000.

32. See <http://www.synagro.com/> (accessed May 11, 2009).

33. Milorganite Advisory Committee, Sept. 22, 1981, MMSD Box 3673.

34. Approval to Enter into License Agreements with Selected Fertilizer Blenders, May 16, 1990; Ralph Hollmon to Commissioners, Nov. 6, 1995; Alan Nees to Jim Hill, Nov. 27, 1995, MMSD Box 9000.

35. Robert C. Cheek to Alan Nees, Mar. 27, 1992, MMSD Box 9000; "New Rules on Organic Foods," *NYT*, Mar. 9, 2000.

36. Don Behm, "Milorganite as Garden Deer Repellent? It Could Happen: Seeking Official Designation," *MJS*, Nov. 19, 2006; Don Behm, "EPA Derails Plans to Market Milorganite as Deer Repellent," *MJS*, Jan. 18, 2009.

37. See, for example, Barry Commoner, *The Closing Circle* (New York: Bantam, 1971), 157–158; Thomas E. Maloney, "Detergent Phosphorus Effect on Algae," *Journal of the Water Pollution Control Federation* 38 (1966): 38–45.

38. D. W. Schindler, "Eutrophication and Recovery in Experimental Lakes: Implications for Lake Management," *Science* 184 (1974): 897–899.

39. *Croplife America Inc. et al. v. City of Madison et al.*, 373 F. Supp. 2d 905 (June 13, 2005).

40. Marie Rohde, "A Greener Future?: Fertilizer Slowdown Has MMSD Pondering Uses for Sludge," *MJS*, Sept. 14, 2006.

41. Lisa Sink, "Brookfield Offers to Share Sewage Sludge," *MJS*, Aug. 18, 2008.

42. Rohde, "A Greener Future?"

43. Comptroller General of the United States, Operation and Maintenance of Municipal Waste Treatment Plants (Washington, DC: Government Printing Office, 1969), 36 p.; Comptroller General of the United States, Continuing Need for Improved Operation and Maintenance of Municipal Waste Treatment Plants (Washington, DC: Government Printing Office, 1977), 75 p.

44. Comptroller General, Operation and Maintenance, 16.

45. Jeremiah F. Reynolds, "Control and Analysis: Activated Sludge System," *Water and Sewage Works* 117 (7) (1970): 251–259, on 251.

46. "Washington," *Water and Sewage Works* 116 (9) (1969): 5A.

47. "Proposed Development of the Sewage Treatment Plant—Operation & Design Branch—Cincinnati Field Investigations Center—to Enhance Pollution Reduction Accomplishments,"

Apr. 12, 1972, Environmental Protection Agency, Office of Enforcement and General Counsel, Cincinnati Field Investigations Center, personal files of James B. Walasek, in possession of author.

48. A. W. West, "Activated Sludge Process Operational Control (Clearance Draft)," Cincinnati Field Investigations Center, Office of Enforcement and General Counsel, Environmental Protection Agency, May 1971, 3–5, personal files of James B. Walasek, in possession of author.

49. Ibid., 1–3; Alfred W. West, "Operational Control Procedures for the Activated Sludge Process: Part I—*Observations*," National Waste Treatment Center, U.S. Environmental Protection Agency, Office of Water Program Operations, May 1974 (Revised No., 1975), 1, personal files of James B. Walasek, in possession of author.

50. Alfred West, "Objectives For Both Parts IV and V Which Will Be Bound as One Pamphlet," Feb. 1975, personal files of James B. Walasek, in possession of author.

51. A. W. West, "Activated Sludge Process Operational Control (Clearance Draft)," Cincinnati Field Investigations Center, Office of Enforcement and General Counsel, Environmental Protection Agency, May 1971, 24.

52. L. Preston, "Editor's Notes: Use More than Your Senses for Process Control," *Operations Forum* 10 (8) (1993): 2.

53. Karen B. Carter, "Wastewater Laboratories: Full Partners in Environmental Protection," *Journal of the Water Pollution Control Federation* 4 (1985): 266–270, on 270.

54. "The Institute—Past, Present and Future," *Institute of Sewage Purification, Journal and Proceedings*, 1949 (Part 2), 166–169.

55. Jeffrey Keefe and Denise Potosky, "Technical Dissonance: Conflicting Portraits of Technicians," in *Between Craft and Science: Technical Work in U.S. Settings*, ed. Stephen R. Barley and Julian E. Orr (Ithaca, NY: Cornell University Press, 1997), 53–81.

56. *History of the Indiana Water Pollution Control Association, 1937–1987*, 50–52, 88–97, WEF Archives, Record Group 4, Box 4.

57. W. R. Hill to Jim Suddreth, Dec. 20, 1982, WEF Archives, Record Group 2, Box 13.

58. Charles H. Jones to Member Association Officers, Aug. 5, 1983, Executive Committee, Background Material, Oct. 1, 1983, WEF Archives, Record Group 2, Box 11.

59. "Executive Summary on Operations Members," Aug. 5, 1983, WEF Archives, Record Group 2, Box 11.

60. Thomas M. F. Doyle to Quincalee Brown, Feb. 5, 1987, WEF Archives, Record Group 2, Box 11, Ex. Dir. Files.

61. "The PWOD Advantage," *Operations Forum* 4 (3) (1987); WEF Archives, Record Group 2, Box 11, Ex. Dir. Files.

62. "Wastewater Operations Competition"; "Sewage Olympics Bolsters Morale of City Employees," *St. Petersburg Times*, May 21, 1988; "'Operations Challenge '88' WPCF Wastewater Competition," WEF Archives, Record Group 2, Box 11.

63. See Rose George, *The Big Necessity: The Unmentionable World of Human Waste and Why It Matters* (New York: Metropolitan Books, 2008), 34–35, for a description of the New York City team.

64. Water Environment Federation, *Operation of Municipal Wastewater Treatment Plants, Manual of Practice No. 11, 5th ed., Vol. II. Liquid Processes* (Alexandria, VA: Water Environment Federation, 1996), 643; Water Environment Federation, *Operation of Municipal Wastewater Treatment Plants, Manual of Practice No. 11, 6th ed., Vol. II: Liquid Processes* (New York: McGraw-Hill, 2008).

65. JA, "Losing Operations Forum," Apr. 1, 2000; TJ, May 23, 2000; MC, "Operations Forum/WET," May 24, 2000; JM, "My Opinion," May 25, 2000; Water Environment Federation, Operation and Maintenance Discussion Forum, <http://www.wef.org/Forums/Thread.cfm?CFApp=13&Thread_ID=854&mc=27> (accessed June 6, 2000).

66. U. Jeppsson, J. Alex, M. N. Pons, H. Spanjers, and P. A. Vanrolleghem, "Status and Future Trends of ICA in Wastewater Treatment: A European Perspective," *Water Science and Technology* 45 (2002): 485–494; Reza Katebi, Michael A. Johnson, and Jacqueline Wilkie, *Control and Instrumentation for Wastewater Treatment Plants* (London: Springer Verlag, 1999).

67. Alvin E. Molvar, Joseph F. Roesler, and Russel H. Babcock, "Instrumentation and Automation Experiences in Wastewater-Treatment Facilities," Municipal Environmental Research Laboratory, U.S. Environmental Protection Agency, EPA 600/2–76–198, Oct.1976. On petroleum, see David F. Noble, *Forces of Production: A Social History of Industrial Automation* (Oxford: Oxford University Press, 1984), 59–66.

68. Ana M. A. Dias and Eugénio C. Ferreira, "Computational Intelligence Techniques for Supervision and Diagnosis of Biological Wastewater Treatment Systems," in *Computational Intelligence Techniques for Bioprocess Modelling, Supervision and Control*, ed. Maria do Carmo Nicoletti and Lakhmi C. Jain, Studies in Computational Intelligence, vol. 218 (New York: Springer, 2009), 127–162; T. Ohtsuki, T. Kawazoe, and T. Masui, "Intelligent Control System Based on Blackboard Concept for Wastewater Treatment Processes," *Water Science and Technology* 37 (1998): 77–85.

69. James R. Getchell, "Instrumentation and Control Systems," in *Fermentation and Biochemical Engineering Handbook: Principles, Process Design, and Equipment*, ed. Henry C. Vogel (Park Ridge, NJ: Noyes Publications, 1983), 401.

70. Gustaf Olsson, "Improving the Operation of Wastewater Treatment Systems by Automatic Control," *Water Quality Research Journal of Canada* 30 (1995): 127–142, on 127.

71. Joseph F. Roesler, Dolloff F. Bishop, and Irwin J. Kugelman, "Current Status of Research in Automation of Wastewater Treatment in the United States," *Progress in Water Technology* 9 (1977): 659–671, on 659, 668.

72. Carmen F. Guarino and Joseph V. Radziul, "Water—Wastewater, I&A, U.S.A.," *Progress in Water Technology* 9 (1977): 35–39; bulk processing quotation on 35.

73. William F. Garber and James J. Anderson, "From the Standpoint of an Operato—What Is Really Needed in the Automation of a Wastewater Treatment Plant," in *Advances in Water Pollution Control*, ed. R. A. Drake (Oxford: Pergamon Press, 1985), 429–442.

74. Olsson, "Improving the Operation," 128.

75. J. Comas, I. Rodriguez-Roda, M. Poch, K. V. Gernaey, C. Rosen, and U. Jeppsson, "Demonstration of a Tool for Automatic Learning and Re-use of Knowledge in the Activated Sludge Process," *Water Science and Technology* 53 (4–5) (2006): 303–311.

76. O. C. Pires, C. Palma, J. C. Costa, I. Moita, M. M. Alves, and E.C. Ferreira, "Knowledge-based Fuzzy System for Diagnosis and Control of an Integrated Biological Wastewater Treatment Process," *Water Science and Technology* 53 (4–5) (2006): 313–320, on 313.

77. Y. H. Hong, M. R. Rosen, and R. Bhamidimarri, "Analysis of a Municipal Wastewater Treatment Plant Using a Neural Network Based Pattern Analysis," *Water Research* 37 (2003): 1608–1618; Ohtsuki, Kawazoe, and Masui, "Intelligent Control System"; M. U. Sanchez, J. Corts, J. Bjar, J. De Garcia, J. Lafuente, and M. Poch, "Concept Formation in WWTP by Means of Classification Techniques: A Compared Study," *Applied Intelligence*, 1997, 147–165.

78. P. A. Paraskevas, I. S. Pantelakis, and T. D. Lekkas, "An Advanced Integrated Expert System for Wastewater Treatment Plants Control," *Knowledge-Based Systems* 12 (1999): 355–361, on 358.

79. G. Olsson, "Lessons Learnt at ICA 2001," *Water Science and Technology* 45 (2002): 1–8.

80. G. Olsson, "Instrumentation, Control and Automation in the Water Industry--State of the Art and New Challenges," *Water Science and Technology* 53 (4–5) (2006): 1–16, on 13.

81. T. S. Moona, Y. J. Kima, J. R. Kimb, J. H. Chaa, D. H. Kima, and C. W. Kima, "Identification of Process Operating State with Operational Map in Municipal Wastewater Treatment Plant," *Journal of Environmental Management* 90 (2009): 772–778.

82. "Wastewater Utilities Respond Slowly to Looming 'Brain Drain' Prospect," *Utility Executive* 8(1) (2005): 1–2.

83. "The Business of Water Pollution," *Environmental Science and Technology* 6 (1972): 974–979, "Envirotech Builds a Waste Control Empire," *Business Week*, Jan. 22, 1972, 54–56; "Municipal and Industrial Waste Water Control Equipment," The Bank of New York Industry Studies, Nov. 1972, in *Implementation of the Federal Water Pollution Control Act (Municipal Construction Grants Program and the State Management Assistance Program)*, Hearings Before the Subcommittee on Oversight and Review of the Committee on Public Works and Transportation, U.S. House of Representatives, Oct. 30, 31; and Nov. 1, 1979, 97–118.

84. "Up and Down with Envirotech," *Forbes*, 117 (7), Apr. 1, 1976, 53; Gene G. Marcial, "Pollution-Control Issues Get Mixed Reaction after a Generally Lackluster Start This Year," *Wall Street Journal*, July 20, 1976, 41; "The Fading Business of Cleansing the Waters," *Business Week*, Mar. 24, 1980, 60; G. L. Van Houtven, S. B. Brunnermeier, and M. C. Buckley, "A Retrospective Assessment of the Costs of the Clean Water Act: 1972 to 1997," U.S. Environmental Protection Agency, Office of Water, 2000, 5.1–5.3.

85. "Up and Down with Envirotech," 53; "The Fading Business," 60.

86. Testimony of Frank P. Sebastian, *Implementation of the Federal Water Pollution Control Act (Concerning the Performance of the Municipal Wastewater Treatment Construction Grants Program)*, Hearings Before the Subcommittee on Investigations and Oversight of the Committee on Public Works and Transportation, U.S. House of Representatives, 97th Congress, July 14, 15, 16, 21, 22, 23, 1981, 109; see also Jeffrey A. Klein and Laurie Doane, "Environmental Control Industry: Municipal Wastewater Equipment Market Review," Kidder, Peabody and Co. Research Dept., Aug. 14, 1979, in *Implementation of the Federal Water Pollution Control Act (Municipal Construction Grants Program and the State Management Assistance Program)*, Hearings Before the Subcommittee on Oversight and Review of the

Committee on Public Works and Transportation, U.S. House of Representatives, Oct. 30, 31; and Nov. 1, 1979, 13, 28.

87. Van Houtven, Brunnermeier, and Buckley, "A Retrospective Assessment," 5.2.

88. [Advertisement for Envirotech Operating Services], *Implementation of the Federal Water Pollution Control Act (Concerning the Performance of the Municipal Wastewater Treatment Construction Grants Program)*, Hearings Before the Subcommittee on Investigations and Oversight of the Committee on Public Works and Trsnsportation, U.S. House of Representatives, 97th Congress, July 14, 15, 16, 21, 22, 23, 1981, 1125.

89. "A man could be operating a gas station one day and if his brother-in-law gets a strategic position with the authority or the city or the municipality, the next day he could be superintendent of a treatment plant." Testimony of Carmen F. Guarino, President, Water Pollution Control Federation. Ibid., 109.

90. Kent Robinson and John Filbert, "OMI," <http://www.alumni.ch2m.com/History/Years/Discipline/omi.html> (accessed Jan. 6, 2009).

91. Peter F. Meiksins, "Scientific Management and Class Relations," *Theory and Society* 13 (1984): 177–209.

92. Emil C. Jensen, "What Is the Best Way to Finance the Nation's Sewage Works Projects?" *Wastes Engineering* 28 (1957): 674–676, 698–699, on 676. A few states had granted franchises to private sewerage systems for municipalities, including New Jersey, Pennsylvania, and Texas. In 1902, there were fewer than fifty in the country. "An Exhibit of Municipal Ownership," *Public Opinion* 32 (1902): 493.

93. "How much it all cost is a difficult and open-ended question," and includes the costs of floating the companies, tax writeoffs, direct cash infusions and write-offs of loans. David Kinnersley, *Coming Clean: The Politics of Water and the Environment* (London: Penguin Books, 1994), 77–79. In 1990, in a curious addendum to the history of the conflict over public and private rights to the activated sludge process, the privatized water and sewerage services company North West Water, which included the Manchester sewage treatment plants where the activated sludge process was first invented, bought the firm Water Engineering, the corporate successor to Activated Sludge Ltd. Graham Searjeant, "North West Water Buys Three Firms," *Times* (London), Nov. 2, 1990. "Water Engineering, Company History," <http://www.waterengineering.co.uk/company-history.asp> (accessed July 1, 2009).

94. Karen J. Bakker, "From Public to Private to . . . Mutual? Restructuring Water Supply Governance in England and Wales," *Geoforum* 34 (2003): 359–374, on 364.

95. David Hall and Emanuele Lobina, "Employment and Profit Margins in UK Water Companies: Implications for Price Regulation Proposals," Public Services International Research Unit, Report 9911-W-UK.doc, Nov. 1999.

96. For a discussion of expert systems and scientific management, see Peter Holden, "'Working-to-Rules': A Case of Taylor-Made Expert Systems," *Interacting with Computers* 1 (1989): 199–219.

97. K. R. Stimson, "Activated Sludge Control Advisor," *Water Science and Technology* 28 (11–12) (1993): 295–302.

98. Bakker, "From Public to Private," 364.

99. Paul Eisenhardt and Andrew Stocking, "Gradual Growth and 17 Long-Term Deals Mark 1997 Water & Wastewater Contract Operations," reprinted from *Public Works Financing*, Mar. 1998, <http://www.waterindustry.org/PWF-1.htm> (accessed Jan. 6, 2009).

100. Bakker "From Public to Private," 370. On retrenchment of water privatization, see also Michael Goldman, "How 'Water for All!' Policy Became Hegemonic: The Power of the World Bank and its Transnational Policy Networks," *Geoforum* 38 (2007): 786–800.

101. Kinnersley, *Coming Clean*, 46.

102. "1992 Needs Survey Report to Congress," U.S. Environmental Protection Agency, Office of Water, Sept. 1993, EPA 832-R-93–002, C-2.

103. John Holusha, "Cities Enlisting Private Companies for Sewage Treatment," *NYT*, May 5, 1996, 18.

104. 40 CFR Section 261.4(a)(1); Memo, Bussard to Warren, Mar. 10, 1997 (RCRA Online #14068); *Response to Congress on Privatization of Wastewater Facilities*, U.S. Environmental Protection Agency, Office of Water, July 1997, EPA 832-R-97–001a.

105. Rick Romell, "Consultants' Advice to Be Sought: Sewer District May Consider Selling Plants." *MJS*, Oct. 7, 1995.

106. "Worker Deal Advances Sewerage Takeover," *MJS*, Jan. 3, 1998; "N.J. Firm to Get Nearly $300 Million over 19 Years," *MJS*, Jan. 6, 1998.

107. "MMSD's Troubled Waters," *MJS*, June 23, 2003; Marie Rohde, "Other Vendors Could Run Sewers, Norquist Warns," *MJS*, Sept. 17, 2002; Marie Rohde, "Improper Maintenance Blamed in Dumping," *MJS*, Sept. 21, 2002; Don Behm, "2 French Firms Vie for MMSD Contract," *MJS*, Sept. 23, 2007.

108. Rohde, "Other Vendors Could Run Sewers"; Rohde, "Sewage Spill Fouls Contract Relations."

109. "MMSD's Troubled Waters," *MJS*, June 23, 2003.

110. "Grading the Sewer System," *MJS*, June 26, 2003.

111. Don Behm, "Contracting Could Save $5 Million, MMSD Says," *MJS*, Nov. 1, 2007.

112. Don Behm, "MMSD Goes with Veolia as Operator," *MJS*, Dec. 4, 2007.

113. The Milwaukee contract was in a direct lineage of the first private operation contract in the United States that had been initiated by Envirotech in 1972. Envirotech was bought by U.S. Filter in 1981. U.S. filter was purchased by Vivendi in 1999. Vivendi became Veolia in 2003.

114. Don Behm, "Sabotage Suspected in Disruption at Milorganite Plant," *MJS*, Feb. 1, 2008.

115. "PWF's 12th Annual Water Outsourcing Report," *Public Works Financing* 225 (2008): 14–17, <http://pwfinance.net/2008_conops_survey.pdf> (accessed Jan. 6, 2009). Eisenhardt and Stocking, "Gradual Growth and 17 Long-Term Deals."

116. Ibid.; *Response to Congress on Privatization of Wastewater Facilities*, U.S. Environmental Protection Agency, Office of Water, July 1997, EPA 832-R-97–001a, July 1997, 37. For more cases of cities canceling operations contracts, see Public Citizen, "Waves of Regret: What Some Cities Have Learned and Other Cities Should Know about Water Privatization Fiascos in the United States," June 2005, <http://www.citizen.org/documents/waves.pdf> (accessed Aug. 5, 2010>. Karen Bakker argues that commercialization is independent of

privatization and commodification. See *An Uncooperative Commodity: Privatizing Water in England and Wales* (Oxford: Oxford University Press, 2003), 6–7.

117. E. F. Schumacher, *Small Is Beautiful* (New York: Harper & Row, 1975); Ray Dinges, *Natural Systems for Water Pollution Control* (New York: Van Nostrand Reinhold, 1982); Nancy Jack Todd and John Todd, *From Eco-Cities to Living Machines: Principles of Ecological Design* (Berkeley: North Atlantic Books, 1993); Carroll Pursell, "The Rise and Fall of the Appropriate Technology Movement in the United States, 1965–1985," *Technology and Culture* 34 (1993): 629–637; Andrew Kirk, "Appropriating Technology: The Whole Earth Catalog and Counterculture Environmental Politics," *Environmental History* 6 (2001): 374–394.

118. Julie Stauffer, *The Water Crisis: Constructing Solutions to Freshwater Pollution* (London: Earthscan Publications, 1998), 75, 77.

119. Sherwood C. Reed, E. Joe Middlebrooks, and Ronald W. Crites, *Natural Systems for Waste Management and Treatment* (New York: McGraw-Hill, 1988), 1.

120. Ibid., 1.

121. The Administrator to Assistant Administrators and Regional Administrators, "EPA Policy on Land Treatment of Municipal Wastewater," Oct. 3, 1977, Appendix A in William J. Jewell and Belford L. Seabrook, *A History of Land Application as a Treatment Alternative*, U.S. Environmental Protection Agency, Technical Report, Apr. 1979, EPA 430/9–79–012, 64–67.

122. See Steven A. Serfling and Dominick Mendola, "Buoyant Contact Surfaces in Waste Treatment Pond," U.S. Patent 4,169,050, Sept. 25, 1979, for a discussion.

123. See <http://www.livingmachines.com/products/livingmachine> (accessed July 9, 2008). Todd trademarked the term "Living Machine" in 1991. U.S. Patent and Trademark Office, Trademark Electronic Search System.

124. Nancy Jack Todd, *A Safe and Sustainable World: The Promise of Ecological Design* (Washington, DC: Island Press, 2005), 18.

125. For a history of the New Alchemy Institute, see the memoir by Todd, *A Safe and Sustainable World*; on relation to back to the land movement, see 9, 13.

126. Robert Gilman, "The Restoration of Waters: An Interview with John and Nancy Jack Todd," *In Context* 25 (1990): 42–47, on 43–44.

127. Gilman, "Restoration of Waters," 44.

128. John H. Todd and Barry Silverstein, U.S. Patent 5,087,353, Feb. 11, 1992. Todd had difficulty getting his patent because of an earlier system of waste treatment patented by two former students of Todd's at San Diego State University. Their system, named "Solar Aquasystems," also received federally financed trials.

129. Gilman, "Restoration of Waters," 45. EEA is now Ecological Engineering Group and markets the technology as "Solar Aquatics Systems."

130. A well-documented account of the controversy was written by a student at the nearby Conway School of Landscape Design. See Jennifer Fraulo, "A History of Ashfield's Wastewater Treatment Facility," <http://www.purplepanthers.com/solaraquatics.htm> (accessed Jan. 9, 2009). See also Ken and Ethel Kipen, "Scapegoating Solar-aquatics," <http://www.purplepanthers.com/rx2.htm> (accessed Feb. 20, 2009) and "Ashfield's Solar-Aquatic Wastewater Treatment Plant," <http://www.purplepanthers.com/ashrxplant.html> (accessed Feb. 20, 2009).

131. Paul Bennet, "Ashfield's Alternative," *Landscape Architecture* 88 (1998): 44–49, on 48.

132. Ibid., 46.

133. Fraulo, "A History of Ashfield's Wastewater Treatment Facility."

134. Ibid.

135. Betsy Calvert, "Sewer Plant Vegetation Harvested," *Union-News* (Springfield, MA), Jan. 11, 2000, B1.

136. Ken and Ethel Kipen, "Scapegoating Solar-aquatics."

137. Virginia Ray, "Plants to Return to Facility to Digest Sewage," *Recorder* (Greenfield, MA), June 9, 2000.

138. Jeremy Dirac, "Colrain—Several Residents of Colrain's Sewer District Aren't Paying Their Sewer Bills and Signed a Letter Demanding to Meet with Sewer Commissioners," *Recorder* (Greenfield, MA), Oct. 31, 2005.

139. For details on one Living Machine installation, see the web site for the Oberlin College Adam Joseph Lewis Center for Environmental Studies, <http://www.oberlin.edu/ajlc/> (accessed Feb. 23, 2010).

140. Melosi, *The Sanitary City*, 369–372.

141. Reed, Middlebrooks, and Crites, *Natural Systems*. Under the auspices of the Water Environment Federation, Reed was responsible for the 2001 *Manual of Practice* that detailed the accepted methods for designing and operating land, water and wetland based treatment systems. This Manual of Practice for "Natural Systems for Wastewater Treatment" contained no reference to Living Machines or similar treatment systems. Todd called members of the environmental engineering profession "gatekeepers." David Riggle and Kevin Gray, "Using Plants to Purify Wastewater," *BioCycle* 40 (1) (1999): 40–41.

142. *The Living Machine® Wastewater Treatment Technology—An Evaluation of Performance and System Cost*, Municipal Technology Branch, U.S. Environmental Protection Agency, Office of Water, 2001, EPA 832-R-01–004.

143. Ibid., 8–7.

144. Ibid., ES-1–2.

145. Ibid., 2–16.

146. Ibid., 8–1.

147. Reed, Middlebrooks, and Crites, *Natural Systems*, 1; "rediscovered" on 2. Reed's later Manual of Practice doesn't define "natural" at all.

148. "Response to Congress on the AEES 'Living Machine' Wastewater Treatment Technology," U.S. Environmental Protection Agency, Office of Water, Apr. 1997, EPA 832-R-97–002, ES-2.h.

149. John Todd, "The Next Transition at OAI: Knowledge Vital to an Age of Information," <http://www.oceanarks.org/abo40_The_Next_Transition_at_OAI.php> (accessed Sept. 21, 2009); John Todd, Erica J. G. Brown, and Erik Wells, "Ecological Design Applied," *Ecological Engineering* 20 (2003): 421–440, on 425.

150. For Worrell Water Technologies, see Dave McNair, "The Tao of Poo: Can Worrell's Green Sewage System Save Water and Planet?," *The Hook* (Charlottesville, VA), June 11, 2009, and <http://www.livingmachines.com/about/history/> (accessed Sept. 21, 2009).

151. James L. Barnard, "Biological Nutrient Removal without Addition of Chemicals," *Water Research* 9 (1975): 485–490. As with the septic tank, patent concerns and disputes have been important in determining the use of this technology as well. The developer of this process, James Barnard, patented it in many parts of the world, but another firm was able to patent a similar process in the United States. Other U.S. scientists working under an EPA grant dedicated their patent to the public. James L. Barnard, "Prejudices, Processes and Patents," in *Proceedings of Second Australian Conference on Biological Nutrient Removal from Wastewater*, 1994, 1–13; *Project Summary. Emerging Technology Assessment of PhoStrip, A/O, and Bardenpho Processes for Biological Phosphorus Removal*, United States Environmental Protection Agency, Water Engineering Research Laboratory, EPA 600 S2–85 008, Mar.1985; *Design of Municipal Wastewater Treatment Plants, Volume II: Chapters 13–20*, WEF Manual of Practice No. 8 (Alexandria, VA: Water Environment Federation, 1992), 956–959.

152. J. B. Neethling, Brian Akke, Mario Benisch, April Gu, Heather Stephens, H. David Stensel, and Rebecca Moore, "Factors Influencing the Reliability of Enhanced Biological Phosphorus Removal," Water Environment Research Foundation, 2005, Report 01-CTS-3.

153. Chad A. Kinney, Edward T. Furlong, Steven D. Zaugg, Mark R. Burkhardt, Stephen L. Werner, Jeffery D. Cahill, and Gretchen R. Jorgensen, "Survey of Organic Wastewater Contaminants in Biosolids Destined for Land Application," *Environmental Science and Technology* 40 (2006): 7207–7215; Chris D. Metcalfe, Shaogang Chu, Colin Judt, Hongxia Li, and Ken D. Oakes, "Antidepressants and Their Metabolites in Municipal Wastewater, and Downstream Exposure in an Urban Watershed," *Environmental Toxicology and Chemistry* 29 (2010): 79–89; O. A. H. Jones, N. Voulvoulis, and J. N. Lester, "Human Pharmaceuticals in Wastewater Treatment Processes," *Critical Reviews in Environmental Science and Technology* 35 (2005): 401–427. For a current bibliography of microcontaminants in sewage, see *Treating Contaminants of Emerging Concern*, U.S. Environmental Protection Agency, Office of Water, EPA 820-R-10–002, <http://epa.gov/waterscience/ppcp/studies/results.html (accessed Mar. 1, 2010)>.

154. Ralph L. Seiler, Steven D. Zaugg, James M. Thomas, and Darcy L. Howcroft, "Caffeine and Pharmaceuticals as Indicators of Waste Water Contamination in Wells," *Ground Water* 37 (1999): 405–410.

155. Margaretha Adolfsson-Erici, Maria Pettersson, Jari Parkkonen, and Joachim Sturve, "Triclosan, a Commonly Used Bactericide Found in Human Milk and in the Aquatic Environment in Sweden," *Chemosphere* 46 (2002): 1485–1489; Chad A. Kinney, Edward T. Furlong, Dana W. Kolpin, Mark R. Burkhardt, Steven D. Zaugg, Stephen L. Werner, Joseph P. Bossio, and Mark J. Benotti, "Bioaccumulation of Pharmaceuticals and Other Anthropogenic Waste Indicators in Earthworms from Agricultural Soil Amended with Biosolid or Swine Manure," *Environmental Science and Technology* 42 (2008): 1863–1870; Chenxi Wu, Alison L. Spongberg, Jason D. Witter, Min Fang, and Kevin P. Czajkowski, "Uptake of Pharmaceutical and Personal Care Products by Soybean Plants from Soils Applied with Biosolids and Irrigated with Contaminated Water," *Environmental Science and Technology*, 2010, 44 (2010): 6157–6161; Richard H. M. M. Schreurs, Edwin Sonneveld, Jenny H. J. Jansen, Willem Seinen, and Bart van der Burg, "Interaction of Polycyclic Musks and UV Filters with the Estrogen Receptor (ER), Androgen Receptor (AR), and Progesterone Receptor (PR) in Reporter Gene Bioassays," *Toxicological Sciences* 83 (2005): 264–272.

156. Patricia Burkhardt-Holm, "Endocrine Disruptors and Water Quality: A State-of-the-Art Review," *International Journal of Water Resources Development* 26 (2010): 477–493.

157. S. G. Schrank, U. Bieling, H. J. José, R. F. P. M. Moreira, and H. Fr. Schröder, "Generation of Endocrine Disruptor Compounds during Ozone Treatment of Tannery Wastewater Confirmed by Biological Effect Analysis and Substance Specific Analysis," *Water Science and Technology* 59 (2009): 31–38.

158. Oliver A. H. Jones, Pat G. Green, Nikolaos Voulvoulis, and John N. Lester, "Questioning the Excessive Use of Advanced Treatment to Remove Organic Micropollutants from Wastewater," *Environmental Science and Technology* 41 (2007): 5085–5089; A. Joss, H. Siegrist, and T. A. Ternes, "Are We about to Upgrade Wastewater Treatment for Removing Organic Micropollutants?," *Water Science and Technology* 57 (2008): 251–255; Yoong K. K. Koh, Tze Y. Chiu, Alan R. Boobis, Mark D. Scrimshaw, John P. Bagnall, Ana Soares, Simon Pollard, Elise Cartmell, and John N. Lester, "Influence of Operating Parameters on the Biodegradation of Steroid Estrogens and Nonylphenolic Compounds during Biological Wastewater Treatment Processes," *Environmental Science and Technology* 43 (2009): 6646–6654.

159. M. Clara, N. Kreuzinger, B. Strenn, O. Gans, and H. Kroiss, "The Solids Retention Time—A Suitable Design Parameter to Evaluate the Capacity of Wastewater Treatment Plants to Remove Micropollutants," *Water Research* 39 (2005): 97–106.

160. "Discussion on Modern Methods of Sewage Disposal," *JRSI* 28 (1908): 1–14, on 9.

161. National Research Council, Committee on Toxicants and Pathogens in Biosolids Applied to Land, *Biosolids Applied to Land: Advancing Standards and Practices* (Washington, DC: National Academy Press, 2002); Yuansong Wei, Renze T. Van Houten, Arjan R. Borger, Dick H. Eikelboom, and Yaobo Fan, "Minimization of Excess Sludge Production for Biological Wastewater Treatment," *Water Research* 37 (2003): 4453–4467; Tania Datta, Yanjie Liu, and Ramesh Goel, "Evaluation of Simultaneous Nutrient Removal and Sludge Reduction Using Laboratory Scale Sequencing Batch Reactors," *Chemosphere* 76 (2009): 697–705.

162. Steingraber, *Living Downstream*.

6   From Sewage to Biotech

1. In the Matter of the Application of Malcolm E. Bergy, John H. Coats, and Vedpal S. Malik, 563 F.2d 1031 (1977), 1038.

2. "Mass Production of Insulin Hinted," *NYT*, June 13, 1978.

3. See the press release from the University of California, Dec. 2, 1977, announcing the earlier experiment on *E. coli* in which the gene for somatistatin was inserted into the bacteria, which used the term "factories" to describe the new bacteria. Appendix H in Herbert W. Boyer, "Recombinant DNA Research at UCSF and Commercial Application at Genentech," an oral history conducted in 1994 by Sally Smith Hughes, Regional Oral History Office, The Bancroft Library, University of California, Berkeley, 2001. As a proof of the concept, Genentech had previously created *E. coli* that produced a human hormone, somatistatin, but this hormone did not have a known therapeutic use. "The First Fruits of Gene-Splicing," *NYT*, Nov. 8, 1982. Eli Lilly licensed the technology and began production in 1983. For the history of humulin, Lilly's synthetic insulin, see Irving S. Johnson, "Human Insulin from Recombinant

DNA Technology," *Science* 219 (1983): 632–637; Gary Walsh, "Therapeutic Insulins and Their Large-Scale Manufacture," *Applied Microbiology and Biotechnology* 67 (2005): 151–159. By 1988, Lilly was producing over $200 million of insulin a year in 10,000-gallon fermentation reactors. Milt Freudenheim, "5 Big Drug Manufacturers Post Higher Nets in Quarter," *NYT*, July 20, 1988. By 2003, when humulin was replaced with a newer recombinant insulin, sales had increased to $1 billion. Eli Lilly and Company, *Annual Report*, 2003, <http://files.shareholder.com/downloads/LLY/715660752x0x221661/1A56BA1D-8E6D-4E1E-A589-83B638F67B0D/English.PDF> (accessed Sept. 3, 2009).

4. Robert Bud, "The Zymotechnic Roots of Biotechnology," *The British Journal for the History of Science* 25 (1992): 127–144.

5. John Kendrew, "Forward," in *A Revolution in Biotechnology*, ed. Jean L. Marx (Cambridge: Cambridge University Press, 1989), n.p.

6. Chas. C. Brown, "Method of Making a Sanitary Investigation of a River," *Science* 21 (1893): 284–285.

7. Edwin Oakes Jordan, "The Relative Abundance of Bacillus Coli Communis in River Water as an Index of the Self-Purification of Streams," *The Journal of Hygiene* 1 (1901): 295–320.

8. Edwin O. Jordan, "Variation in Bacteria," *Proceedings of the National Academy of Sciences of the United States of America* 1 (1915): 160–164. See William Burrows, "Biographical Memoir of Edwin Oakes Jordan, 1866–1936," *National Academy of Sciences Biographical Memoirs* 20 (1938): 195–228.

9. Alfred MacConkey, "Lactose-Fermenting Bacteria in Faeces," *The Journal of Hygiene* 5 (1905): 333–379. C.-E. A. Winslow, I. J. Kligler and W. Rothberg, "Studies on the Classification of the Colon typhoid Group of Bacteria with Special Reference to Their Fermentative Reactions," *The Journal of Bacteriology* 4 (1919): 429–503.

10. For the relation between these early studies of bacterial variation and the rise of *E. coli* as a model for genetic studies in the 1930s and 1940s, see William C. Summers, "From Culture as Organism to Organism as Cell: Historical Origins of Bacterial Genetics," *Journal of the History of Biology* 24 (1991): 171–190.

11. Sally Smith Hughes, "Making Dollars out of DNA: The First Major Patent in Biotechnology and the Commercialization of Molecular Biology, 1974–1980," *Isis* 92 (2001): 541–575.

12. Carl Zimmer, *Microcosm:* E. coli *and the New Science of Life* (New York: Pantheon Books, 2008), 160–170.

13. *Diamond v. Chakrabarty*, 447 U.S. 303 (1980).

14. Ananda M. Chakrabarty, "Microorganisms Having Multiple Compatible Degradative Energy-generating Plasmids and the Preparation Thereof," U.S. Patent 4,259,444, Mar. 31, 1981.

15. Daniel J. Kevles, "Patents, Protections, and Privileges: The Establishment of Intellectual Property in Animals and Plants," *Isis* 98 (2007): 323–331, on 324. Glen Bugos and Daniel J. Kevles, "Plants as Intellectual Property: American Law, Policy, and Practice in World Context," *Osiris*, 2nd Ser., 7 (1992): 119–148.

16. Brief for the Respondent, *Diamond v. Chakrabarty*, 1979 U.S. Briefs 136; Brief for the Petitioner, 19 U.S. Briefs 136.

17. In the Matter of the Application of Malcolm E. Bergy, John H. Coats, and Vedpal S. Malik, 563 F. 2d 1031 (1977); 596 F 2nd 952 (1979).

18. *563 F. 2d 1031 (1977), 1033.*

19. In the Matter of the Application of Ananda M. Chakrabarty, 571 F.2d 40 (1978).

20. The Patent Office appealed *Bergy* to the Supreme Court, where it was vacated and remanded. *Parker v. Bergy*, 438 U.S. 932 (1978). It was combined with *Chakrabarty* and ruled on again by the U.S. Court of Customs and Patent Appeals, in 596 F 2nd 952 (1979), which repeats much of 563 F. 2d 1031 quoted previously, but applied to both *Chakrabarty* and *Bergy*. The Patent Office appealed to the U.S. Supreme Court. On Bergy's decision to withdraw, see Linda Greenhouse, "Commerce Cases Taken by Justices," *NYT*, Jan.15, 1980.

21. *596 F.2d 952 (1979), 977.*

22. *563 F.2d 1031 (1977), 1037–1038.*

23. Arguments on Final Hearing, 131. *Union Solvents vs. Guaranty Trust*, Equity no. 802, District Court, Delaware. NARA-Mid Atlantic RG 21, Box 272.

24. 19 USPQ 178 (1932).

25. *563 F.2d 1031 (1977), 1038.*

26. *563 F.2d 1031 (1977), 1038, and 596 F.2d 952 (1979), 997.*

27. Philip Leder and Timothy A. Stewart, "Transgenic Non-human Mammals," U.S. Patent 4,736,866, Apr. 12, 1988.

28. 447 U.S. 303 (1980), 309. Estimated from a total of 2,355 patents through Aug. 17, 2010, listed within U.S. Patent and Trademark Office classification codes 800/8–20 (Multicellular Living Organisms, Non-human Animals). Some patents are listed in multiple subclasses, and some do not claim the animal itself. There were 689 unique mouse patents alone.

29. Thomas D. Kiley, "Genentech Legal Counsel and Vice President, 1976–1988, and Entrepreneur," an oral history conducted in 2000 and 2001 by Sally Smith Hughes, Regional Oral History Office, The Bancroft Library, University of California, Berkeley, 2002, 39–40. Genentech went on to patent a number of bacteria as well. See, for example, David V. Goeddel and Sidney Pestka, "Microbial Production of Mature Human Leukocyte Interferon K and L," U.S. Patent 4,810,645, Mar. 7, 1989, in which *E. coli* and other transformed microbes are patented.

30. Antonio Villaverde, "Editorial: Old Bugs for New Tasks; the Microbial Offer in the Proteomics Era," *Microbial Cell Factories* 1 (2002): 1, <http://www.microbialcellfactories.com/content/1/1/4>.

31. Jessica C, Zweers, Imrich Barák, Dörte Becher, Arnold J. M. Driessen, Michael Hecker, Vesa P. Kontinen, Manfred J. Saller, L'udmila Vavrová, and Jan Maarten van Dijl, "Towards the Development of *Bacillus subtilis* as a Cell Factory for Membrane Proteins and Protein Complexes," *Microbial Cell Factories* 7 (2008): 10, <http://www.microbialcellfactories.com/content/7/1/10>.

32. Lidia Westers, Helga Westers, Geeske Zanen, Haike Antelmann, Michael Hecker, David Noone, Kevin M. Devine, Jan Maarten van Dijl, and Wim J. Quax, "Genetic or Chemical Protease Inhibition Causes Significant Changes in the *Bacillus subtilis* Exoproteome," *Proteomics* 8 (2008): 2704–2713; P. L. Foley and M. L. Shuler, "Considerations for the Design and Construction of a Synthetic Platform Cell for Biotechnological Applications," *Biotechnology and Bioengineering* 105 (2010): 26–36; Yaramah M. Zalucki, Christopher E. Jones, Preston S. K. Ng, Benjamin L. Schulz, and Michael P. Jennings, "Signal Sequence Non-

optimal Codons Are Required for the Correct Folding of Mature Maltose Binding Protein," *Biochimica et Biophysica Acta-Biomembranes* 1798 (2010): 1244–1249.

33. M. Jenzsch, R. Simutis, and A. Lübbert, "Optimization and Control of Industrial Microbial Cultivation Processes," *Engineering in Life Sciences* 6 (2006): 117–124, on 117.

34. Dennis G. Kleid, "Scientist and Patent Agent at Genentech," an oral history conducted in 2001 and 2002 by Sally Smith Hughes for the Regional Oral History Office, The Bancroft Library, University of California, Berkeley, 2002, 37.

35. Fred A. Middleton, "First Chief Financial Officer at Genentech, 1978–1984," an oral history conducted in 2001 by Glenn E. Bugos for the Bancroft, Regional Oral History Office, The Bancroft Library, University of California, Berkeley, 2002, 10.

36. Kleid oral history, 36; Daniel G. Yansura, "Senior Scientist at Genentech," an oral history conducted in 2001 and 2002 by Sally Smith Hughes for the Regional Oral History Office, The Bancroft Library, University of California, Berkeley, 2002, 78.

37. Middleton oral history, 10.

38. Yansura oral history, 77.

39. David V. Goeddel, Dennis G. Kleid, Francisco Bolivar, Herbert L. Heyneker, Daniel G. Yansura, Roberto Crea, Tadaaki Hirose, Adam Kraszewski, Keiichi Itakura, and Arthur D. Riggs, "Expression in *Escherichia coli* of Chemically Synthesized Genes for Human Insulin," *Proceedings of the National Academy of Sciences of the United States of America* 76 (1979): 106–110; Irving S. Johnson, "Human Insulin from Recombinant DNA Technology," *Science* 219 (1983): 632–637.

40. Brian McNeil and Linda M. Harvey, "Fermentation: An Art from the Past, a Skill for the Future," in *Practical Fermentation Technology* (Chichester, UK: Wiley, 2008), 1–2, on 2.

41. V. Vojinovi, J. M. S Cabral, and L. P. Fonseca, "Real-Time Bioprocess Monitoring Part I: In situ Sensors," *Sensors and Actuators B* 114 (2006): 1083–1091, on 1084.

42. B. Lennox, G. A. Montague, H. G. Hiden, G. Kornfeld, and P. R. Goulding, "Process Monitoring of an Industrial Fed-Batch Fermentation," *Biotechnology and Bioengineering* 74 (2001): 125–135, on 125.

43. Konstantin B. Konstantinov and Toshiomi Yoshida, "Knowledge-based Control of Fermentation Processes," *Biotechnology and Bioengineering* 39 (1992): 479–486, on 479.

44. Michio Sugeno, "An Introductory Survey of Fuzzy Control," *Information Sciences* 36 (1985): 59–83; Suteaki Shioya, Kazuyuki Shimizu, and Toshiomi Yoshida, "Knowledge-based Design and Operation of Bioprocess Systems," *Journal of Bioscience and Bioengineering* 87 (1999): 261–266.

45. Bernd Hitzmann, Andreas Lübbert, and Karl Schügerl, "An Expert System Approach for the Control of a Bioprocess. I: Knowledge Representation and Processing," *Biotechnology and Bioengineering* 39 (1992): 33–43, on 33.

46. Eleftherios T. Papoutsakis, "Engineering Solventogenic Clostridia," *Current Opinion in Biotechnology* 19 (2008): 420–429; Thaddeus Chukwuemeka Ezeji, Nasib Qureshi, and Hans Peter Blaschek, "Bioproduction of Butanol from Biomass: From Genes to Bioreactors," *Current Opinion in Biotechnology* 18 (2007): 220–227.

47. *2010 Ethanol Industry Outlook*, Renewable Fuels Association, February 2010, <http://www.ethanolrfa.org/pages/annual-industry-outlook> (accessed Sept. 19, 2010). The number

of plants increased from fewer than 10 in 1980, to 50 in 1999, and 189 in 2010 (<http://www.eia.doe.gov/kids/energy.cfm?page=tl_ethanol>,[accessed Feb. 25, 2011]).

48. Jim Robbins, "Yellowstone's Microbial Riches Lure Eager Bioprospectors," *NYT*, Oct. 14, 1997.

49. Alan T. Bull, "Biotechnology, the Art of Exploiting Biology," in *Microbial Diversity and Bioprospecting*, ed. Alan T. Bull (Washington, DC: ASM Press, 2004), 3–10, on 3.

50. World Resources Institute, as cited by Cori Hayden, *When Nature Goes Public: The Making and Unmaking of Bioprospecting in Mexico* (Princeton, NJ: Princeton University Press, 2003), 53.

51. See, for example, Walter V. Reid et al., eds., *Biodiversity Prospecting: Using Genetic Resources for Sustainable Development* (Washington, DC: World Resources Institute, 1993).

52. Jack Ralph Kloppenburg, *First the Seed: The Political Economy of Plant Biotechnology, 1492–2000*, 2nd ed. (Madison: University of Wisconsin Press, 2005), 335–344. Hayden, *When Nature Goes Public*, 63–66.

53. Kloppenburg, *First the Seed*, 286–289.

54. Hayden, *When Nature Goes Public*. This paragraph is based on Hayden's work, although she does not use the multiple contradictions to frame her research.

55. On bioprospecting in the ocean, see Stefan Helmreich, *Alien Ocean: Anthropological Voyages in Microbial Seas* (Berkeley: University of California Press, 2009).

56. Quoted in Helmreich, *Alien Ocean*, 125–126.

57. Environmental Protection Agency, 40 CFR Part 455, Pesticide Chemicals Manufacturing Category Effluent Limitations Guidelines, Pretreatment Standards, and New Source Performance Standards, [FRL-4105–1], RIN 2040-AB32, 57 FR 12560, Apr. 10, 1992; for data on the activated sludge plant, see NPDES Permit Rationale & Development of Proposed Effluent Limitation, Monsanto Company–Muscatine, Iowa, Permit No. 70–00–1-02, available through the Enforcement and Compliance History Online (ECHO) system, facility 110018869474, at EPA's ENVIROFACTS web site, <http://www.epa.gov/envirofw/> (accessed Feb. 25, 2010). These data are from the late 1990s and early 2000s rather than the time of the initial experiments on biodegradation; Terry M. Balthazor and Laurence E. Hallas, "Glyphosate-Degrading Microorganisms from Industrial Activated Sludge," *Applied and Environmental Microbiology* 51 (1986): 432–434; Laurence E. Hallas, William J. Adams, and Michael A. Heitkamp, "Glyphosate Degradation by Immobilized Bacteria: Field Studies with Industrial Wastewater Effluent," *Applied and Environmental Microbiology* 58 (1992):1215–1219; Michael A. Heitkamp, William J. Adams, and Laurence E. Hallas, "Glyphosate Degradation by Immobilized Bacteria: Laboratory Studies Showing Feasibility for Glyphosate Removal from Waste Water," *Canadian Journal of Microbiology* 38 (1992): 921–928.

58. John E. Franz, "N-Phosphonomethyl-Glycine, Phytotoxicant Compositions," U.S. Patent 3,799,758, Mar. 26, 1974.

59. David Barboza, "The Power of Roundup," *NYT*, Aug. 2, 2001.

60. Barboza, "The Power of Roundup."

61. The molecular basis for resistance has only recently been elucidated. Todd Funke, Huijong Han, Martha L. Healy-Fried, Markus Fischer, and Ernst Schönbrunn, "Molecular Basis for the Herbicide Resistance of Roundup Ready Crops," *Proceedings of the National Academy of Sciences* 103 (2006): 13010–13015.

62. Gerard F. Barry, Ganesh M. Kishore, Stephen R. Padgette, and William C. Stallings, "Glyphosate-tolerant 5-enolpyruvylshikimate-3-phosphate synthases," United States Patent 5,633,435, May 27, 1997; Hallas, Adams, and Heitkamp, "Glyphosate Degradation."

63. <http://www.gmo-compass.org/eng/agri_biotechnology/gmo_planting/342.genetically_modified_soybean_global_area_under_cultivation.html> (accessed Sept. 8, 2009). "Adoption of Genetically Engineered Crops in the U.S.," Economic Research Service, United States Department of Agriculture, <http://www.ers.usda.gov/Data/BiotechCrops/> (accessed Sept. 16, 2010).

64. Monsanto, *Annual Report, 2008*, <http://www.monsanto.com/investors/Documents/Pubs/2008/annual_report.pdf>.

65. M. L. Zapiola, C. K. Campbell, M. D. Butler, and C. A. Mallory-Smith, "Escape and Establishment of Transgenic Glyphosate-resistant Creeping Bentgrass *Agrostis stolonifera* in Oregon, USA: A 4-year Study," *Journal of Applied Ecology* 45 (2008): 486–494; Lidia S. Watrud, E. Henry Lee, Anne Fairbrother, Connie Burdick, Jay R. Reichman, Mike Bollman, Marjorie Storm, George King, and Peter K. Van de Water, "Evidence for Landscape-Level, Pollen-mediated Gene Flow from Genetically Modified Creeping Bentgrass with CP4 EPSPS as a Marker," *Proceedings of the National Academy of Sciences* 101 (2004): 14533–14538. The USDA was sued and must now conduct an environmental impact statement prior to any additional field trials and consider listing genetically modified creeping bentgrass as a "noxious weed." *International Center for Technology Assessment et al. v. Mike Johanns et al. and The Scotts Company*, 473 F. Supp. 2d 9 (Feb. 5, 2007).

66. S. I. Warwick, M.-J. Simard, A. Légère, H. J. Beckie, L. Braun, B. Zhu, P. Mason, G. Séguin-Swartz, and C. N. Stewart, "Hybridization between *Brassica napus* L. and Its Wild Relatives: *B. rapa* L., *Raphanus raphanistrum* L., *Sinapis arvensis* L., and *Erucastrum gallicum* (Willd.) O. E. Schulz," *Theoretical and Applied Genetics* 107 (2003): 528–539. S. I. Warwick, A. Légère, M.-J. Simard, and T. James, "Do Escaped Transgenes Persist in Nature? The Case of an Herbicide Resistance Transgene in a Weedy *Brassica rapa* Population," *Molecular Ecology* 17 (2008): 1387–1395. See also C. Neal Stewart Jr., Matthew D. Halfhill, and Suzanne I. Warwick, "Transgene Introgression from Genetically Modified Crops to Their Wild Relatives," *Nature Reviews Genetics* 4 (2004): 806–817.

7 Conclusion

1. "Address of Mr. F. O. Ward," General Congress of Hygiene at Brussels, First Meeting, Sept. 20, 1852, reprinted in *Transactions of the Sanitary Institute* 2 (1880): 267–287, on 268, 271.

2. See, for instance, Charles Richson, *The Observance of the Sanitary laws, Divinely appointed, in the Old Testament Scriptures, sufficient to ward off Preventable Diseases from Christians as well as Israelites: A Sermon Preached in the Cathedral, Manchester, on Sunday Morning, April 30th, 1854*, Knowsley Pamphlet Collection, 21, <http://www.jstor.org/stable/60101325> (accessed May 27, 2009). Pamela K. Gilbert notes that during this period "the confluence of epidemic disease, sanitary theory, and the development of medical mapping . . . led to new representations of the city as itself a body." *Mapping the Victorian Social Body* (Albany: State University of New York Press, 2004), 110.

3. Leo Marx, *The Machine in the Garden: Technology and the Pastoral Ideal in America* (Oxford: Oxford University Press, [1964] 2000), 23.

4. Marx, *The Machine in the Garden*, 312. See also Melville's story "Tartarus of Maids," where he writes of industrial machines as "inflexible iron animals" being fed and served by "tame" attendants.

5. Brief for Complainant-Appellant, *Cameron Septic Tank Co. v. Village of Saratoga Springs et al.*, 1908, Box 1100, Record Group 276, U.S. Court of Appeals, 2nd Circuit, National Archives and Records Administration, Northeast Region, New York, NY, 69.

6. *Cameron v. Saratoga Springs*, 159 F. 453 (1908), 463. This quotation is the judge's rephrasing of Cameron's arguments.

7. Brief for Complainant-Appellant, *Cameron Septic Tank Co. v. Village of Saratoga Springs et al.*, 178.

8. Robert Gilman, "The Restoration of Waters: An Interview with John and Nancy Jack Todd," *In Context* 25 (1990): 42–47.

9. John Todd, "The New Alchemists," in *Design Outlaws on the Ecological Frontier*, ed. Chris Zelov (Philadelphia: Knossus Publishing, 1997), 164–175, on 172.

10. John H. Todd and Barry Silverstem, "Solar Aquatic Apparatus for Wastewater Treatment," U.S. Patent 5,087,353, Feb. 11, 1992.

11. John Todd, "Ecological Engineering, Living Machines, and the Visionary Landscape," in *Ecological Engineering for Wastewater Treatment*, 2nd ed., ed. Carl Etnier and Björn Guterstan (Boca Raton: Lewis Publishers, 1997), 113–122, on 113–114. For discussion of another kind of "living machine," see Clay McShane and Joel A. Tarr, *The Horse in the City: Living Machines in the Nineteenth Century* (Baltimore: Johns Hopkins University Press, 2007).

12. Gilman, "The Restoration of Waters," 45–46.

13. Todd, "The New Alchemists," 172.

14. Ibid., 172.

15. Richard White, *The Organic Machine: The Remaking of the Columbia River* (New York: Hill & Wang), 108.

16. Donna Haraway, "A Cyborg Manifesto: Science, Technology, and Socialist-Feminism in the Late Twentieth Century," in *Simians, Cyborgs and Women: The Reinvention of Nature* (New York: Routledge, 1991), 149–181.

17. White, *The Organic Machine*, 109, 111–112.

18. Arthur John Martin, "The Bio-Aeration of Sewage," *Minutes of Proceedings of the Institution of Civil Engineers* 217 (1924): 96–205, on 166.

19. For a discussion of the wild and wilderness, see William Cronon, "The Trouble with Wilderness; or, Getting Back to the Wrong Nature," in *Uncommon Ground: Rethinking the Human Place in Nature*, ed. William Cronon (New York: W. W. Norton & Co., 1995), 69–90.

20. *Trout Unlimited v. Lohn*, 559 F 3d 646, Mar. 16, 2009.

21. Department of Commerce, National Oceanic and Atmospheric Administration, National Marine Fisheries Service, "Interim Policy on Artificial Propagation of Pacific Salmon under the Endangered Species Act," Apr. 5, 1993, 58 FR 17573. See the analysis by Patti Goldman, "The Current Attack on the Salmon Listings: Alsea Valley Alliance and Its Implications," Earthjustice Legal Defense Fund, Seattle, Washington, n.d., 12, <http://www.earthjustice.org/sites/default/files/library/references/Salmon-20Listing-20Paper.pdf> (accessed Feb. 24, 2011).

For a history of salmon in the Pacific Northwest, see White, *Organic Machine*; Joseph E. Taylor III, *Making Salmon: An Environmental History of the Northwest Fisheries Crisis* (Seattle: University of Washington Press, 1999); and Jim Lichatowich, *Salmon without Rivers: A History of the Pacific Salmon Crisis* (Washington, DC: Island Press, 1999). For consideration of similar issues in populations of Western cutthroat trout, see *American Wildlands et al. v. Gale Norton et al.*, 193 F. Supp. 2d 244 (2002).

22. Audio Recording of Oregon State Legislature, Joint Committee on Stream Restoration and Species Recovery, July 8, 1999, 3:00 PM, Tape 138A, transcription by author.

23. Jonathan Brinckman, "Clubbing of Salmon Unleashes Outrage," *The Oregonian* (Portland), Mar. 4, 2000.

24. *Alsea Valley Alliance v. Evans*, 161 F Supp. 2nd 1154, Sept. 10, 2001, 1163.

25. Goldman, "The Current Attack on the Salmon Listings," 14.

26. *Oregon Trollers Association v. Carlos M. Gutierrez*, 452 F.3d 1104, July 6, 2006; *Jack Marincovich et al. v. Conrad C. Lautenbacher, Jr. et al.*, 553 F. Supp. 2d 1237, Mar. 31, 2008; *Alsea Valley Alliance et al. v. Conrad C. Lautenbacher*, U.S. Dist. Lexis 60203, Aug. 14, 2007; *Modesto Irrigation District et al. v. Donald L Evans et al.*, U.S. Dist. Lexis 30304, May 12, 2004.

27. Department of Commerce, National Oceanic and Atmospheric Administration, National Marine Fisheries Service, "Policy on the Consideration of Hatchery-Origin Fish in Endangered Species Act Listing Determinations for Pacific Salmon and Steelhead, Part III," 70 FR 37204, June 28, 2005.

28. *Alsea Valley Alliance et al. v. Conrad C. Lautenbacher et al.*, U.S. Dist. Lexis 60203, 2007; 66 ERC (BNA) 1108; 37 ELR 20218, Aug. 14, 2007. The judge was Michael R. Hogan.

29. *Trout Unlimited et al. v. Lohn*, U.S. Dist. Lexis 42858, 2007; 65 ERC (BNA) 1633, June 13, 2007, 54–55, 70.

30. *Trout Unlimited v. Lohn*, 559 F 3d 646, Mar. 16, 2009.

31. USPO Patent #634,423.

32. Department of Commerce, National Oceanic and Atmospheric Administration, "Policy on the Consideration of Hatchery-Origin Fish in Endangered Species Act Listing Determinations for Pacific Salmon and Steelhead," 70 FR 37204, June 28, 2005, 37214.

33. White, *Organic Machine*, 111.

34. Bruno Latour, *We Have Never Been Modern*, trans. Catherine Porter (Cambridge, MA: Harvard University Press, 1993).

35. *Time* magazine ranked all of these, along with the microbial fuel cell, as among the fifty best inventions of 2009 (<http://www.time.com/time/specials/packages/completelist/0,29569,1934027,00.html> [accessed Mar. 7, 2011]). On microbial fuel cells, see, for example, S. Venkata Mohan, S. Veer Raghavulu, and P. N. Sarma, "Biochemical Evaluation of Bioelectricity Production Process from Anaerobic Wastewater Treatment in a Single Chambered Microbial Fuel Cell (MFC) Employing Glass Wool Membrane," *Biosensors and Bioelectronics* 23 (2008): 1326–1332.

36. The transgenic Atlantic salmon was patented by Choy L. Hew and Garth L. Fletcher, "Transgenic Salmonid Fish Expressing Exogenous Salmonid Growth Hormone," U.S. Patent 5,545,808, Aug. 13, 1996. For a discussion of the AquAdvantage salmon, see Dennis Doyle

Takahashi Kelso, "The Migration of Salmon from Nature to Biotechnology," in *Engineering Trouble: Biotechnology and Its Discontents*, ed. Rachel A. Schurman and Dennis Doyle Takahashi Kelso (Berkeley: University of California Press, 2003), 84–110.

37. Andrew Pollack, "Panel Leans in Favor of Engineered Salmon," *NYT*, Sept. 20, 2010. In response the California state assembly has recently passed a bill requiring genetically modified fish to carry a label. Susan Carpenter, "Genetically Engineered Salmon Must Be Labeled, California Assembly Bill Says," May 6, 2011, <http://latimesblogs.latimes.com/green space/2011/05/genetically-engineered-salmon.html> (accessed May 13, 2011).

38. Nick Bingham, "Slowing Things Down: Lessons from the GM Controversy," *Geoforum* 39 (2008): 111–122.

Selected Bibliography

Archives

Bancroft Library, University of California, Berkeley
    Regional Oral History Office
City Archives, Grand Rapids, Michigan
    Environmental Protection Department, Wastewater Treatment Plant
Community Archives, Vigo County Public Library, Terre Haute, IN
    Commercial Solvents Corporation
Manchester Archives, Manchester, UK
    Manchester Rivers Committee Minutes
Milwaukee County Historical Society
    Daniel Hoan Papers
Milwaukee Metropolitan Sewerage District
    Records Office
Milwaukee Public Library, Humanities Department
    Local History Collection
National Archives
    College Park, MD
        Patented Case Files, United States Patent Office, RG 241
    Philadelphia, PA
        U.S. Court of Appeal Records, RG 276
        U.S. District Court Records, RG 21
    New York, NY
        U.S. Court of Appeals Records, RG 276
Turfgrass Information Center, Michigan State University
    Noer/Milorganite® MMSD Collection
University Archives, University of Illinois at Urbana-Champaign
    Edward Bartow Papers

Short Course, Conference and Institute Programs and Proceedings
Water Environment Federation Archives, Alexandria, VA
Water Resources Center Archives, University of California, Berkeley
    California Bureau of Sanitary Engineering Papers

Newspapers, Professional Journals, and Magazines

American Society of Civil Engineers, Transactions

Applied and Environmental Microbiology

Association of Managers of Sewage Disposal Works, List of Members and Proceeding; Institute of Sewage Purification, Journal and Proceedings; Water Pollution Control

Biotechnology and Bioengineering

Chicago Daily Tribune; Chicago Tribune

Engineering News

Engineering Record; Engineering News-Record

Greenkeepers' Reporter

Incorporated Association of Municipal and County Engineers, Proceedings

Institution of Civil Engineers, Minutes of Proceedings

Milwaukee Journal

Milwaukee Sentinel; Milwaukee Journal Sentinel

Municipal Engineering

National Greenkeeper

New York Times

Operations Forum

Sanitary Record

Sewage Works Journal; Sewage and Industrial Wastes; Journal of the Water Pollution Control Federation; Water Environment Research

Society of Chemical Industry, Journal

Surveyor and Municipal and County Engineer

Sanitary Institute, Transactions; Journal of the Sanitary Institute; Journal of the Royal Sanitary Institute

Times (London)

Water Environment and Technology

Water Research

Water Works and Sewerage; Water and Sewerage Works

Patents

George E. Waring Jr., "Sewering and Draining Cities," U.S. Patent 236,740, Jan. 18, 1881.

Charles E. Avery, "Manufacture of Lactates," U.S. Patent 330,815, Nov. 17, 1885.

John C. Pennington, "Retting-Bath," U.S. Patent 509,396, Nov. 28, 1893.

Carl Wehmer, "Process of Making Citric Acid," U.S. Patent 515,033, Feb. 20, 1894.

Donald Cameron and Frederick James Commin, "Improvements in the Treatment of Sewage, and in Apparatus therefor," British Patent 21,142, Apr. 25, 1896.

Donald Cameron, Frederick J. Commin, and Arthur J. Martin, "Apparatus for Treating Sewage," U.S. Patent 634,429, Oct. 3, 1899.

Joseph T. Wood, "Bate," U.S. Patent 638,828, Dec. 12, 1899.

Karl Imhoff, "Sewage Treatment Apparatus," U.S. Patent 924,664, June 15, 1909.

Leslie C. Frank, "Process of Purifying Sewage or Other Wastes and Apparatus Therefor," U.S. Patent 1,139,024, May 11, 1915.

Walter Jones and Jones & Attwood, Ltd., "Improvements in and Connected with the Purification of Sewage and Analogous Liquids," British Patents 22,736 and 22,737, Dec. 20, 1915.

C. H. Nordell, "Aerating Apparatus," U.S. Patent 1,208,821, Dec. 19, 1916.

Walter Jones, "Purification of Sewage and Analogous Liquids," U.S. Patents 1,247,540, 1,247,542, and 1,247,543, Nov. 20, 1917.

Walter Jones, "Apparatus for the Purification of Sewage and Analogous Liquids," U.S. Patent 1,282,587, Oct. 22, 1918.

Walter Jones, "Purification of Sewage and Analogous Liquids," U.S. Patent 1,286,017, Nov. 26, 1918.

Charles Weizmann, "Production of Acetone and Alcohol by Bacteriological Processes," U.S. Patent 1,315,585, Sept. 9, 1919.

Walter Jones, "Purification of Sewage and Other Liquids," U.S. Patent 1,341,561, May 25, 1920.

Francis Richard O'Shaughnessy, "Improvements in the Treatment of Sewage Liquor and like Impure Liquids," British Patent 218,399, June 27, 1924.

Fritz Simmer, "Process for Biological Purification of Waste Water," U.S. Patent 1,751,459, Mar. 18, 1930.

Samuel C. Prescott and Kisaku Morikawa, "Production of Butyl and Isopropyl Alcohols," U.S. Patent 1,933,683, Nov. 7, 1933.

Paul Zigerli, "Purification of Sewage or Waste Liquors," U.S. Patent 2,158,954, May 16, 1939.

E. B. Mallory, "Process of Controlling the Purification of Sewage," U.S. Patent 2,154,132, Apr. 11, 1939.

Carl Shelly Miner Jr. and Bernard Wolnak, "Fermentation Activation," U.S. Patent 2,544,273, Mar. 6, 1951.

Thomas R. Wood, Murray Hill, and David Hendlin, "Process for Production of Vitamin B12," U.S. Patent 2,595,499, May 6, 1952.

Steven A. Serfling and Dominick Mendola, "Buoyant Contact Surfaces in Waste Treatment Pond," U.S. Patent 4,169,050, Sept. 25, 1979.

Ananda M. Chakrabarty, "Microorganisms Having Multiple Compatible Degradative Energy-Generating Plasmids and the Preparation Thereof," U.S. Patent 4,259,444, Mar. 31, 1981.

John H. Todd and Barry Silverstein, "Solar Aquatic Apparatus for Treating Waste," U.S. Patent 5,087,353, Feb. 11, 1992.

Choy L. Hew and Garth L. Fletcher, "Transgenic Salmonid Fish Expressing Exogenous Salmonid Growth Hormone," U.S. Patent 5,545,808, Aug. 13, 1996.

Gerard F. Barry, Ganesh M. Kishore, Stephen R. Padgette, and William C. Stallings, "Glyphosate-Tolerant 5-Enolpyruvylshikimate-3-Phosphate Synthases," U.S. Patent 5,633,435, May 27, 1997.

Government Documents

U.K.

Preliminary Report of the Commission Appointed to Inquire Into the Best Mode of Distributing the Sewage of Towns, and Applying it to Beneficial and Profitable Uses, (London: Eyre and Spottiswoode, 1858) [2372].

Sewage of Towns. Second Report of the Commission Appointed to Inquire Into the Best Mode of Distributing the Sewage of Towns and Applying it to Beneficial and Profitable Uses, 1861 [2882].

Sewage of Towns. Third Report and Appendices of the Commission Appointed to Inquire Into the Best Mode of Distributing the Sewage of Towns and Applying it to Beneficial and Profitable Uses, 1865 [3472].

First Report of The Commissioners Appointed in 1868 to Inquire Into The Best Means of Preventing the Pollution of Rivers. (Mersey and Ribble Basins.) Vol. I. Report and Plans, 1870 [C. 37].

Royal Commission on Metropolitan Sewage Discharge. *First Report of the Commissioners,* 1884 [C. 3842].

Royal Commission on Metropolitan Sewage Discharge. *Second and Final Report of the Commissioners,* 1884 [C. 4253]

Royal Commission on Metropolitan Sewage Discharge. *Minutes of Evidence Taken Before the Commission, From May 1884 to October 1884 Together with a Selection from the Appendices, and a Digest of the Evidence, Vol. II.,* 1885 [C. 4253-I]

Royal Commission on Sewage Disposal. *Interim Report of the Commissioners Appointed in 1898 to Inquire and Report What Methods of Treating and Disposing of Sewage (Including Any Liquid from Any Factory or Manufacturing Process) May Properly be Adopted,* 3 vols. 1901, 1902 [Cd. 685, 686, 686-1].

Royal Commission on Sewage Disposal. *Second Report of the Commissioners Appointed in 1898 to Inquire and Report What Methods of Treating and Disposing of Sewage May Properly Be Adopted,* 1902 [Cd. 1178].

Royal Commission on Sewage Disposal. *Fifth Report of the Commissioners Appointed to Inquire and Report What Methods of Treating and Disposing of Sewage (Including Any Liquid from Any Factory or Manufacturing Process) May Properly Be Adopted. Methods of Treating and Disposing of Sewage. Report*, 1908 [Cd. 4278].

Royal Commission on Sewage Disposal. *Final Report of the Commissioners Appointed to Inquire and Report What Methods of Treating and Disposing of Sewage (Including any Liquid from any Factory or Manufacturing Process) May Properly Be Adopted. General Summary of Conclusions and Recommendations*, 1915 [Cd. 7821].

United States

State and Local

Annual Budget Report to the Board of Trustees, The Sanitary District of Chicago

Sewage Gas (Illinois)

Digester (Indiana)

Wm. Ripley Nichols and George Derby. "Sewerage; Sewage; The Pollution of Streams; The Water-Supply of Towns. A Report to the State Board of Health of Massachusetts." *Fourth Annual Report of the State Board of Health of Massachusetts*, January 1873, 19–132.

"Water Supply and Sewerage." *Nineteenth Annual Report of the State Board of Health of Massachusetts*. Boston, Wright & Potter Printing Company, 1888.

Experimental Investigations by the State Board of Health of Massachusetts, upon the Purification of Sewage by Filtration and by Chemical Precipitation and Upon the Intermittent Filtration of Water. Made at Lawrence, Mass., 1888–1890, Part II of Report on Water Supply and Sewerage. Boston: Wright and Potter, 1890.

Annual Report, Milwaukee Sewerage Commission

Report of the Annual Missouri Water and Sewerage Conference

Proceedings, Annual Meeting, New Jersey Sewage Works Association

Proceedings of the Oklahoma Water and Sewage Short Course

Annual Report, Ohio Conference on Sewage Treatment

Official Bulletin, North Dakota Water and Sewage Works Conference

Proceedings of the Texas Water Utilities Short School

Federal

The Taylor and Other Systems of Shop Management. Hearings Before Committee of the House of Representatives to Investigate the Taylor and Other Systems of Shop Management under Authority of H Res 90, vol. 3, 1912.

Comptroller General of the United States. Operation and maintenance of municipal waste treatment plants. Washington, DC: Government Printing Office, 1969.

Comptroller General of the United States. Continuing need for improved operation and maintenance of municipal waste treatment plants. Washington, DC: Government Printing Office, 1977.

Implementation of the Federal Water Pollution Control Act (Municipal Construction Grants Program and the State Management Assistance Program). Hearings Before the Subcommittee on Oversight and Review of the Committee on Public Works and Transportation, U.S. House of Representatives, Oct. 30, 31; and Nov. 1, 1979.

Implementation of the Federal Water Pollution Control Act (Concerning the Performance of the Municipal Wastewater Treatment Construction Grants Program). Hearings Before the Subcommittee on Investigations and Oversight of the Committee on Public Works and Transportation, U.S. House of Representatives, 97th Congress, July 14, 15, 16, 21, 22, 23, 1981.

Department of Commerce, National Oceanic and Atmospheric Administration, "Policy on the Consideration of Hatchery-Origin Fish in Endangered Species Act Listing Determinations for Pacific Salmon and Steelhead," 70 FR 37204, June 28, 2005, 37214.

Court Cases

Boulton v. Bull, 2 H Blackstone 463 (1795).

Hornblower v. Boulton 8 Term Reports 95 (1799).

O'Reilly v. Morse, 56 U.S. 62 (1854).

Le Roy v. Tatham, 55 U.S. 156 (1853).

Cochrane v. Deener, 94 U.S. 780 (1877).

Cameron Septic Tank Co. v. Village of Saratoga Springs et al., 151 F. 242 (1907).

Cameron Septic Tank Co. v. Village of Saratoga Springs et al., 159 F. 453 (1908).

Guaranty Trust Co. Of New York et al. v. Union Solvents Corporation, 54 F. 2d 400 (1931).

Union Solvents Corporation v. Guaranty Trust Company, 61 F. 2d 1041 (1932).

Ex parte Prescott, 19 USPQ 178 (1932)

Guaranty Trust Co. et al. v. Union Solvents Corporation, 3 F. Supp. 572 (1933).

Guthard et al. v. Sanitary Dist. of Chicago, 8 F. Supp. 329 (1934).

City of Milwaukee v. Activated Sludge Inc., 69 F. 2d 577 (1934).

Sanitary District of Chicago v. Activated Sludge Inc., 90 F. 2d 727 (1937).

Activated Sludge Inc. et al. v. Sanitary Dist. of Chicago, 64 F. Supp. 25 (1946).

In the Matter of the Application of Malcolm E. Bergy, John H. Coats, and Vedpal S. Malik 563 F. 2d 1031 (1977).

In the Matter of the Application of Ananda M. Chakrabarty, 571 F. 2d 40 (1978).

In the Matter of the Application of Malcolm E. Bergy, John H. Coats, and Vedpal S. Malik; In the Matter of the Application of Ananda M. Chakrabarty 596 F. 2d 952 (1979).

Diamond v. Chakrabarty, 447 U.S. 303 (1980).

Alsea Valley Alliance v. Evans, 161 F. Supp. 2nd 1154 (2001).

Trout Unlimited v. Lohn, 559 F. 3d 646 (2009).

Books

Bailey-Denton, J. *Sewage Disposal. Ten Years' Experience (Now Fourteen Years) in Works of Intermittent Downward Filtration, Separately and in Combination with Surface Irrigation; With Notes on the Practice and Results of Sewage Farming.* London: E & FN Spon, 1885.

Burns, Robert Scott. *Outlines of Modern Farming, Vol. V. Utilisation of Town Sewage.—Irrigation.—Reclamation of Waste Land,* 2nd ed. London: Virtue Brothers, & Co., 1865.

Buswell, Arthur M. *The Chemistry of Water and Sewage Treatment.* New York: The Chemical Catalog Company, 1928.

Clark, H. W., and Stephen DeM. Gage. *A Review of Twenty-one years' Experiments upon the Purification of Sewage at the Lawrence Experiment Station,* reprinted from the 40th Annual Report of the State Board of Health. Boston: Wright & Potter Printing Co., State Printers, 1909.

Collings, Gilbeart H. *Commercial Fertilizers: Their Sources and Use,* 5th ed. New York: McGraw-Hill, 1955.

Corfield, W. H. *The Treatment and Utilisation of Sewage.* London: Macmillan and Co., 1887.

Curtis, George Tickner. *A Treatise on the Law of Patents for Useful Inventions in the United States of America.* Boston: Charles C. Little and James Brown, 1849.

Dibdin, W. J. *The Purification of Sewage and Water,* 3rd ed. London: Sanitary Publishing Company, 1897.

Dinges, Ray. *Natural Systems for Water Pollution Control.* New York: Van Nostrand Reinhold, 1982.

Dunbar, Professor Dr. *Principles of Sewage Treatment,* trans. H. T. Calvert. London: Charles Griffin & Co., 1908.

Etnier, Carl, and Björn Guterstan. *Ecological Engineering for Wastewater Treatment,* 2nd ed. (Boca Raton: Lewis Publishers 1997)

Ferme, George. *Local Board Sewage Farming: A Letter to Clare Sewell Read, Esq., M.P.* (London: Daldy, Isbister & Co., 1876)

Fowler, Gilbert J. *Sewage Works Analyses.* New York: Wiley, 1902.

Fowler, Gilbert J. *An Introduction to the Biochemistry of Nitrogen Conservation.* London: E. Arnold & Co., 1934.

Frankland, Percy, and Mrs. Percy Frankland. *Micro-organisms in Water: Their Significance, Identification and Removal.* London: Longmans, Green, and Co., 1894.

Fuller, George W. *Sewage Disposal.* New York: McGraw-Hill, 1912.

Granville, A. B. *The Great London Question of the Day; or, Can Thames Sewage Be Converted Into Gold?* London: Edward Stanford, 1865.

Hansen, Emil Chr. *Practical Studies in Fermentation, Being Contributions to the Life History of Micro-organisms,* trans. Alex K. Miller. London: E & FN Spon, 1896.

James, Robert. *The Grant and Validity of British Patents for Inventions.* London: John Murray, 1903.

Jones, Lieut.-Col. Alfred S., and H. Alfred Roechling. *Natural and Artificial Sewage Treatment.* London: E & FN Spon, 1902.

Kinnicutt, Leonard P., C.-E. A. Winslow, and R. Winthrop Pratt. *Sewage Disposal*. New York: Wiley, 1910.

Lawes, Edward. *The Act for Promoting the Public Health: with Notes, an Analytical Index, and (By Way of Appendix) the Nuisances Removal and Diseases Prevention Act, 1848*. London: Shaw and Sons, 1849.

Letheby, Henry. *The Sewage Question*. London: Baillière, Tindall and Cox, 1872

Martin, Arthur J. *The Sewage Problem: A Review of the Evidence Collected by the Royal Commission on Sewage Disposal*. London: The Sanitary Publishing Company, 1905.

Martin, Arthur J. *The Activated Sludge Process*. London: MacDonald and Evans, 1927.

Martin, Arthur J. *The Work of the Sanitary Engineer; a Handbook for Engineers, Students, and Others Concerned with Public Health*. London: MacDonald and Evans, 1935.

Metcalf, Leonard, and Harrison P. Eddy. *American Sewerage Practice, vol. III, Disposal of Sewage*, 2nd ed. New York: McGraw-Hill, 1916.

Pearse, Langdon, ed. *Modern Sewage Disposal: Anniversary Book of the Federation of Sewage Works Associations*. New York: Federation of Sewage Works Associations, 1938.

Poore, George Vivian. *Essays on Rural Hygiene*. London: Longmans, Green, and Co., 1893.

Porter, J. Edward. *The Activated Sludge Process of Sewage Treatment: A Bibliography of the Subject*, 2nd ed. Rochester, NY: General Filtration Company, 1921.

Rafter, George W. "Sewage Irrigation," *Water-Supply and Irrigation Papers of the United States Geological Survey No. 3*, 1897.

Rafter, George. "Sewage Irrigation, Part II," *Water-Supply and Irrigation Papers of the United States Geological Survey No. 22*, 1899.

Reed, Sherwood C., E. Joe Middlebrooks, and Ronald W. Crites. *Natural Systems for Waste Management and Treatment*. New York: McGraw-Hill, 1988.

Report of the Judges Appointed by the Royal Agricultural Society of England to Adjudicate the Prizes in the Sewage Farm Competition, 1879. London: William Clowes, 1880.

Rideal, Samuel. *Sewage and the Bacterial Purification of Sewage*. New York: Wiley, 1900.

Robinson, Henry. *Sewerage and Sewage Disposal*. London: Biggs and Co., 1905.

Schumacher, E. F. *Small Is Beautiful*. New York: Harper & Row, 1975.

Slater, J. W. *Sewage Treatment, Purification, and Utilization*. London: Whittaker & Co., 1888.

Some Studies in Biochemistry by Some Students of Dr. Gilbert J. Fowler. Bangalore: Phoenix Printing House, 1924.

Thudichum, George. *The Bacterial Treatment of Sewage. A Handbook for Councillors, Engineers, and Surveyors*. London: The Councillor and Guardian Offices, n.d.

Todd, Nancy Jack, and John Todd. *From Eco-Cities to Living Machines: Principles of Ecological Design*. Berkeley: North Atlantic Books, 1993.

Utilization of Sewage Sludge as Fertilizer. Manual of Practice No. 2. Federation of Sewage Works Associations, 1946.

Wilson, George. *A Handbook of Hygiene and Sanitary Science*, 2nd ed. Philadelphia: Lindsay and Blakiston, 1873.

Winslow, C.-E. A., and Earle B. Phelps. "Investigations on the Purification of Boston Sewage Made at the Sanitary Research Laboratory and Sewage Experiment Station of the Massachusetts Institute of Technology; with History of the Sewage-Disposal Problem." United States Geological Survey. *Water Supply and Irrigation Paper no. 185*, 1906.

Winslow Taylor, Frederick. *Principles of Scientific Management*. New York: Harper and Brothers, 1911.

Secondary Sources

Abbott, Andrew. *The System of Professions: An Essay on the Division of Expert Labor*. Chicago: University of Chicago Press, 1988.

Allen, Michelle. *Cleansing the City: Sanitary Geographies in Victorian London*. Athens: Ohio University Press, 2008.

Allenby, Braden R., and Deanna J. Richards, eds. *The Greening of Industrial Ecosystems*. Washington: National Academy Press, 1994.

Apple, Rima D. "Patenting University Research: Harry Steenbock and the Wisconsin Alumni Research Foundation." *Isis* 80 (1989): 375–394.

Aulie, Richard P. "The Mineral Theory." *Agricultural History* 48 (1974): 369–382.

Bakker, Karen. *An Uncooperative Commodity: Privatizing Water in England and Wales*. Oxford: Oxford University Press, 2003.

Bakker, Karen J. "From Public to Private to . . . Mutual? Restructuring Water Supply Governance in England and Wales." *Geoforum* 34 (2003): 359–374.

Barley, Stephen R., and Julian E. Orr, eds. *Between Craft and Science: Technical Work in U.S. Settings*. Ithaca, NY: Cornell University Press, 1997.

Benidickson, Jamie. *The Culture of Flushing: A Social and Legal History of Sewage*. Vancouver: University of British Columbia Press, 2007.

Bennett, Stuart. "'The Industrial Instrument—Master of Industry, Servant of Management': Automatic Control in the Process Industries, 1900–1940." *Technology and Culture* 32 (1991): 69–81.

Bennett, Stuart. "The Use of Measuring and Controlling Instruments in the Chemical Industry in Great Britain and the USA during the Period 1900–1939." In *Determinants in the Evolution of the European Chemical Industry, 1900–1939*, ed. A. S. Travist, Harm G. Schroter, Ernst Homburg, and Peter J. T. Morris, 215–237. Dordrecht: Kluwer Academic Publishers, 1998.

Bent, Stephen A., Richard L. Schwaab, David G. Conlin, and Donald D. Jeffery. *Intellectual Property Rights in Biotechnology Worldwide*. New York: Stockton Press, 1987.

Berkes, Fikret. *Sacred Ecology: Traditional Ecological Knowledge and Resource Management*. Philadelphia: Taylor & Francis, 1999.

Bingham, Nick. "Slowing Things Down: Lessons from the GM Controversy." *Geoforum* 39 (2008): 111–122.

Boons, Frank, and Jennifer Howard-Grenville, eds. *The Social Embeddedness of Industrial Ecology*. Cheltenham: Edward Elgar, 2009.

Braverman, Harry. *Labor and Monopoly Capital: The Degradation of Work in the Twentieth Century*. New York: Monthly Review Press, 25th anniversary ed., 1998.

Brock, William H. *Justus von Liebig: The Chemical Gatekeeper*. Cambridge, UK: Cambridge University Press, 1997.

Brock, William H. *William Crookes (1832–1919) and the Commercialization of Science*. Aldershot: Ashgate, 2008.

Buchanan, R. A. "Institutional Proliferation in the British Engineering Profession, 1847–1914." *The Economic History Review* 38 (1985): 42–60.

Bud, Robert. "The Zymotechnic Roots of Biotechnology." *The British Journal for the History of Science* 25 (1992): 127–144.

Bud, Robert. *The Uses of Life: A History of Biotechnology*. Cambridge, UK: Cambridge University Press, 1993.

Bugos, Glen, and Daniel J. Kevles. "Plants as Intellectual Property: American Law, Policy, and Practice in World Context." *Osiris*, 2nd Ser., 7 (1992): 119–148.

Bull, Alan T., ed. *Microbial Diversity and Bioprospecting*. Washington, DC: ASM Press, 2004.

Burchardt, Jeremy. *Paradise Lost: Rural Idyll and Social Change in England Since 1800*. London: I. B. Tauris, 2002.

Burkhardt-Holm, Patricia. "Endocrine Disruptors and Water Quality: A State-of-the-Art Review." *International Journal of Water Resources Development* 26 (2010): 477–493.

Calhoun, Daniel H. *The American Civil Engineer: Origins and Conflict*. Cambridge, MA: Technology Press, Massachusetts Institute of Technology, 1960.

Castree, Noel. *Nature*. London: Routledge, 2005.

Ceccatti, John Simmons. "Science in the Brewery: Pure Yeast Culture and the Transformation of Brewing Practices in Germany at the End of the 19th Century," PhD dissertation, University of Chicago, 2001.

Coombs, E. P. *Activated Sludge Ltd.—The Early Years*. Bournemouth: C. Roy Coombs, 1992.

Cooper, Gregory J. *The Science of the Struggle for Existence: On the Foundations of Ecology*. Cambridge, UK: Cambridge University Press 2003.

Cottrell, P. L. "Resolving the Sewage Question: Metropolis Sewage & Essex Reclamation Company, 1865–81." In *Cities of Ideas: Civil Society and Urban Governance in Britain 1800–2000*, ed. Robert Colls and Richard Rodger, 67–95. Aldershot: Ashgate, 2004.

Cronon, William. *Nature's Metropolis: Chicago and the Great West*. New York: W. W. Norton, 1991.

Cronon, William. "The Trouble with Wilderness; or, Getting Back to the Wrong Nature." In *Uncommon Ground: Rethinking the Human Place in Nature*, ed. William Cronon, 69–90. New York: W. W. Norton & Co., 1995.

Demain, Arnold L., and Nadine A. Solomon. *Manual of Industrial Microbiology and Biotechnology*. Washington D.C.: American Society for Microbiology, 1986.

Dickens, Peter. *Reconstructing Nature: Alienation, Emancipation and the Division of Labour*. London: Routledge, 1996.

Dierig, Sven, Jens Lachmund, and J. Andrew Mendelsohn, eds. *Science and the City*, *Osiris 18*. Chicago: University of Chicago Press, 2003.

Donnelly, James. "Consultants, Managers, Testing Slaves: Changing Roles for Chemists in the British Alkali Industry, 1850–1920." *Technology and Culture* 35 (1994): 100–128.

Duffy, John. *The Sanitarians: A History of American Public Health*. Urbana: University of Illinois Press, 1990.

Eisenberg, R. S. "Public Research and Private Development: Patents and Technology Transfer in Government-Sponsored Research." *Virginia Law Review* 82 (1996): 1663–1727.

Etnier, Carl, and Björn Guterstan. *Ecological Engineering for Wastewater Treatment*, 2nd ed. Boca Raton, FL: Lewis Publishers, 1997.

Etzkowitz, Henry, and Loet Leydesdorff. *Universities and the Global Knowledge Economy: A Triple Helix of University–Industry–Government Relations*. London: Pinter, 1997.

Encyclopedia of Bioprocess Technology: Fermentation, Biocatalysis, and Bioseparation. New York: Wiley, 1999.

Falkusa, Malcolm. "The Development of Municipal Trading in the Nineteenth Century." *Business History* 19 (1997): 134–161.

Foster, John Bellamy. *Marx's Ecology: Materialism and Nature*. New York: Monthly Review Press, 2000.

Foss-Mollan, Kate. *Hard Water: Politics and Water Supply in Milwaukee, 1870–1995*. West Lafayette IN: Purdue University Press, 2001.

George, Rose. *The Big Necessity: The Unmentionable World of Human Waste and Why It Matters*. New York: Metropolitan Books, 2008.

Giddens, Anthony. *The Consequences of Modernity*. Stanford: Stanford University Press, 1990.

Gilbert, Pamela K. *Mapping the Victorian Social Body*. Albany: State University of New York Press, 2004.

Glamann, Kristoff. "The Scientific Brewer: Founders and Successors during the Rise of the Modern Brewing Industry." In *Enterprise and History: Essays in Honour of Charles Wilson*, ed. D. C. Coleman and Peter Mathias, 186–198. Cambridge, UK: Cambridge University Press, 2006.

Goddard, Nicholas. "A Mine of Wealth"? The Victorians and the Agricultural Value of Sewage." *Journal of Historical Geography* 22 (1996): 274–290.

Goldblith, Samuel A. *Pioneers in Food Science Volume I. Samuel Cate Prescott: M.I.T. Dean and Pioneer Food Technologist*. Trumbull, CT: Food & Nutrition Press, 1993.

Goldman, Joanne Abel. *Building New York's Sewers: Developing Mechanisms of Urban Management*. West Lafayette, IN: Purdue University Press, 1997.

Graedel, T. E., and B. R. Allenby. *Industrial Ecology*, 2nd ed. Englewood Cliffs: Prentice Hall, 2003.

Haber, Samuel. *The Quest for Authority and Honor in the American Professions, 1750–1900*. Chicago: University of Chicago Press, 1991.

Halliday, Stephen. *The Great Stink of London: Sir Joseph Bazalgette and the Cleansing of the Victorian Metropolis*. Stroud: Sutton Publishing, 1999.

Hamlin, Christopher. "Providence and Putrefaction: Victorian Sanitarians and the Natural Theology of Health and Disease." *Victorian Studies* 28 (1985): 381–411.

Hamlin, Christopher. *What Becomes of Pollution? Adversary Science and the Controversy on the Self-Purification of Rivers in Britain, 1850–1900.* New York: Garland Publishing Inc., 1987.

Hamlin, Christopher. "Muddling in Bumbledom: On the Enormity of Large Sanitary Improvements in Four British Towns, 1855–1885." *Victorian Studies* 32 (1988): 55–83.

Hamlin, Christopher. "William Dibdin and the Idea of Biological Sewage Treatment." *Technology and Culture* 29 (1988): 189–218.

Hamlin, Christopher. *Public Health and Social Justice in the Age of Chadwick: Britain, 1800–1854.* Cambridge, UK: Cambridge University Press, 1998.

Hamlin, Christopher. "The City as a Chemical System? The Chemist as Urban Environmental Professional in France and Britain, 1780–1880." *Journal of Urban History* 33 (2007): 702–728.

Hayden, Cori. *When Nature Goes Public: The Making and Unmaking of Bioprospecting in Mexico.* Princeton: Princeton University Press, 2003.

Haraway, Donna. "A Cyborg Manifesto: Science, Technology, and Socialist-Feminism in the Late Twentieth Century. In *Simians, Cyborgs and Women: The Reinvention of Nature*, 149–181. New York: Routledge, 1991.

Hassan, John. *A History of Water in Modern England and Wales.* Manchester: Manchester University Press, 1998.

Helmreich, Stefan. *Alien Ocean: Anthropological Voyages in Microbial Seas.* Berkeley: University of California Press, 2009.

Hughes, Sally Smith. "Making Dollars Out of DNA: The First Major Patent in Biotechnology and the Commercialization of Molecular Biology, 1974–1980." *Isis* 92 (2001): 541–575.

Hynes, H. B. N. *Biology of Polluted Waters.* Liverpool: Liverpool University Press, 1963.

Isenmann, Ralf. "Industrial Ecology: Shedding More Light on its Perspective of Understanding Nature as Model." *Sustainable Development* 11 (2003): 143–158.

Jenkins, David, Michael G. Richard, and Glen T. Daigger. *Manual on the Causes and Control of Activated Sludge Bulking, Foaming, and Other Solids Separations Problems*, 3rd ed. Boca Raton, FL: CRC Press, 2004.

Jewell, William J., and Belford L. Seabrook. *A History of Land Application as a Treatment Alternative.* U.S. Environmental Protection Agency, Technical Report, April 1979, EPA 430/9–79–012.

Johns, Adrian. *Piracy: the Intellectual Property Wars from Gutenberg to Gates.* Chicago: The University of Chicago Press, 2009.

Johnson, Irving S. "Human Insulin from Recombinant DNA Technology." *Science* 219 (1983): 632–637.

Johnson, Jim [Bruno Latour]. "Mixing Humans and Nonhumans Together: the Sociology of a Door-closer." *Social Problems* 35 (1998): 298–310.

Jones, O. A. H., N. Voulvoulis, and J. N. Lester. "Human Pharmaceuticals in Wastewater Treatment Processes." *Critical Reviews in Environmental Science and Technology* 35 (2005): 401–427.

Kellett, J. R. "Municipal Socialism, Enterprise and Trading in the Victorian City." *Urban History* 5 (1978): 36–45.

Kelso, Dennis Doyle Takahashi. "The Migration of Salmon from Nature to Biotechnology." In *Engineering Trouble: Biotechnology and Its Discontents*, ed. Rachel A. Schurman and Dennis Doyle Takahashi Kelso, 84–110. Berkeley: University of California Press, 2003.

Kevles, Daniel J. "Patents, Protections, and Privileges: The Establishment of Intellectual Property in Animals and Plants." *Isis* 98 (2007): 323–331.

Kinnersley, David. *Coming Clean: The Politics of Water and the Environment*. London: Penguin Books, 1994.

Kirk, Andrew. "Appropriating Technology: The Whole Earth Catalog and Counterculture Environmental Politics." *Environmental History* 6 (2001): 374–394.

Kleinman, Daniel L. *Impure Cultures: University Biology and the World of Commerce*. Madison: University of Wisconsin Press, 2003.

Kloppenburg, Jack Ralph. *First the Seed: The Political Economy of Plant Biotechnology, 1492–2000*, 2nd ed. Madison: University of Wisconsin Press, 2005.

Landes, David S. *The Unbound Prometheus: Technological Change and Industrial Development in Western Europe from 1750 to Present*, 2nd ed. Cambridge, UK: Cambridge University Press, 2003.

Latour, Bruno. *We Have Never Been Modern*, trans. Catherine Porter. Cambridge, MA: Harvard University Press, 1993.

Laurent, John. "Science, Society and Politics in Late Nineteenth-Century England: A Further Look at Mechanics' Institutes." *Social Studies of Science* 14 (1984): 585–619.

Layton, Edwin. *The Revolt of the Engineers: Social Responsibility and the American Engineering Profession*, reissued. Baltimore: The Johns Hopkins University Press, 1986.

Leavitt, Judith Walzer. *The Healthiest City: Milwaukee and the Politics of Health Reform*. Princeton: Princeton University Press, 1982.

Lichatowich, Jim. *Salmon Without Rivers: A History of the Pacific Salmon Crisis*. Washington, DC: Island Press, 1999.

Lifset, Reid. "A Metaphor, a Field, and a Journal." *Journal of Industrial Ecology* 1 (1997): 1–3.

Lucier, Paul. *Scientists and Swindlers: Consulting on Coal and Oil in America, 1820–1890*. Baltimore: The Johns Hopkins University Press, 2008.

Paul Lucier. "The Professional and the Scientist in Nineteenth-Century America." *Isis* 100 (2009): 699–732.

Luckin, Bill. *Pollution and Control: A Social History of the Thames in the Nineteenth Century*. Bristol: Adam Hilger, 1986.

Marsh, Jan. *Back to the Land: The Pastoral Impulse in Victorian England from 1880 to 1914*. London: Quartet, 1983.

McCracken, Robert A., and Dennis Sebian. "Lawrence Experiment Station: Birthplace of Environmental Research in America." *The Diplomate* 24 (2) (1988): 12–18.

McGarey, Barbara M., and Annette C. Levey, "Patents, Products, and Public Health: An Analysis of the CellPro March-In Petition." *Berkeley Technology Law Journal* 14 (1999): 1095–1116.

McNeil, Brian, and Linda M. Harvey. "Fermentation: An art from the Past, a Skill for the Future." In *Practical Fermentation Technology*, ed. Brian McNeil and Linda M. Harvey, 1–2. Chichester, UK: Wiley, 2008.

McShane, Clay, and Joel A. Tarr. *The Horse in the City: Living Machines in the 19th Century*. Baltimore: Johns Hopkins University Press, 2007.

Melosi, Martin V. *The Sanitary City: Urban Infrastructure in America from Colonial Times to the Present*. Baltimore: The Johns Hopkins University Press, 2000.

Meiksins, Peter. "The 'Revolt of the Engineers' Reconsidered." *Technology and Culture* 29 (1988): 219–246.

Meiksins, Peter F. "Scientific Management and Class Relations: A Dissenting View." *Theory and Society* 13 (1984): 177–209.

National Research Council, Committee on Toxicants and Pathogens in Biosolids Applied to Land. *Biosolids Applied to Land: Advancing Standards and Practices*. Washington, DC: National Academy Press, 2002.

Nelson, Daniel, ed. *Scientific Management Since Taylor*. Columbus: Ohio State University Press, 1992.

Noble, David F. *America by Design: Science, Technology and the Rise of Corporate Capitalism*. New York: Oxford University Press, 1979.

Noble, David F. *Forces of Production: A Social History of Industrial Automation*. Oxford University Press, Oxford, 1984.

Nuwer, Michael. "From Batch to Flow: Production Technology and Work-Force Skills in the Steel Industry, 1880–1920." *Technology and Culture* 29 (1988): 808–838.

Olsson, G. "Instrumentation, control and automation in the water industry–state of the art and new challenges." *Water Science and Technology* 53 (4–5) (2006): 1–16.

O'Rourke, Maureen A. "Toward a Doctrine of Fair Use in Patent Law." *Columbia Law Review* 100 (2000): 1177–1250.

Otero, José Manuel, Gianni Panagiotou, and Lisbeth Olsson. "Fueling Industrial Biotechnology Growth with Bioethanol." *Advances in Biochemical Engineering/Biotechnology* 108 (2007): 1–40.

Platt, Harold L. *Shock Cities: The Environmental Transformation and Reform of Manchester and Chicago*. Chicago: University of Chicago Press, 2005.

Purcell, Carroll. "The Rise and Fall of the Appropriate Technology Movement in the United States, 1965–1985." *Technology and Culture* 34 (1993): 629–637.

Pursell, Carroll. *Technology in Postwar America: A History*. New York: Columbia University Press, 2007.

Radford, Gail. "From Municipal Socialism to Public Authorities: Institutional Factors in the Shaping of American Public Enterprise." *The Journal of American History* 90 (2003): 863–890.

Rawlings, D. E., and B. D. Johnson, eds. *Biomining*. Berlin: Springer, 2007.

Reich, Leonard S. *The Making of American Industrial Research: Science and Business at GE and Bell, 1876–1926*. Cambridge, UK: Cambridge University Press, 1985.

Reid, Walter V., et al., eds. *Biodiversity Prospecting: Using Genetic Resources for Sustainable Development*. Washington, DC: World Resources Institute, 1993.

Reid, David. *Paris Sewers and Sewermen: Realities and Representations*. Cambridge, MA: Harvard University Press, 1991.

Reinharz, Jehuda. *Chaim Weizmann: The Making of a Zionist Leader*. Oxford: Oxford University Press, 1985.

Rodgers, Daniel T. *Atlantic Crossings: Social Politics in a Progressive Age*. Cambridge, MA: Harvard University Press, 1998.

Scarpino, Philip V. *Great River: An Environmental History of the Upper Mississippi, 1890–1950*. Columbia: University of Missouri Press, 1986.

Schivelbusch, Wolfgang. *The Railway Journey: The Industrialization of Time and Space in the Nineteenth Century* Berkeley: University of California Press, 1986.

Schrepfer, Susan R., and Philip Scranton, eds. *Industrializing Organisms: Introducing Evolutionary History*. New York: Routledge, 2004.

Scott, James. *Seeing Like a State: How Certain Schemes to Improve the Human Condition Have Failed*. New Haven: Yale University Press, 1998.

Servos, John W. "Engineers, Businessmen, and the Academy: The Beginnings of Sponsored Research at the University of Michigan." *Technology and Culture* 37 (1996): 721–762.

Shapin, Steven, and Barry Barnes. "Science, Nature and Control: Interpreting Mechanics' Institutes." *Social Studies of Science* 7 (1977): 31–74.

Shapiro-Shapin, Carolyn G. "'A Really Excellent Scientific Contribution': Scientific Creativity, Scientific Professionalism, and the Chicago Drainage Case, 1900–1906." *Bulletin of the History of Medicine* 71 (1997): 385–411.

Sherman, Brad, and Lionel Bently. *The Making of Modern Intellectual Property Law: The British Experience, 1760–1911*. Cambridge, UK: Cambridge University Press, 1999.

Sigsworth, E. M. "Science and the Brewing Industry, 1850–1900." *The Economic History Review* 17 (1965): 536–550.

Sinclair, Bruce. *A Centennial History of the American Society of Mechanical Engineers*. Toronto: University of Toronto Press, 1980.

Slaton, Amy E. *Reinforced Concrete and the Modernization of American Building, 1900–1930*. Baltimore: The Johns Hopkins University Press, 2001.

Stanbridge, H. H. *History of Sewage Treatment in Britain*, 12 vols. Maidstone, UK: The Institute of Water Pollution Control, 1976, 1977.

Stauber, John, and Sheldon Rampton. *Toxic Sludge Is Good For You!* Monroe, ME: Common Courage Press, 1995.

Stauffer, Julie. *The Water Crisis: Constructing Solutions to Freshwater Pollution*. London: Earthscan Publications, 1998.

Steinberg, Theodore. *Nature Incorporated: Industrialization and the Waters of New England*. Amherst: University of Massachusetts Press, 1991.

Steingraber, Sandra. *Living Downstream: An Ecologist Looks at Cancer and the Environment*. Reading, MA: Addison-Wesley, 1997.

Stevenson, Lloyd G. "Science Down the Drain: On the Hostility of Certain Sanitarians to Animal Experimentation, Bacteriology and Immunology." *Bulletin of the History of Medicine* 29 (1955): 1–26.

Stewart, C. Neal Jr., Matthew D. Halfhill, and Suzanne I. Warwick. "Transgene Introgression from Genetically Modified Crops to Their Wild Relatives." *Nature Reviews Genetics* 4 (2004): 806–817.

Summers, William C. "From Culture as Organism to Organism as Cell: Historical Origins of Bacterial Genetics." *Journal of the History of Biology* 24 (1991): 171–190.

Tarr, Joel A. *The Search for the Ultimate Sink: Urban Pollution in Historical Perspective*. Akron: University of Akron Press, 1996.

Taylor, Joseph E. III. *Making Salmon: An Environmental History of the Northwest Fisheries Crisis*. Seattle: University of Washington Press, 1999.

Teich, Mikulas. "Fermentation Theory and Practice: the Beginnings of Pure Yeast Cultivation and English Brewing, 1883–1913." *History of Technology* 8 (1983): 117–133.

Thackrey, Arnold, ed. *Private Science: Biotechnology and the Rise of the Molecular Sciences*. Philadelphia: University of Pennsylvania Press, 1998.

Thelan, David P. *The New Citizenship: Origins of Progressivism in Wisconsin, 1885–1900*. Columbia: University of Missouri Press, 1972.

Tomes, Nancy. *Gospel of Germs: Men, Women, and the Microbe in American Life*. Cambridge, MA: Harvard University Press, 1998.

Treating Contaminants of Emerging Concern. U.S. Environmental Protection Agency, Office of Water, EPA 820-R-10–002, <http://epa.gov/waterscience/ppcp/studies/results.html>.

Tropea, Joseph L. "Rational Capitalism and Municipal Government: The Progressive Era." *Social Science History* 13 (1989): 137–158.

Van de Poel, Ibo. "The Bugs Eat the Waste: What Else Is There to Know? Changing Professional Hegemony in the Design of Sewage Treatment Plants." *Social Studies of Science* 38 (2008): 605–634.

Vernon, Keith. "Pus, Sewage, Beer and Milk: Microbiology in Britain, 1870–1940." *History of Science* 28 (1990): 289–325.

Vernon, Keith. "Microbes at Work. Micro-organisms, the D.S.I.R. and Industry in Britain, 1900–1936." *Annals of Science* 51 (1994): 593–613.

Warren, Charles E. *Biology and Water Pollution Control*. Philadelphia: Saunders, 1971.

Water Environment Federation *Activated Sludge, Manual of Practice OM-9*. Alexandria, VA: Water Environment Federation, 2002.

Water Environment Federation. *Operation of Municipal Wastewater Treatment Plants, Manual of Practice No. 11, 6th ed., 3 vols*. Alexandria, VA: Water Environment Federation, 2007.

Water Environment Federation. *Automation of Wastewater Treatment Facilities, Manual of Practice No. 21, 3rd ed*. Alexandria, VA: Water Environment Federation, 2007.

Water Environment Federation. *Design of Municipal Wastewater Treatment Plants, Manual of Practice No. 8, 5th ed., 3 Vols*. Alexandria, VA: Water Environment Federation, 2009.

White, Richard. *The Organic Machine: The Remaking of the Columbia River*. New York: Hill & Wang, 1995.

Whitston, Kevin. "The Reception of Scientific Management by British Engineers, 1890–1914." *The Business History Review* 71 (1997): 207–229.

Williams, Raymond. *The Country and the City*. New York: Oxford University Press, 1973.

Wilson, Alan. "Technology and Municipal Decision-Making: Sanitary Systems in Manchester 1868–1910," PhD thesis, University of Manchester, 1990.

Worboys, Michael. *Spreading Germs: Disease Theories, and Medical Practice in Britain, 1865–1900*. Cambridge: Cambridge University Press, 2000.

Worster, Donald. *Nature's Economy: A History of Ecological Ideas*, 2nd ed. Cambridge, UK: Cambridge University Press, 1994.

Wylie, J. C. *Fertility From Town Wastes*. London: Faber and Faber Ltd., 1955.

Young, Robert M. *Darwin's Metaphor: Nature's Place in Victorian Culture*. Cambridge, UK: Cambridge University Press, 1985.

Zimmer, Carl. *Microcosm: E. coli and the New Science of Life*. New York: Pantheon Books, 2008.

Index

Urban and Industrial Environments

Series editor: Robert Gottlieb, Henry R. Luce Professor of Urban and Environmental Policy, Occidental College

Kerry H. Whiteside, *Precautionary Politics: Principle and Practice in Confronting Environmental Risk*

Ronald Sandler and Phaedra C. Pezzullo, eds., *Environmental Justice and Environmentalism: The Social Justice Challenge to the Environmental Movement*

Julie Sze, *Noxious New York: The Racial Politics of Urban Health and Environmental Justice*

Robert D. Bullard, ed., *Growing Smarter: Achieving Livable Communities, Environmental Justice, and Regional Equity*

Ann Rappaport and Sarah Hammond Creighton, *Degrees That Matter: Climate Change and the University*

Michael Egan, *Barry Commoner and the Science of Survival: The Remaking of American Environmentalism*

David J. Hess, *Alternative Pathways in Science and Industry: Activism, Innovation, and the Environment in an Era of Globalization*

Peter F. Cannavò, *The Working Landscape: Founding, Preservation, and the Politics of Place*

Paul Stanton Kibel, ed., *Rivertown: Rethinking Urban Rivers*

Kevin P. Gallagher and Lyuba Zarsky, *The Enclave Economy: Foreign Investment and Sustainable Development in Mexico's Silicon Valley*

David N. Pellow, *Resisting Global Toxics: Transnational Movements for Environmental Justice*

Robert Gottlieb, *Reinventing Los Angeles: Nature and Community in the Global City*

David V. Carruthers, ed., *Environmental Justice in Latin America: Problems, Promise, and Practice*

Tom Angotti, *New York for Sale: Community Planning Confronts Global Real Estate*

Paloma Pavel, ed., *Breakthrough Communities: Sustainability and Justice in the Next American Metropolis*

Anastasia Loukaitou-Sideris and Renia Ehrenfeucht, *Sidewalks: Conflict and Negotiation over Public Space*

David J. Hess, *Localist Movements in a Global Economy: Sustainability, Justice, and Urban Development in the United States*

Julian Agyeman and Yelena Ogneva-Himmelberger, eds., *Environmental Justice and Sustainability in the Former Soviet Union*

Jason Corburn, *Toward the Healthy City: People, Places, and the Politics of Urban Planning*

JoAnn Carmin and Julian Agyeman, eds., *Environmental Inequalities Beyond Borders: Local Perspectives on Global Injustices*

Louise Mozingo, *Pastoral Capitalism: A History of Suburban Corporate Landscapes*

Gwen Ottinger and Benjamin Cohen, eds., *Technoscience and Environmental Justice: Expert Cultures in a Grassroots Movement*

Samantha MacBride, *Recycling in the United States: Progress, Promise, and the Politics of Diversion*

Andrew Karvonen, *Politics of Urban Runoff: Nature, Technology, and the Sustainable City*

Daniel Schneider, *Hybrid Nature: Sewage Treatment and the Contradictions of the Industrial Ecosystem*